The effects of UV radiation in the marine environment

Decreasing concentrations of ozone in the stratosphere radically influence the effects of UV radiation on the environment. This book provides a comprehensive review of UV radiation effects specifically in the marine environment. A multidisciplinary approach is adopted to discuss all aspects from physical, chemical and biological perspectives.

The book begins by describing the attenuation of UV radiation in the atmosphere and seawater, outlining the photochemical reactions involved and highlighting the role that such chemistry can play in influencing the biogeochemical cycling of various elements. The deleterious consequences of such radiation on organisms are discussed, from viruses and bacteria through phytoplankton and zooplankton to fish and mammals. The strategies adopted by these organisms to mitigate such harmful repercussions and a synthesis of the UV-induced response at a community level are also considered.

The book is aimed at researchers and graduate students in photobiology, photochemistry and environmental science. It will also be useful as a supplementary text for courses in oceanography, climatology and ecology.

T0185823

CAMBRIDGE ENVIRONMENTAL CHEMISTRY SERIES

Series Editors:
P. G. C. Campbell, *Institut National de la Recherche Scientifique,
Université du Québec, Québec, Canada*
R. M. Harrison, *School of Chemistry, University of Birmingham, UK*
S. J. de Mora, *IAEA Marine Environment Laboratory, Monaco*

Other books in the series:
A. C. Chamberlain *Radioactive Aerosols*
M. Cresser and A. Edwards *Acidification of Freshwaters*
M. Cresser, K. Killham and A. Edwards *Soil Chemistry and its Applications*
R. M. Harrison and S. J. de Mora *Introductory Chemistry for the
Environmental Sciences* Second Edition
S. J. de Mora *Tributyltin: Case Study of an Environmental Contaminant*
T. D. Jickells and J. E. Rae *Biogeochemistry of Intertidal Sediments*

The effects of UV radiation in the marine environment

Edited by

STEPHEN DE MORA
IAEA Marine Environment Laboratory, Monaco

SERGE DEMERS
ISMER, Rimouski, Québec

and

MARIA VERNET
University of California at San Diego

CAMBRIDGE
UNIVERSITY PRESS

CAMBRIDGE UNIVERSITY PRESS
Cambridge, New York, Melbourne, Madrid, Cape Town, Singapore, São Paulo

Cambridge University Press
The Edinburgh Building, Cambridge CB2 2RU, UK

Published in the United States of America by Cambridge University Press, New York

www.cambridge.org
Information on this title: www.cambridge.org/9780521632188

First published 2000
This digitally printed first paperback version 2005

A catalogue record for this publication is available from the British Library

Library of Congress Cataloguing in Publication data

The effects of UV radiation in the marine environment / edited by Stephen de Mora, Serge
    Demers, and Maria Vernet.
        p. cm. – (Cambridge environmental chemistry series)
    Includes index.
    ISBN 0 521 63218 8
    1. Ultraviolet radiation – Environmental aspects. 2. Marine ecology. I. De Mora, S. J. II.
    Demers, Serge, 1951– . III. Vernet, Maria. IV. Series.
QH543.6.E45 2000

577.7–dc21 99-15231 CIP

ISBN-13 978-0-521-63218-8 hardback
ISBN-10 0-521-63218-8 hardback

ISBN-13 978-0-521-02095-4 paperback
ISBN-10 0-521-02095-6 paperback

Contents

Contributors

Authors indicated with an asterisk at the heads of chapters are corresponding authors.

Charles R. Booth, Biospherical Instruments Inc., 5340 Riley Street, San Diego, CA 92110-2621, USA

Robert P. Bukata, Aquatic Ecosystem Conservation Branch, National Water Research Institute, 867 Lakeshore Road, Burlington, Ontario L7R 4A6, Canada

Stephen J. de Mora, International Atomic Energy Agency, Marine Environment Laboratory, 4 Quai Antoine ler, BP 800, MC 98012, Monaco

Serge Demers, Institut des Sciences de la Mer de Rimouski, 310 allée des Ursulines, Rimouski, Québec G5L 3A1, Canada

Susana B. Díaz, Consejo Nacional de Investigaciones Cientificas y Tecnicas, Ruta 3 y Malvinas Arg., 9410 Ushuaia, Tierra del Fuego, Argentina

Wade H. Jeffrey, Center for Environmental Diagnostics and Bioremediation, University of West Florida, 11000 University Parkway, Pensacola, FL 32514, USA

John H. Jerome, Aquatic Ecosystem Conservation Branch, National Water Research Institute, 867 Lakeshore Road, Burlington, Ontario L7R 4A6, Canada

Jason P. Kase, Center for Environmental Diagnostics and Bioremediation, University of West Florida, 11000 University Parkway, Pensacola, FL 32514, USA

David J. Kieber, State University of New York, College of Environmental Science and Forestry, Chemistry Department, 1 Forestry Drive, Syracuse, New York 13210, USA

Kenneth Mopper, Chemistry Department, Washington State University, Pullman, WA 99164-4630, USA

John H. Morrow, Biospherical Instruments Inc., 5340 Riley Street, San Diego, CA 92110-2621, USA

Behzad Mostajir, Institut des Sciences de la Mer de Rimouski, 310 allée des Ursulines, Rimouski, Québec G5L 3A1, Canada

Patrick J. Neale, Environmental Research Center, Smithsonian Institution, PO Box 28, Edgewater, MD 21037, USA

Suzanne Roy, Institut des Sciences de la Mer de Rimouski, 310 allée des Ursulines, Rimouski, Québec G5L 3A1, Canada

Maria Vernet, Marine Research Division, Scripps Institution of Oceanography, La Jolla, CA 92093-0218, USA

Warwick F. Vincent, Département de biologie, Laval University, Sainte-Foy, Québec G1K 7P4, Canada

Robert F. Whitehead, Institut des Sciences de la Mer de Rimouski, 310 allée des Ursulines, Rimouski, Québec G5L 3A1, Canada

Steven W. Wilhelm, Department of Microbiology, University of Tennessee, Knoxville, TN 37996, USA

Craig E. Williamson, Department of Earth and Environmental Sciences, 31 Williams Drive, Lehigh University, Bethlehem, PA 18015-3188, USA

Horacio E. Zagarese, Universidad Nacional del Comahue, Centro Regional Universitario Bariloche, U.P. Universidad, 8400 San Carlos de Bariloche, Argentina

Preface

Insofar as UV radiation (UVR) is a natural component of solar radiation, the marine environment has always been exposed to UVR. However, anthropogenic influences via stratospheric ozone depletion have caused an increase in the UVR flux to the earth's surface in recent years. In this way, the effects of enhanced UVR on the marine environment can be considered to be a new problem. Moreover, the relative change in the UVR flux is greatest in the polar regions, which are most susceptible to 'ozone hole' formation in the spring, arguably a very sensitive time for marine organisms. Recognition of the Antarctic ozone hole initially prompted considerable research aimed at phytoplankton, owing to their importance in primary production. Subsequent investigations were extended to consider both the microbial loop and higher trophic levels. Elucidating responses at the ecosystem level remains the ongoing challenge in this field.

This book is aimed at researchers and the postgraduate student market, providing a state of the art review of UVR effects in the marine environment. A multidisciplinary approach is adopted such that basic properties of the relevant physics, chemistry and biology are included. The fate and effects of UVR are treated as a continuum, with the underlying intent to try to follow the pathway and fate of a UV photon. The progression considers firstly optical properties and the attenuation of UVR in the atmosphere and seawater. Then the wavelength dependence of absorption by both molecules and organisms is explained. Next, photochemical reactions are described, highlighting the role that such chemistry can play in influencing the biogeochemical cycling of several elements. Two chapters discuss the deleterious consequences of such radiation and the strategies adopted by organisms to mitigate harmful repercussions. Thereafter, the text considers the effects of UVR on organisms, following the stepwise progression through the trophic levels from viruses and bacteria through phytoplankton and zooplankton to fish. Finally, a brief synthesis of the UV-induced response at a community level is presented.

This book has evolved from a workshop entitled 'The Effects of UV Radiation on Various Ecosystems at Different Latitudes' held in Ensenada, Mexico, in September 1996. Many of the authors here were invited speakers at this meeting. Financial support for the workshop from the Inter-American Institute for Global Change is gratefully acknowledged. Finally, we wish to thank the numerous referees for commenting on individual chapters.

Stephen de Mora
Serge Demers
Maria Vernet

1

○ ○ ○ ○ ○ ○ ○ ○ ○ ○ ○ ○ ○ ○ ○ ○ ○ ○ ○ ○

Enhanced UV radiation – a new problem for the marine environment

Robert F. Whitehead, Stephen J. de Mora* and Serge Demers

1.1 Introduction

UV irradiance at the earth's surface is intimately related to stratospheric ozone. This gas tends to be concentrated in the lower stratosphere (hence the notion of an ozone layer) and is primarily responsible for the absorption of solar UV radiation (UVR). UVR has been recognised for many years (e.g. Worrest, Dyke & Thomson, 1978; Worrest et al., 1981; Calkins, 1982) as a potential stress for organisms in a variety of environments and as a factor in biogeochemical cycling (Zepp, Callaghan & Erickson, 1995). The trend in recent years of an intensifying, but periodic, anthropogenic-induced decline in stratospheric ozone concentrations with concurrent enhanced UV-B radiation is quite alarming. Altered solar radiation regimes can potentially upset established balances in marine ecosystems and thus presents a new problem. Most attention has been given to the 'ozone hole' over Antarctica that has been recorded annually since the 1980s. However, recent observations have confirmed measurable ozone losses over other regions, including the development of an Arctic ozone hole. The major factor responsible for the destruction of the ozone layer is anthropogenic emissions of chlorofluorocarbons (CFCs). These gases, having no natural sources, are non-toxic and inert in the troposphere, but are photolysed in the stratosphere, thereby releasing reactive chlorine atoms that catalytically destroy ozone. Other anthropogenic contributions to ozone depletion may include global changes in land use and the increased emission of nitrogen dioxide as a result of fertiliser applications (Bouwman, 1998). Paradoxically, the anthropogenic emissions of greenhouse gases that tend to cause a temperature increase at the earth's surface also produce a decrease in stratospheric temperatures. This decrease in stratospheric temperatures leads to enhanced formation of polar stratospheric clouds and may serve to increase ozone loss in polar regions (Salawitch, 1998; Shindell, Rind & Lonergan, 1998).

The understanding of the atmospheric chemistry involved in ozone depletion has greatly expanded since the link to CFCs was first proposed in 1974 (Molina & Rowland, 1974). A worldwide network of ozone observation stations has documented continuing ozone reductions over many areas of the globe. The effects have been especially pronounced in the Antarctic region, where an ozone hole, characterised by the depletion of 60% or more of the ozone, opens up each spring over an area that is now slightly larger than the size of Canada (Smith *et al.*, 1992). In the Arctic and into the North Temperate Zone, the ozone layer diminished by 15% to 20% during the 1991–2 winter[1]. The increases in atmospheric carbon dioxide anticipated over the next 50 years should lead to stratospheric cooling, thereby accelerating the destruction of stratospheric ozone and perhaps leading to an Arctic ozone hole as severe as that over Antarctica (Austin, Butchart & Shine, 1992). The latest Environmental Canada (Wardle *et al.*, 1997) report indicates ozone loss over the Arctic of up to 45% during the spring of 1997 in response to atmospheric conditions that may be indicative of changes due to stratospheric cooling (Mühler *et al.*, 1997; Wardle *et al.*, 1997). The magnitude of ozone destruction is predicted to increase over the next century despite international efforts to reduce the usage and emission of CFCs in accordance with the Montreal Protocol (Shindell *et al.*, 1998).

Regardless of the cause, the decrease in stratospheric ozone concentrations provokes an increase of UV-B radiation in the wavelength range 280 to 320 nm (Crutzen, 1992; Smith *et al.*, 1992; Kerr & McElroy, 1993). For example, an annual increase in UV-B of up to 35% has been observed in Canada for the winter–spring period during 1989–93 (Kerr & McElroy, 1993). The UV-B wave band represents less that 0.8 % of the total energy reaching the surface of the earth but is responsible for almost half of the photochemical effects in the aquatic and marine environments. Although not widely recognised due to a lack of field measurements, biologically effective levels of solar UVR penetrate water columns to significant depths: at least 30 m for UV-B (280–320 nm) and 60 m for UV-A (320–400 nm) (Smith & Baker, 1979; Holm-Hansen, Lubin & Helbing, 1993). Even in highly productive lakes and coastal regions, UVR can penetrate to at least 20 m (Kirk, 1994b; Scully & Lean, 1994) and this penetration increases as stratospheric ozone declines (Smith *et al.*, 1992).

---

[1] A number of articles on Arctic ozone and atmospheric chemistry can be found in *Science* (1993) **261**.

The environmental impact of this rise in solar UV-B has recently become a source of much concern and speculation in public as well as scientific literature.

Solar UV-B radiation is known to have a wide range of harmful effects, generally manifested as reduced productivity, on freshwater and marine organisms, including bacterioplankton and phytoplankton (Vincent & Roy, 1993; Cullen & Neale, 1994; Booth et al., 1997). Analogous studies on zooplankton and on the early life history stages of fishes indicate that exposure to relatively low levels of UV-B also deleteriously affects these groups (Holm-Hansen et al., 1993). All plant, animal and microbial groups appear to be susceptible to UV-B, but to a highly variable extent that depends on the individual species and its environment (Vincent & Roy, 1993). In addition, UV-B may have significant effects on community structure that are not apparent through studies based on individual species or trophic levels (e.g. Bothwell, Sherbot & Pollock, 1994; Vernet et al., 1994).

This chapter provides an introduction to some fundamental aspects of the behaviour of solar radiation in the atmosphere and water column. The fate of photons is also considered in respect of basic photochemistry and photobiology. The introduction is intended to form a basis for the understanding of the relationships amongst anthropogenic-related changes in the atmosphere, changes in solar radiation and the new problems they present to marine ecosystems. Subsequent chapters elucidate effects on specific biological structures and organisms, trophic-level interactions, photochemical reactions and biogeochemical cycling.

1.2   The solar spectrum and the nature of light

The effect of solar radiation on chemical and biological processes in the marine environment depends on both intensity and spectral distribution. There is significant natural variability in the factors that attenuate solar radiation and UV-B in both the atmosphere and the ocean. At the edge of the earth's atmosphere, the solar energy reaching a surface perpendicular to the radial direction from the sun is approximately $1394\,\mathrm{W\,m^{-2}}$ and has a spectrum characterised as UV (UV-C 200–280 nm, UV-B 280–320 nm, UV-A 320–400 nm), photosynthetically available radiation (PAR 400–700 nm) and infrared (IR > 700 nm) (Figure 1.1). The energy characteristic of each wavelength is determined by the relationship:

$$E = hc/\lambda \tag{1.1}$$

where E is the energy in joules, h is the Planck constant, c is the speed of light, and λ is the wavelength in metres. When dealing with biological and chemical systems, the most commonly used unit is the mole photon (also called an Einstein) which contains N photons (where N is Avogadro's number $= 6.023 \times 10^{23}$). The radiant energy of 1 mole photon is defined by:

$$E_{\text{(mole photon)}} = Nhc/\lambda = 1.19629 \times 10^8 \text{ J}/\lambda \qquad (1.2)$$

Thus, the energy of a mole photon varies inversely with wavelength (Figure 1.2). For example, the energy of 1 mole photon of 300 nm light is 398 kJ. In contrast, the energy of 1 mole photon of 700 nm light is only 171 kJ. The large increase in energy with decreasing wavelength has important chemical and biological implications when one is considering systems under changing solar spectral distributions.

1.3   Attenuation of solar energy

1.3.1   Attenuation in the atmosphere
In general terms, the relative solar spectral distribution outside the

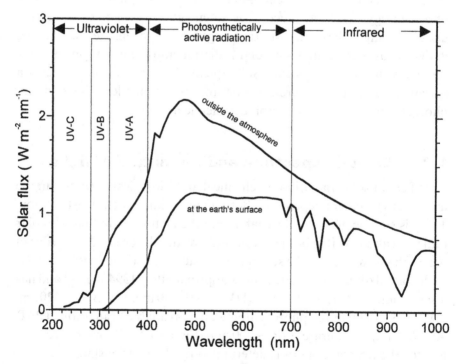

Figure 1.1.   General characteristics of solar radiation outside the atmosphere and at the earth's surface.

atmosphere comprises 51% in the IR region, 41% in the visible (PAR) region and 8% in the UV region (Figure 1.3). Passing through the atmosphere, the radiation is subject to scattering and absorption which reduces its intensity by ~35% before it reaches the earth's surface. As a result, the spectral distribution at the earth's surface differs from that experienced at the edge of the atmosphere and is a combination of direct and diffuse radiation. The amount of scattering and absorption is a function of the atmospheric composition (gases and particles) and the pathlength of the photons through the atmosphere. Thus, given a uniform atmospheric composition the spectral distribution and intensity would still vary as a function of solar zenith angle (i.e. time of day, season and latitude). A typical solar spectral distribution for a low latitude (30° N) site on a sunny day with the sun at zenith is composed of about 43% IR, 52% PAR and 5% UV radiation. For the same location with a zenith angle of 60° or 79°, the distribution changes to about 45% IR, 52% PAR and 3% UV radiation or 53% IR, 46% PAR and 1% UV, respectively (Figure 1.3). The reason for the larger relative reduction at the UV end of the spectrum

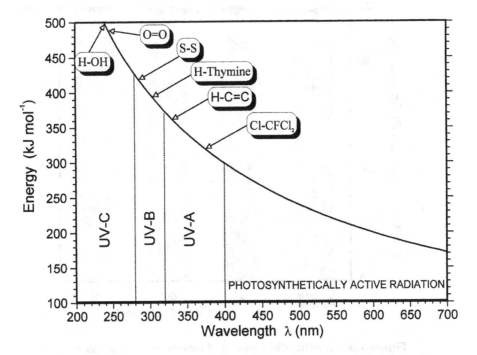

Figure 1.2. The inverse relationship between energy per mole and wavelength of solar radiation. Bond dissociation energies of some important biomolecular bonds are indicated by the location of the arrows on the curve.

with increasing atmospheric pathlength is two-fold:

1. *Enhancement of scattering:* Scattering in the atmosphere is inversely proportional to the fourth power of the wavelength and is therefore more effective in the UV region. Scattering may redirect a photon's path away from the earth such that it is lost back to space or may enhance the probability of absorption due to longer pathlengths.

2. *Enhancement of absorption:* UV radiation <320 nm is strongly absorbed by ozone and to some extent by oxygen (Figure 1.4). Longer pathlengths effectively increase the total ozone encountered by a photon and thereby enhances the probability of absorption.

1.3.1.1 Absorbance of UV and the ozone cycle

The strong reduction in UVR (<320 nm) reaching ground level (Figure 1.1) is due primarily to absorption by ozone and oxygen. Although ozone is a trace gas in the atmosphere (maximum concentration ~8 parts per

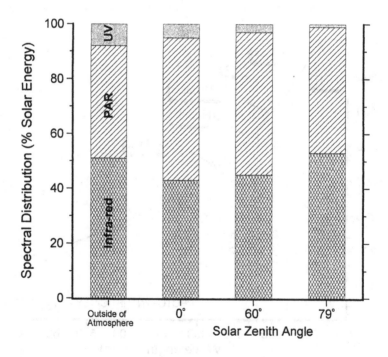

Figure 1.3. Spectral distributions of wavelength regions as a percentage of total solar radiation. Bars represent solar radiation outside the atmosphere and at the earth's surface (30° N) for three solar zenith angles. Atmospheric attenuation causes the largest relative reduction at the UV end of the spectrum.

million by volume (ppmv) at $\sim 35\,km$ altitude), the attenuation of UV by ozone is orders of magnitude higher than that of oxygen. The absorption of UV with enough energy to break the $O\!\!=\!\!O$ bond ($\Delta H = 494\,kJ\,mol^{-1}$ requires $\lambda < 240\,nm$) is the first step in the production of ozone (O_3):

$$O_2 + h\nu(\lambda < 240\,nm) \rightarrow O + O \qquad [1.1]$$

where ν is the wave frequency.

The O atoms released may then react with O_2 to form O_3:

$$2(O + O_2 + M \rightarrow O_3 + M) \qquad [1.2]$$

where M is a collision chaperone that absorbs excess energy but is itself unreactive. Net ozone production is $3\,O_2 \rightarrow 2\,O_3$. Ozone can be destroyed by direct photolysis:

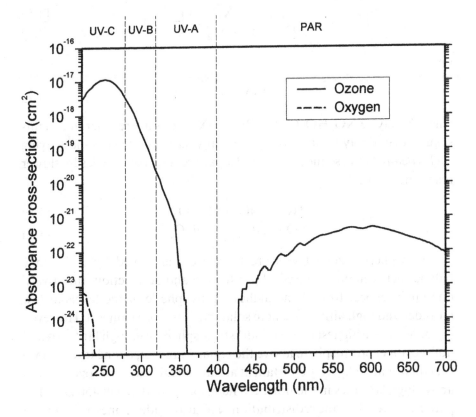

Figure 1.4. Spectral characteristics of the absorbance cross-sections of oxygen (O_2) and ozone (O_3) at 298 K. Whereas wavelengths in the UV-C and UV-B regions are strongly absorbed by O_3, UV-A and PAR are little affected. (Data from Inn & Tanaka, 1953; Molina & Molina, 1986.)

$$O_3 + h\nu(\lambda < 410\,\text{nm}) \rightarrow O + O_2 \qquad\qquad [1.3]$$

or by recombination with O:

$$O_3 + O \rightarrow 2O_2 \qquad\qquad [1.4]$$

Net ozone destruction is thus $2O_3 \rightarrow 3O_2$.

However, at all times the concentration of oxygen far exceeds that of ozone and the recombination reaction is slower than production. If pure oxygen reactions were the only mechanism for ozone production and destruction, the ozone layer would be approximately twice as thick as is currently observed. Thus, other destruction reactions are necessary to explain natural ozone levels.

The rate of recombination is greatly enhanced by catalytic cycles of the general form involving a free radical, X:

$$O_3 + X \rightarrow XO + O_2 \qquad\qquad [1.5]$$
$$O + XO \rightarrow O_2 + X \qquad\qquad [1.6]$$

or

$$O_3 + XO \rightarrow 2O_2 + X \qquad\qquad [1.7]$$

where X may be NO, HO, Cl, I or Br. The X species are regenerated in this sequence and may be involved in as many as 100 000 ozone-destroying cycles before being sequestered into less active reservoir species by slower reactions such as:

$$HO + NO_2 \rightarrow HNO_3 \qquad\qquad [1.8]$$
$$XO + NO_2 \rightarrow XONO_2 \qquad\qquad [1.9]$$

Stratospheric ozone levels are therefore maintained by a dynamic balance between photochemical production and destruction. Intuitively, one might expect to find the highest stratospheric ozone levels at low latitudes and high altitudes where solar irradiance is strongest. However, ozone levels are highest in the middle stratosphere over high latitudes and not the upper stratosphere above the equator. In fact, ozone levels above the equator are relatively constant at about 260 DU[2] whereas ozone levels above high latitudes in the northern hemisphere may reach 450 DU. The pattern is a result of the redistribution of high altitude ozone-rich air from the tropics to lower altitudes in the polar regions (Figure 1.5).

[2] 100 Dobson units, DU, are equivalent to an ozone layer 1 mm thick at 0 °C and 1 atm pressure.

The natural O_3 cycle can be perturbed by interactions with anthropogenic compounds, most notably CFCs (WMO, 1995). CFCs were first produced in the 1930s and were heralded as non-toxic, non-flammable compounds with a wide variety of uses as refrigerants, propellants for aerosol cans, cleaning compounds for electronic parts and blowing agents for foam manufacturing. Over the 50 years since the introduction of CFCs, their concentrations in the atmosphere, in general, have shown a steady increase, with a corresponding decrease in stratospheric ozone (Figure 1.6). Mario Molina and F. Sherwood Rowland first proposed their role in the destruction of atmospheric ozone in 1974. They shared the 1995 Nobel Prize for Chemistry with Paul Crutzen for their work in this field. The ozone hole over Antarctica was first reported in 1985 and led to work that has firmly established the link between CFCs and ozone depletion.

CFCs are quite stable and inert in the troposphere. They have long residence times in the atmosphere and are mixed into the stratosphere, attaining notable concentrations. Once in the stratosphere, CFCs are exposed to UV radiation of sufficient energy to break the carbon–chlorine bonds. The released chlorine can then attack O_3 in the following reaction sequence:

$$Cl + O_3 \rightarrow ClO + O_2 \qquad [1.10]$$
$$ClO + ClO \rightarrow Cl_2O_2 \qquad [1.11]$$
$$Cl_2O_2 + hv \rightarrow Cl + ClOO \qquad [1.12]$$

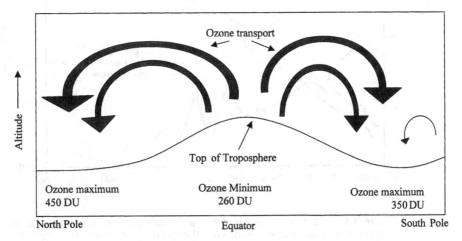

Figure 1.5. Generalised atmospheric redistribution of O_3 without the influence of O_3 depletion. Highest O_3 production occurs over the equator and tropics, but atmospheric circulation transports the O_3 produced there towards the poles, giving rise to an O_3 maximum at higher latitudes. (Adapted from Stolarski, 1988.)

$$ClOO + O \rightarrow ClO + O_2 \tag{1.13}$$
$$2 \times (Cl + O_3 \rightarrow ClO + O_2) \tag{1.14}$$

giving a net destruction of $2\,O_3 \rightarrow 3\,O_2$.

These gas phase reactions can occur anywhere in the stratosphere, however, the rates are not sufficiently fast to explain the large ozone hole that has been observed in the spring over Antarctica since the early 1980s. In the gas phase reactions, reactive chlorine species (Cl, ClO) can be removed from the ozone destruction cycle and transformed into non-reactive reservoir chlorine compounds (HOCl, $ClONO_2$) by reactions [1.8] and [1.9]. A rapid conversion of reservoir chlorine into reactive chlorine is necessary to explain the ozone hole over Antarctica. The mechanism for this rapid conversion is heterogeneous (gas–solid) reactions catalysed on the surface of polar stratospheric clouds (PSCs). PSCs are composed largely of condensed nitric acid, which also reduce atmospheric NO_2 concentrations. Low NO_2 concentrations extend the life of reactive chlorine species by reducing the importance of reaction [1.9] in the gas

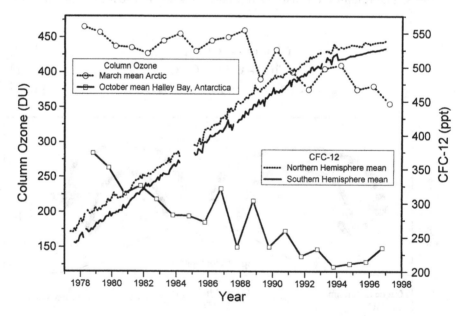

Figure 1.6. Comparison of the decrease in springtime stratospheric ozone over the Arctic and the Antarctic with the build-up of chlorofluorocarbon (CFC)12 in the northern and southern hemispheres. Natural atmospheric chlorine concentrations are relatively constant ($\cong 0.6$ p.p.b.v.) whereas anthropogenic sources have steadily increased since the introduction of CFCs. ppt, parts per trillion. (CFC data from Elkins, NOAA; Antarctic ozone from British Antarctic Survey; Arctic ozone from Environment Canada.)

phase. Surface-catalysed gas–solid reactions on PSCs causing the ozone hole are:

$$ClONO_{2(g)} + HCl_{(s)} \rightarrow Cl_{2(g)} + HNO_{3(s)} \qquad [1.15]$$
$$ClONO_{2(g)} + H_2O_{(s)} \rightarrow HOCl_{(g)} + HNO_{3(s)} \qquad [1.16]$$

where subscript g denotes gas and s denotes solid.

The strong polar vortex that surrounds Antarctica during the winter reduces atmospheric exchange with lower latitudes and allows the build-up of stratospheric Cl_2 and HOCl. With the increase of solar radiation in the spring, the build-up of Cl_2 and HOCl released from the gas–solid reactions is photolysed and reactive Cl atoms are released to drive the catalytic ozone destruction. The process continues until the PSCs dissipate later in the spring due to stratospheric warming. Spring also brings a weakening of the polar vortex, which allows air with higher ozone concentration to invade from lower latitudes together with the advection of air with low ozone air away from Antarctica. An ozone hole of similar magnitude is not common over the Arctic because of a weaker polar vortex and higher average stratospheric temperatures that reduce PSCs formation in the early spring when there is sufficient sunlight for catalytic ozone destruction to occur. However, climatic changes in response to global warming may induce cooler stratospheric temperatures producing more PSCs (Pawson & Naujokat, 1997) and a stronger polar vortex over the Arctic (Chubachi, 1997). These conditions were observed in the winter/spring of 1995–6 and 1996–7 (Fioletov *et al.*, 1997; Mühler *et al.*, 1997), but it is too soon to determine whether this is a developing pattern or a transient event.

Outside Antarctica, ozone reductions are less dramatic but are still significant (Stolarski *et al.*, 1992). The mechanism for ozone reduction at low and mid-latitudes is more equivocal than that for Antarctica (Pyle, 1997). Mass exchange of low ozone air across the polar vortex boundary could affect the global ozone budget. Heterogeneous gas–solid reactions might also occur outside the polar vortex. A ubiquitous layer of sulfate aerosols, partly of volcanic origin, may provide the necessary surfaces for the gas–solid reactions to occur (Hoffman & Solomon, 1989; Solomon *et al.*, 1996). Atmospheric measurements indicate that the increase sulfate aerosols after the 1991 eruption of Mt Pinatubo did correlate with higher levels of reactive chlorine species and lower levels of nitrogen oxides (McCormick, Thomason & Trepte, 1995). Both observations suggest that sulfate aerosols may be sites of heterogeneous reactions. However, the relative roles of chemical impacts of the sulfate layer and mass transport across the vortex boundary are still unresolved (Shepherd, 1997).

1.3.1.2 Effects on UV intensity and UV/PAR ratios

The absorbance cross-section of ozone increases by two orders of magnitude between 320 nm and the peak value at ~ 250 nm (Figure 1.4). A reduction in the ozone encountered by incoming solar radiation, either through ozone loss or reduced solar zenith angles, would first of all result in increased UV-B radiation (Madronich, 1991; Kerr & McElroy, 1993) (Figure 1.7a,b). In principle, UV-C would also increase, but absorption in this region is so efficient that a tremendous reduction in ozone is required in order for significant amounts to reach the earth's surface. As can be seen in Figure 1.4, UV-A is not affected by ozone absorption and PAR absorption is negligible compared to UV-B. The reduction of atmospheric O_3 causes two significant changes to ground-level UV-B, namely a shift in the spectrum towards shorter, more energetic wavelengths and an increase in intensity of the whole UV-B band (Figure 1.7a,b). Although the increase in integrated UV fluence due to ozone depletion in Antarctica (Figure 1.7b) is less than a factor of 5, the change in the ratio of low ozone irradiance to high ozone irradiance increases exponentially with decreasing UV-B wavelengths (Figure 1.7a). The shift in spectrum and the lack of significant effects in the UV-A and PAR regions also produces a change in the ratios of UV-B to UV-A and PAR. The change in the ratio of UV-B to PAR has been shown to be well correlated with ozone levels (Smith *et al.*, 1992). These changes in ratios are important to biological processes, as damage is usually associated with the shorter more energetic wavelengths whereas repair and photosynthesis require longer wavelengths (Cullen, Neale & Lesser, 1992).

1.3.1.3 Effects of clouds and particles

Atmospheric clarity and cloud cover can have a significant effect on UV radiation at ground level (Booth *et al.*, 1997). Atmospheric radiation modelling has shown that stratus clouds can reduce erythema-weighted UV radiation by up to 75% compared to clear sky conditions (350 DU O_3 and a solar zenith angle of 55°) (Tsay & Stamnes, 1992). This model also showed that cirrus clouds, stratospheric and tropospheric particles attenuated UV-B by 12%, 6%, and 5%, respectively. When ozone levels were reduced to 260 DU, stratus clouds still reduced UV-B levels to 56% less than in clear sky, 350 DU O_3 conditions. Under partly cloudy conditions, however, cumulus clouds have been shown to enhance total sky UV-B irradiance by up to 30% over short periods (Mimms &

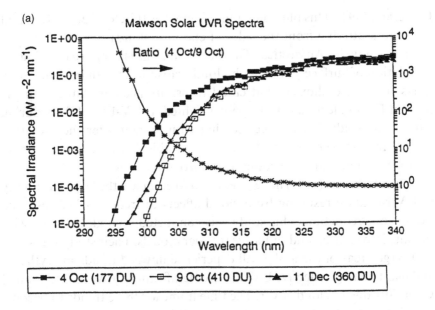

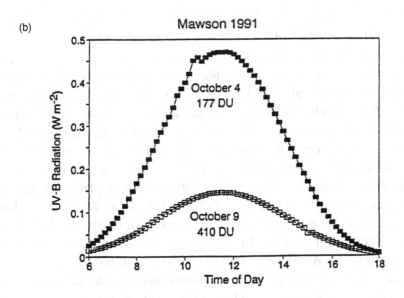

Figure 1.7. (a) The influence of the Antarctic ozone hole on the spectrum of solar radiation reaching the earth's surface at Mawson Station. Spectra are shown for the same solar angle but with varying amounts of column O_3. The loss of O_3 allows shorter, more energetic wavelengths to penetrate the atmosphere altering the ratios of UV-B to UV-A and to PAR. (b) The effect of the Antarctic ozone hole on the energy of UV-B radiation impacting marine environments. An approximate five-fold increase in UV-B is shown for a day under the ozone hole (Roy *et al.*, 1994).

Frederick, 1994). This phenomenon is known as the cloud edge effect and is due to scattering from the sides of cumulus clouds.

In a study in Antarctica, Gautier *et al.* (1994) reported a higher correlation of surface UV-B with cloud transmittance than with ozone levels. They also showed that cloud transmittance correlated with the ratio of DNA-effective UV (heavily weighted in UV-B) to UV-A during part of the study. This suggested that cloud might alter the radiation spectrum; however, the authors believed this effect would be small compared to alterations produced by ozone reductions.

Lubin & Jensen (1995) have attempted to compare the large variability in UV radiation resulting from cloud affects to trends in UV resulting from ozone reduction. This study was conducted on a global scale with satellite information and radiative transfer models. Their study suggested that large areas of the globe will experience upward trends in erythema and plant damage-effective UV that are significant relative to interannual variability due to cloud cover. The time frame for these trends to become significant over the background variability is 10 to 100 years, depending on location, from the onset of ozone reduction (*circa* 1980).

1.3.1.4 Season and latitude

Season and latitude play an important role in the attenuation of solar radiation by controlling solar zenith angles and thereby the thickness of the air column through which radiation must pass before reaching the ground (Madronich, 1993). For example, a change in solar zenith angle from 55° to 75° implies a doubling of air mass. Solar radiation, including UV-B, is consequently highest and least variable in the tropics and decreases at higher latitudes (Figure 1.8). The combination of higher solar angles and the global distribution of ozone (excluding depletion) enhances the attenuation and the truncation of shorter UV-B wavelengths such that latitudes above 55° never experience the intensity or integrated daily dose common between 0° and 30°. Indeed, even under severe ozone depletion, UV irradiance in Antarctic is less than that prevailing at the equator. However, high latitude ecological systems have developed under less intense UV regimes and may not be readily adaptable to large increases in intensity or shifts in the spectrum (Weiler & Penhale, 1994).

1.3.2 Attenuation in the water column

1.3.2.1 Surface reflection

The attenuation of solar radiation by the water column begins when light strikes the water surface. The penetration of light into the water column

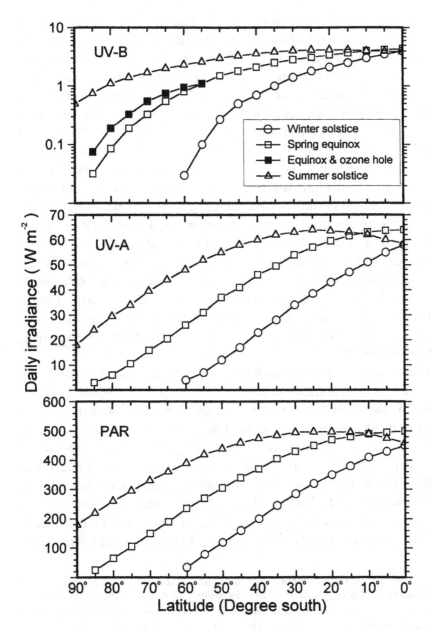

Figure 1.8. The influence of latitude and season on the solar energy at the earth's surface. Highest total energy and highest UV energy are at the equator, but the largest seasonal variability occurs at high latitudes. PAR, photosynthetically available radiation. (Adapted from Holm-Hansen *et al.*, 1993.)

can be diminished by reflection at the surface. For direct solar radiation and a smooth water surface, the percentage of light reflected can be calculated with the angle of the incident light (90° − solar elevation) and the relative refractive index of water. The computed reflectance as a percentage of incoming radiation for such conditions is given in Table 1.1. However, due to atmospheric effects, the solar radiation at the sea surface is composed of both direct and diffuse (i.e. sky) radiation. Theoretical calculations indicate that if all the radiation reaching the water surface were diffuse radiation, then approximately 6.6% would be reflected (Jerlov, 1976). Actual measurements (including a small contribution due to backscatter out of the water) show that diffuse radiation is reflected to an extent between 6% and 11% (Neumann & Pierson, 1966). Diffuse radiation is present to some extent in all solar radiation at ground level due to atmospheric scattering and its effect on reflection can be seen even under clear sky conditions (Table 1.1; Campbell & Aarup, 1989). On the other hand, there is still a direct component of solar radiation under heavily overcast skies and thus some variations in reflection due to changing solar elevations can be expected for all sky conditions (Neumann & Pierson, 1966).

Thus far, the assumption has been made that the sea surface is perfectly smooth, but observations demonstrate that this is seldom the case. Wind-roughened surfaces influence reflection because of changes in the angle of incidence for incoming light due to waves and changes in the nature of the air–sea interface through the production of bubbles and white-caps. Higher wind speeds tend to increase reflectance of direct solar radiation from high solar elevations, but the effect is minimal (Jerlov, 1976; Preisendorf & Mobley, 1986). At low solar elevations ($< 20°$), however, increasing wind speed significantly reduces direct reflectance by producing lower average incidence angles (Figure 1.9). The reflectance of

Table 1.1. Percentage reflectance of direct and global radiation from a smooth water surface as a function of solar elevations (h), angle of incidence $= 90° - h$

Solar elevation (h)	90°	60°	50°	40°	30°	20°	10°	5°
% Reflectance of direct radiation	2.0	2.1	2.4	3.4	5.9	13.3	34.9	58.3
% Reflectance of global radiation (direct + diffuse) clear sky	3	3	3	4	6	12	27	42

Data from Jerlov, 1976.

diffuse radiation is also reduced with increasing wind speed from 6.6% at $0\,\mathrm{m\,s^{-1}}$ to 4.7% at $20\,\mathrm{m\,s^{-1}}$ (Preisendorf & Mobley, 1986). White-caps and surface bubbles tend to increase all reflectance, but their lifetimes are relatively short and their influence is minor under most conditions (Kirk, 1994a).

The influence of air–sea transmittance on the spectral quality of underwater light is even more equivocal and relatively little work has been done in this area. The percentage of total (global) radiation that is composed of sky (diffuse) radiation is influenced by solar elevation, atmospheric clarity and cloud cover. On average as a manifestation of the inverse relationship between scattering and wavelength, the violet end of the spectrum contains a higher percentage of sky radiation than the red end of the spectrum. A wavelength dependence for reflectance of solar radiation could therefore be expected, particularly at low solar elevations, due to differences in the reflectance of direct and diffuse light (Jerlov, 1976). UV at ground level exhibits a high ratio of diffuse to direct radiation regardless of sky conditions and thus reflection of these wavelengths is relatively invariant with changes in solar angles (Jerlov,

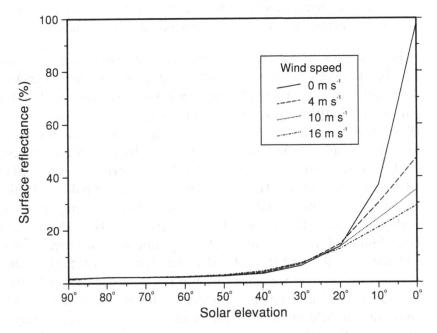

Figure 1.9. Percentage reflectance of direct radiation as a function of solar elevation for various wind speeds. (Data from Austin, 1974.)

1976). Although this differential reflection of wavelengths could influence the spectral quality of underwater light, especially at high latitudes with lower average sun elevations, the extent of its influence is difficult to determine and is probably negligible in comparison with spectral variability in incident solar radiation (Campbell & Aarup, 1989).

1.3.2.2 Scattering and absorption

Once solar radiation has entered the water column, it is subject only to two physical processes: scattering and absorption. These two processes are dependent on the optical properties of the water column, which are functions of the scattering and absorption by water itself and that of particles and dissolved substances in the water. Optical properties can be divided into two classes as follows (Preisendorf, 1976):

1. Inherent properties are independent of the incident light field, but are spectrally dependent. They include index of refraction, coefficients for attenuation, absorption and scattering, and the volume scattering function.
2. Apparent properties such as irradiance attenuation depend on the inherent properties and the radiance distribution of the light field. Consequently irradiance attenuation depends on numerous factors and extrapolation among water bodies is not unambiguous.

For radiation in the solar spectrum, irradiance attenuation by water itself is little affected by the dissolved inorganic salts or gases in seawater. Therefore, the inherent optical properties of pure seawater itself will not vary with salinity but are slightly influenced by temperature and pressure. (Pure seawater refers to seawater containing all the common inorganic salts but lacking any suspended particles or dissolved organic constituents.) Both pure water and pure seawater have high absorptivity at the red and IR end of the spectrum. However, contrary to earlier published values, recent work has shown that the attenuation of UV wavelengths by water itself is very low, being about $0.01\,\mathrm{m^{-1}}$ (Kirk, 1994a). Therefore, the contribution of irradiance attenuation by pure seawater itself serves to shift the spectral distribution of underwater light towards the blue end of the spectrum relative to the solar spectrum (for more details, see Chapter 2).

The variation in irradiance attenuation for seawater experiencing a constant light field can thus be attributed to variations in scattering and absorption (i.e. inherent optical properties) due to dissolved organic constituents together with living and non-living particles (Figure 1.10). Reviews of aquatic radiative transfer are available (Preisendorf, 1976;

Mobley, 1994; Kirk, 1994a). As in the atmosphere, scattering can influence the rate at which photons are absorbed, but a scattered photon is still able to interact within the aquatic ecosystem unless it is backscattered out of the water column. Although, by definition, attenuation considers a scattered photon to be lost from the main beam, only absorption actually removes a photon from the system. From an irradiance measurement standpoint, the scalar irradiance (E_o) will always be higher than downwelling irradiance (E_d) due to the influence of scattering. The most relevant optical parameter for the evaluation of vertical penetration of radiation into the water column is the vertical attenuation coefficient (K_d). If the water column is optically homogeneous, downwelling irradiance can be related to K_d by:

$$E_{d\lambda}(z) = E_{d\lambda}(0)e^{-K_{d\lambda}z} \tag{1.3}$$

where $E_{d\lambda}(z)$ is downwelling irradiance at λ at depth z, $E_{d\lambda}(0)$ is downwelling irradiance at λ just below the surface and $K_{d\lambda}$ is vertical attenuation coefficient of λ at z in m^{-1}. The value of K_d is a result of all the

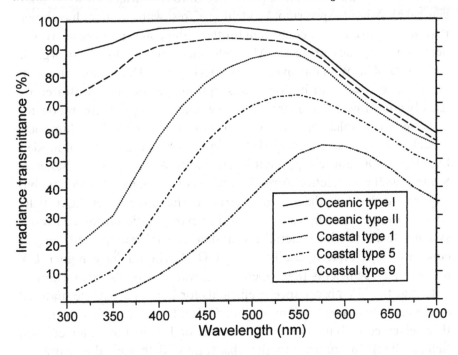

Figure 1.10. Percentage transmittance per metre for downward irradiance in various types of marine waters as classified by Jerlov (1976). Differences in transmittance can be attributed to variations in dissolved and suspended material in the various water types. (Data from Jerlov, 1976.)

radiative transfer processes in the water column and is determined mainly by the inherent optical properties (Kirk, 1994a). By using field analyses and computer modelling, Kirk (1994a) has defined an empirical relationship for the value of K_d in natural waters:

$$K_d = 1/\mu_o(a^2 + G(\mu_o)ab)^{1/2} \qquad (1.4)$$

where μ_o is the cosine of the refracted solar beam just below the surface, a is the absorption coefficient, b is the scattering coefficient and $G(\mu_o)$ is a coefficient whose value is determined by the shape of the scattering phase function, which specifies the angular distribution of scattering.

The UV absorbance coefficient for natural waters on has been shown to be highly dependent on the concentration of dissolved organic material (DOM), also known as humic material, yellow substance, gelbstoff, and chromophoric DOM (Smith & Baker, 1979; Kirk, 1994b). The characteristic absorbance spectra for DOM are very low at the red end of the spectrum and rise exponentially with decreasing wavelength in the UV. Particulate material with humic substances adsorbed onto the surface show the same general absorbance spectrum. Modelling of freshwater systems has shown that the UV absorbance coefficient for the water column can be fairly well determined by measuring DOM levels (measured as dissolved organic carbon (DOC)) (Williamson *et al.*, 1996). However, DOC concentrations and their relative contribution to absorption coefficients in seawater are usually much lower than in freshwater systems and a predictive model for seawater based solely on DOC concentrations is not feasible. Numerical models for radiative transfer (Mobley *et al.*, 1993) and empirical models based on DOC and chlorophyll levels (Smith & Baker, 1981; Baker & Smith, 1982) are effective at predicting average optical parameters, but require incident irradiance to determine the underwater light field. Examination of the empirical models and extrapolation from freshwater systems leads to the generalisation that coastal waters with higher concentrations of terrestrially derived DOC should have higher UV absorbance coefficients than open ocean water. It is also important to note that the UV absorbance coefficients for living particulate material can be equal to or greater than that of DOC and, therefore, phytoplankton themselves contribute to the attenuation of UV in the water column (self-shading). In contrast to the absorption coefficient, the scattering contribution to K_d can be reasonably estimated for both fresh and seawater from turbidity measurements (Kirk, 1985).

Once K_d has been measured or estimated for a particular water body, it can be used with equation (1.3) to estimate the E_d at any depth. The value of K_d is a convenient parameter for the comparison of UV penetration in different water bodies, but is difficult to visualise. Alternatively, the depth of water at which 10% of surface UV remains can be calculated ($2.3/K_d$) and provides a more easily visualised parameter for comparing water bodies (Figure 1.11). Because K_d is dependent upon wavelength, the spectral distribution of underwater light changes with depth. This is considered in more detail in Chapter 2. In general, the UV end of the spectrum is attenuated much faster than PAR, especially in coastal waters with high DOC. As a result, UV irradiance will often fall to immeasurable intensities at much shallower depths than PAR. Physical features, such as the rate and depth of vertical mixing in the water column, can thus influence the light regimes to which planktonic organisms are exposed, together with the distribution and concentration of photochemical reactants and products.

1.4 Effect of light on biomolecules and processes

Given that a photon's journey into the water column is not unimpeded, what reactions/responses are provoked due to the energy input of a photon? Two basic principles that apply to resultant processes are:

1. Only a photon that is absorbed by a substance can produce an effect.
2. Each photon absorbed by a substance can activate only one molecule.

The first principle is intuitive in that the energy of the photon must be transferred to the substance for an effect to occur. The second principle requires that a photon be removed from the system after absorption. However, it does not imply that every absorbed photon will produce a change. For example, the energy transferred to a biomolecule by a photon may be deactivated through non-radiative internal conversions returning the molecule to its original state without causing a change in the biosystem. Photochemical processes are quantified by the quantum yield ($\boldsymbol{\Phi}$):

$$\boldsymbol{\Phi} = \frac{\text{moles of substance reacted or produced}}{\text{moles of photons absorbed}}$$

Quantum yields can be defined for any type of photochemical process, such as phosphorescence, fluorescence or direct photolysis. Quantum yields for these primary photochemical processes have a maximum of 1 (i.e. 100% of the molecules absorbing a photon will produce the same reaction). Likewise, the maximum quantum yield for direct photochemical production is 2 if the original compound is photolysed into identical products (i.e. Cl_2). In other words, the absorption of one photon can cleave only one bond, thus producing a maximum of two products. However, secondary photochemical processes can have much larger quantum yields. For example, the direct photolysis of molecular chlorine can produce two chlorine atoms ($\Phi_{max\ for\ Cl\ production} = 2$):

$$Cl_2 + hv \rightarrow Cl + Cl$$

However, the released atomic chlorine can then react with another molecule, such as ozone, to start a catalytic chain reaction. Thus, this secondary photochemical effect, namely the destruction of ozone, can have a quantum yield much greater than unity. Unfortunately, quantum yields for photochemical processes are not readily derived from chemical structures and must be determined experimentally.

1.4.1   Bond energies

The possible consequences of an absorbed photon can be discerned by comparing the energy of a photon mole for various wavelengths to the dissociation energy of various chemical bonds (Figure 1.2). This figure clearly demonstrates that the dissociation of O_2 in the atmosphere requires $\lambda < 240$ nm. In addition, the figure also shows that UV wavelength

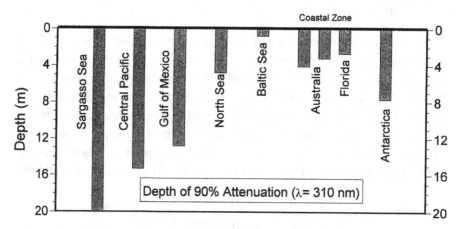

Figure 1.11.   Depth of 90% attenuation ($2.3/K_d$) of UV (305 nm) radiation for various water bodies. Large differences in attenuation can be attributed to the varying inherent optical properties of surface waters. (Data from Kirk, 1994a.)

energy is required to break bonds commonly found in many biomolecules. However, for a molecular bond to be broken the substance containing that bond must absorb light at the appropriate wavelength. This can be clearly seen in the case of CFCs. Although wavelengths into the UV-A region have sufficient energy to cleave a chlorine atom from a CFC (Figure 1.2), CFCs display no absorption for wavelengths longer than 290 nm (Molina & Rowland, 1974). In fact, absorption in CFCs falls rapidly for wavelengths longer than 200 nm and thus CFC photolysis in the atmosphere is limited to altitudes above 20 km, well into the stratosphere.

1.4.2 Production of radicals

A substance may absorb light of insufficient energy to cause dissociation of any molecular bonds. In this case, the energy absorbed activates a molecule to an excited state. If the energy is sufficient, as is more likely the case with UV wavelengths, then the absorbed energy can cause the transition of an electron from one molecular orbital to a higher energy orbital (Figure 1.12). The electronically excited molecule then can return to the ground state via one of four mechanisms:

1. Through internal conversion (i.e. non-radiative transfer).
2. Emission of a photon (fluorescence).
3. Intersystem crossing followed by phosphorescence or non-radiative transfer.
4. By interaction with another molecule. Photoinduced change in biological systems is usually determined by the quantum yield of this mechanism.

In photosynthetic organisms or aquatic environments where oxygen concentrations are high, excited molecules usually interact with another molecule via one of two photodynamic effects. Both mechanisms cause photo-oxidative stress to the organisms and are capable of initiating damage to most cellular constituents. In type I processes, the excited or photosensitised molecule ($P*$) reacts with a substrate molecule (S) to produce two free radicals:

$$P* + S \rightarrow P^{-\cdot} + S^{+\cdot}$$

These radicals can react with O_2 to produce a repaired P, an oxidised S and the superoxide anion:

$$P^{-\cdot} + O_2 \rightarrow P + O_2^{-\cdot}$$

$$S^{+\cdot} + O_2 \rightarrow S^{2+} + O_2^{-\cdot}$$

The superoxide radical produced is highly reactive with many biomolecules.

In type II mechanisms, the photosensitised molecule collides with ground state triplet O_2 to produce singlet O_2:

$$^3P_1 + {}^3O_2 \rightarrow {}^1P_0 + {}^1O_2$$

or

$$^1P_1 + {}^3O_2 \rightarrow {}^3P_1 + {}^1O_2$$

O_2 is unusual in that its triplet state is of lower energy than the singlet state, and thus singlet O_2 is more reactive than the ground state triplet. The highly reactive singlet O_2 can subsequently then react with other molecules by chemically combining (chemical quenching) or transferring the excitation energy (physical quenching).

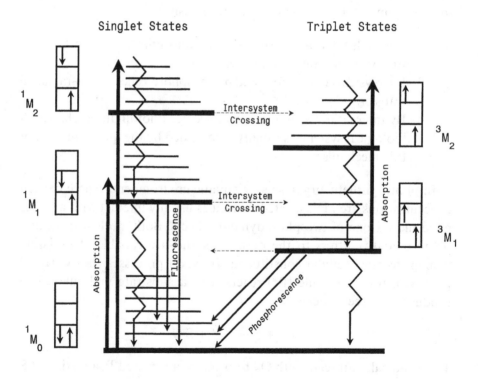

Figure 1.12. Jaborski diagram of molecular excitement due to the absorption of a photon and the various pathways for returning to ground state.

1.4.3 UV in photosynthesis

Absorption of solar radiation by chlorophyll and accessory photosynthetic pigments is the first step in photosynthetic carbon fixation. An extensive review of light reactions in photosynthesis can be found in Kirk (1994a) and Falkowski & Raven (1997). The energy absorbed enables photosynthesis to move electrons through a redox span of 1.25 V from water to reduced carbon. This reduction of CO_2 can be summarised by:

$$2H_2O + hv \rightarrow O_2 + 4H^+ + 4e^-$$
$$CO_2 + 4H^+ + 4e^- \rightarrow (CH_2O) + H_2O$$

The movement of electrons through the system involves two photochemical reactions, one each in photosystems (PS) I and II. As explained below, the fixation of one mole of carbon dioxide involves the capture of at least 8 mole photons, 4 mole photons at each photosystem. The energy required for the reduction of one mole of carbon dioxide is 482 kJ. The energy of 4 mole photons of 700 nm light is more than sufficient to initiate the reaction $4 \times 171 = 684$ kJ. As the quantum yield of photosynthesis seldom exceeds 0.36 the minimum energy available from 700 nm light is $171 \times 0.36 = 61.56$ kJ. Thus, the minimum mole photon required for photosynthesis is $684/61.56 = 8$. Absorption of higher energy, shorter wavelength light does not lower this minimum. Shorter wavelength light can move chlorophyll to higher excited states (second excited state) than does long wavelength light, but the second excited state quickly loses its excess energy as heat and returns to the first excited state from which the primary reaction of photosynthesis proceeds.

If each chlorophyll molecule acted independently, a significant amount of time would be required to absorb 8 mole photons and so power photosynthesis. Biophysical research has led to the concept of photosynthetic units consisting of 200–300 light-harvesting antenna chlorophyll bound in pigment–protein complexes along with other accessory pigments and a reaction centre chlorophyll. The antenna chlorophyll are able to transfer energy from absorbed photons to the reaction centre chlorophyll, which is then photo-oxidized to supply electrons for photosynthesis. The energy absorbed by an antenna chlorophyll passes efficiently to the reaction centre by isoenergic radiationless transfer or inductive resonance. The donor in the singlet excited state passes its energy to the acceptor. The acceptor moves to the singlet excited state and the donor returns to ground state. This process requires considerable overlap in the emission and absorbance spectra of the donor and acceptor, respectively. It also requires that the donor and acceptor be in close proximity.

The reaction centres of PS I and PS II are conceptually connected in the Z scheme of photosynthesis (Figure 1.13). PS I and PS II act independently, as exemplified by following an electron pathway through the system. PS II contains chlorophyll with a maximum absorbance at 680 nm. Upon absorption of light energy, P680 is photo-oxidized to P680* and loses an electron to the primary electron acceptor, phaeophytin. This step raises the oxidation–reduction potential of the electron about 2.0 V. The electron is passed down the electron transport chain by plastoquinone, cytochrome complex, and plastocyanin. Plastocyanin can then donate the electron to PS I. The energy released through the electron transport chain between PS I and PS II allows the production of ATP from ADP (photophosphorylation).

After donating the electron, P680* needs to be reduced back to P680 before it can absorb another photon to repeat the cycle. The electron to reduce P680* comes from water, but P680* is not a sufficiently powerful oxidant to remove the electron. Thus, P680* is reduced by an intermediate species known as Z, which is believed to be an amino acid within a chlorophyll protein. In turn, the electron given up by Z is replaced by an electron from the oxidation of water. The water-oxidation reaction produces four electrons, but the PS II electron chain can utilise only one electron at a time. The problem is believed to be overcome by a type of charge accumulator in the PS II reaction centre called S states.

The charge accumulator begins at S_0. After a photon causes the photo-oxidation of P680, S advances one step to S_1 by supplying the replacement electron. After a second photon has been absorbed, the S_1 state donates an electron and moves to S_2. The third photon causes the movement S_2 to S_3. After the fourth photon, S_4 has released four electrons and is a sufficiently strong oxidant to complete the water oxidation reaction. The four electrons removed to form O_2 and $4H^+$ from water are used to reset S_4 to S_0 and the cycle begins again. The exact chemical nature of the charge accumulator is not known, but is believed to be a protein containing four manganese atoms. It has been observed that oxygen evolution is not possible without four manganese atoms for every P680 molecule in PS II. Additionally, chloride and calcium ions are essential in the function of the water oxidation cycle, but their exact roles are not clear.

At the primary PS I chlorophyll, absorbed light energy photo-oxidizes P700 to P700*, releasing an electron to the acceptor A_0, and again raises the oxidation–reduction potential of the electron. From X, the electron may be transferred to ferredoxin and can then be used for the production of NADPH from $NADP^+$. NADPH is transferred to the Calvin cycle where reduction of CO_2 occurs. Alternatively, the electron from X can be passed into the electron chain between PS II and PS I, resulting in cyclic electron flow. Thus, the electron necessary to reduce P700* back to P700 can come from cyclic transport or PS II.

Although solar radiation is separated into different regions by wavelength, these distinctions are arbitrarily defined and in themselves have no photobiological relevance. From the standpoint of photosynthesis, which depends on solar radiation to function, it is perhaps more realistic to consider the solar spectrum at the earth's surface as a continuum from $\sim 290\,nm$ to $800\,nm$. The action spectra of photosynthesis shows response throughout this range and some studies have shown that light down to at

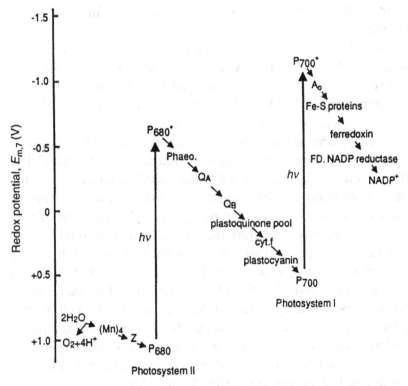

Figure 1.13. Generalised Z scheme of photosynthesis showing the connected but independent PS I and II. Electrons for the system are derived from water oxidation at PS II that also produces the oxygen evolved from photosynthesis. The eventual reduction of CO_2 is carried out in conjunction with the NADPH supplied to the Calvin cycle. Phaeo, phaeophytin; Q_A, Q_B, plastoquinone-binding proteins; cyt., cytochrome; A_0, acceptor of electrons from P700; FD, ferredoxin; $E_{m,7}$, midpoint redox potential scale at pH7, such that the $E_{m,7}$ for the oxidation of water by a one-electron reaction is $+0.8\,V$. (From Kirk, 1994a.)

least 300 nm can be utilised for photosynthesis (for references, see Holm-Hansen *et al.*, 1993). Furthermore, accessory pigments are able to absorb light at various wavelengths and funnel the captured energy to chlorophyll for photosynthesis. The question of whether light of any wavelength is beneficial or detrimental is one of cost/benefit. If the damage cost of absorbing light of a particular wavelength or region outweighs the benefit it provides in terms of fuel for photosynthesis or other metabolic functions, then it will be not be useful to the organism. This might be manifested as a threshold value above which a wavelength or region has deleterious effects. The threshold value can be defined in terms of total dose or dose rate. Helbing *et al.* (1992) posed this question for UVR intensity (dose rate) effects on photoinhibition for Antarctic phytoplankton. Their results suggested that there is a threshold dose rate of between 5 and 10 W m^{-1} for UVR (down to 305 nm) before the onset of photoinhibition.

Whereas UV damage to photosystems appears mainly to affect PS II, PS I seems much less sensitive (Vincent & Roy, 1993). In addition to disruption of photosynthesis, the UV-induced degradation of PS II affects cellular concentrations of ATP resulting in changes in metabolic regulation (Holm-Hansen *et al.*, 1993). The primary target for UV-B damage to PS II has not been determined and both the acceptor and donor sides of electron transport have been suggested (Barbato *et al.*, 1995). The acceptor side mechanisms are not well understood, but may involve the overreduction of quinone acceptors in conjunction with the production of singlet oxygen that attacks reaction centre proteins (Vassiliev *et al.*, 1994). Donor side damage may result from the production of abnormally stable oxidising radicals due to impairment of electron transport from the manganese cluster (Renger *et al.*, 1989). These authors concluded that the primary damage was to the function of water oxidation. The deterioration was assumed to be caused by structural changes to the D1/D2 polypeptide matrix, but a specific target was not located. Barbato *et al.* (1995) detected degradation products of the D1 protein after UV-B irradiation and concluded that the donor side in the presence of manganese from the water oxidation system was responsible for the cleavage.

In a study of coastal phytoplankton communities, Vassiliev *et al.* (1994) found decreases in PS II photochemical energy conversion efficiency with increased PAR; however, no significant differences were found when samples from coastal waters were treated with enhanced UV-B or screened to exclude UV-B. Results from open ocean communities showed that treatments which excluded UV-A and/or UV-B enhanced PS II efficiency. They concluded that for both communities oversaturation by PAR was an overriding factor for reduction in PS II efficiency and damage to the reaction centres was reversible and appeared to be completely repaired overnight. However, repair of reaction centres requires substantial rates of proteins synthesis (Vincent & Roy, 1993), which may be compromised by UV stress.

As PS I appears to be relatively insensitive to UV radiation, most of the repair resources will be used at PS II. Although UV photoinhibition at PS II is thought to be mediated by a photosensitiser different from that in PAR photoinhibition, the repair process is probably the same. Damaged reaction centres are built *de novo* in a process that requires several hours. The repair process involves the resynthesis of proteins, which require considerable transcription of nucleic acids. Concurrent damage to DNA and RNA by UVR may thus interfere with the rebuilding process.

1.4.4 UV in biogeochemical cycles

Biogeochemical cycles refer to the complex interactions of biological, chemical and physical processes that determine the flux of material and energy within the environment. Global biogeochemical cycles involve the mechanisms that serve to transport and transform material within and among the terrestrial, atmospheric and oceanic pools. As such, the effects of increased UV-B radiation on one pool may provoke perturbations in the other pools (Zepp *et al.*, 1995). For example, it has been suggested that the diminution of marine primary productivity due to increased UV-B may reduce the ocean's ability to sequester carbon via the 'biological pump' (Häder & Worrest, 1991; UNEP, 1991). This would in turn moderate the oceanic uptake of atmospheric CO_2 and thus diminish the ocean's role as a sink for increased anthropogenic CO_2 emissions. It should be noted, however, that the factors governing control of biogeochemical fluxes are complex, and simple extrapolations of cause and effect from one part of the cycle are rarely sufficient (Cullen & Neale, 1994).

UVR may also have direct effects on marine biogeochemical cycles, mainly through interactions with DOM (Miller, 1994). Direct photodegradation of DOM has been shown to produce CO, CO_2, OCS and other low molecular weight organic compounds (Kieber, McDaniel & Mopper, 1989; Valentine & Zepp, 1993; Miller & Zepp, 1995). In addition, the photobleaching of DOM may allow deeper penetration of UV-B into the water column. The transformation of relatively recalcitrant DOM into more labile low molecular weight compounds may serve to increase food availability to bacterioplankton, which constitute a major component of the marine microbial food web (for a review, see Moran & Zepp, 1997). Whether an increased food source can outweigh the direct detrimental effects of UV-B on bacterioplankton and increase bacterial production is still unresolved (Arrigo & Brown, 1996).

Other marine photochemical processes involving UV-B (see Chapters 4 and 5) which have important consequences for biogeochemical cycles are metal redox reactions (e.g. Faust, 1994) and the production of reactive species such as hydroxyl radical, superoxide anion, hydrogen peroxide and singlet oxygen (for reviews, see Blough & Zepp, 1995; Blough, 1997). These reactive oxygen species are closely linked to the interactions with DOM and sunlight. They are potent oxidants that can play a role in a variety of chemical reactions as well as increasing the extracellular oxidative stress to organisms (Dahl, 1994). In addition to the influence on redox reactants, solar radiation can directly affect metal species through photoreduction. The oxidation state of metals in aquatic environments has a direct influence on their environmental mobility and bioavailability and toxicity to organisms. Primary production in several areas of the ocean is thought to be limited owing to low levels of bioavailable iron and manganese. Photoreduction of these metals transforms colloidal oxide–hydroxides into more labile (and bioavailable) dissolved species, and may therefore stimulate phytoplankton productivity (Wells *et al.*, 1991).

1.5   Summary

Due to stratospheric ozone depletion caused by anthropogenic activity, ground level UV-B radiation is increasing to levels that are significantly higher than background variability for many parts of the globe. This presents a new problem for the marine environment, the ramifications of which are only now being explored. Detrimental effects at the molecular and cellular levels have been clearly demonstrated for some marine organisms, but the resulting changes in community structures, ecosystems or biogeochemical cycles still remain poorly defined. The ability to predict these changes will be based on the understanding of a wide variety of physical, chemical and biological phenomena. As such, multidisciplinary studies with inputs ranging from physics to biochemistry are needed to answer the questions. Subsequent chapters explore in detail photochemical and photobiological processes, and examine the effect of UVR on organisms at different trophic levels.

References

Arrigo, K. R. & Brown, C. W. (1996). Impact of chromophoric dissolved organic matter on UV inhibition of primary productivity in the sea. *Marine Ecology Progress Series*, **140**, 207–16.

Austin, J., Butchart, N. & Shine, K. P. (1992). Possibility of an arctic ozone hole in a doubled-CO_2 climate. *Nature*, **360**, 221–2.

Austin, R. (1974). The remote sensing of spectral radiance from below the ocean surface. In *Optical Aspects of Oceanography*, ed. N. Jerlov & E. Nielsen, pp. 317–44. Academic Press, London.

Baker, K. S. & Smith, R. C. (1982). Bio-optical classification and model of natural waters. 2. *Limnology and Oceanography*, **27**, 500–9.

Barbato, R., Frizzo, A., Friso, G., Rigoni, F. & Giacometti, G. M. (1995). Degradation of the D1 protein of photosystem-II reaction centre by ultraviolet-B radiation requires the presence of functional manganese on the donor side. *European Journal of Biochemistry*, **227**, 723–9.

Blough, N. (1997). Photochemistry in the sea-surface microlayer. In *The Sea Surface and Global Change*, ed. P. Liss & R. Duce, pp. 383–424. Cambridge University Press, Cambridge.

Blough, N. V. & Zepp, R. (1995). Reactive oxygen species in natural waters. In *Active Oxygen in Chemistry*, ed. C. S. Foote, J. S. Valentine, A. Greenberg & J. F. Liebman, pp. 280–333. Chapman & Hall, New York.

Booth, C., Morrow, J., Coohill, T., Cullen, J. J., Frederick, J., Häder, D.-P., Holm-Hansen, O., Jeffery, W., Mitchell, D., Neale, P. J., Sobolev, I., Leun, J. v. d. & Worrest, C. (1997). Impacts of solar UVR on aquatic microorganisms. *Photochemistry and Photobiology*, **65**, 252–69.

Bothwell, M. L., Sherbot, D. M. J. & Pollock, C. M. (1994). Ecosystem response to solar ultraviolet radiation: influence of trophic-level interactions. *Science*, **265**, 97–100.

Bouwman, A. (1998). Nitrogen oxides and tropical agriculture. *Nature*, **392**, 866–7.

Calkins, J. (ed.) (1982). *The Role of Solar Ultraviolet Radiation in Marine Ecosystems*. Plenum Press, New York.

Campbell, J. & Aarup, T. (1989). Photosynthetically available radiation at high latitudes. *Limnology and Oceanography*, **34**, 1490–9.

Chubachi, S. (1997). Annual variation of total ozone at Syowa Station, Antarctica. *Journal of Geophysical Research*, **102**, 1349–54.

Crutzen, P. J. (1992). Ultraviolet on the increase. *Nature*, **356**, 140–5.

Cullen, J. J. & Neale, P. J. (1994). Ultraviolet radiation, ozone depletion, and marine photosynthesis. *Photosynthesis Research*, **39**, 303–20.

Cullen, J. J., Neale, P. J. & Lesser, M. (1992). Biological weighting function for the inhibition of phytoplankton photosynthesis by ultraviolet radiation. *Science*, **258**, 646–50.

Dahl, T. A. (1994). Examining the role of singlet oxygen in photosensitized cytotoxicity. In *Aquatic and Surface Photochemistry*, ACS Symposium Series, ed. G. Helz, R. G. Zepp & D. Crosby, pp. 241–57. Lewis Publishers, Boca Raton, FL.

Falkowski, P. G. & Raven, J. (1997). *Aquatic Photosynthesis*. Blackwell Science, Malden.

Faust, B. C. (1994). A review of the photochemical redox reactions of iron(III) species in the atmospheric, oceanic, and surface waters: influences on geochemical cycles and oxidant formation. In *Aquatic and Surface Photochemistry*, ACS Symposium Series, ed. G. R. Helz, R. G. Zepp & D. G. Crosby, pp. 3–38. Lewis Publishers, Boca Raton, FL.

Fioletov, V. E., Kerr, J. B., Wardle, D. I., Davies, J., Hare, E. W., McElroy, C. T. & Tarasick, D. W. (1997). Long-term ozone decline over the Canadian Artic to early 1997 from ground-based and balloon observations. *Geophysical Research Letters*, **24**, 2705–8.

Gautier, C., He, Y., Yang, S. & Lubin, D. (1994). Role of clouds and ozone on spectral ultraviolet-B radiation and biologically active UV dose over Antarctica. In *Ultraviolet Radiation in Antarctica: Measurements and Biological Effects*, Antarctic Research Series, ed. C. S. Weiler & P. A. Penhale, pp. 83–91. American Geophysical Union, Washington, DC.

Häder, D.-P. & Worrest, R. (1991). Effects of enhanced solar ultraviolet radiation on aquatic ecosystems. *Photochemistry and Photobiology*, **53**, 717–25.

Helbing, E. W., Villafañe, V., Ferrario, M. & Holm-Hansen, O. (1992). Impact of natural ultraviolet radiation on rates of photosynthesis and on specific marine phytoplankton species. *Marine Ecology Progress Series*, **80**, 89–100.

Hoffman, D. J. & Solomon, S. (1989). Ozone destruction though heterogeneous chemistry following the eruption of El Chichón. *Journal of Geophysical Research*, **D94**, 5029–41.

Holm-Hansen, O., Lubin, D. & Helbing, E. W. (1993). Ultraviolet radiation and its effects on organisms in aquatic environments. In *Environmental UV Photobiology*, ed. A. R. Young, L.O. Björn, J. Moan & W. Nultsch, pp. 379–425. Plenum Press, New York.

Inn, E. & Tanaka, Y. (1953). Absorption coefficients of ozone in the ultraviolet and visible regions. *Journal of the Optical Society of America*, **43**, 870–3.

Jerlov, N. (1976). *Marine Optics*. Elsevier Oceanography Series 14. Elsevier Scientific Publishing, New York.

Kerr, J. B. & McElroy, C. T. (1993). Evidence for large upward trends of ultraviolet-B radiation linked to ozone depletion. *Science*, **262**, 1032–4.

Kieber, D. J., McDaniel, J. A. & Mopper, K. (1989). Photochemical source of biological substrates in seawater: implications for geochemical carbon cycling. *Nature*, **341**, 637–9.

Kirk, J. T. O. (1985). Effect of suspenoids (turbidity) on penetration of solar radiation in aquatic ecosystems. *Hydrobiologia*, **125**, 195–208.

Kirk, J. T. O. (1994a). *Light and Photosynthesis in Aquatic Ecosystems*. Cambridge University Press, Cambridge.

Kirk, J. T. O. (1994b). Optics of UV-B radiation in natural waters. *Ergebnisse der Limnologie (Archiv für Hydrobiologie. Beihefte)*, **43**, 1–16.

Lubin, D. & Jensen, E. H. (1995). Effects of clouds and stratospheric ozone depletion on ultraviolet radiation trends. *Nature*, **377**, 710–13.

Madronich, S. (1991). Implications of recent total atmospheric ozone measurements for biologically active radiation reaching the earth's surface. *Geophysical Research Letters*, **18**, 2269–72.

Madronich, S. (1993). The atmosphere and UV-B radiation. In *Environmental UV Photobiology*, ed. A. R. Young, L. O. Björn, J. Moan & W. Nultsch, pp. 1–39. Plenum Press, New York.

McCormick, M. P., Thomason, L. W. & Trepte, C. R. (1995). Atmospheric effects of the Mt Pinatubo eruption. *Nature*, **373**, 399–404.

Miller, W. L. (1994). Recent advances in the photochemistry of natural dissolved organic matter. In *Aquatic and Surface Photochemistry*, ACS Symposium Series, ed. G. R. Helz, R. G. Zepp & D. G. Crosby, pp. 111–28. Lewis, Boca Raton, FL.

Miller, W. L. & Zepp, R. G. (1995). Photochemical production of dissolved inorganic carbon from terrestrial organic matter: significance to the oceanic carbon cycle.

Geophysical Research Letters, **22**, 417–20.

Mimms, F. M. & Frederick, J. E. (1994). Cumulus clouds and UV-B. *Nature*, **371**, 291.

Mobley, C. (1994). *Light and Water*. Academic Press, New York.

Mobley, C., Gentili, B., Gordon, R., Jin, Z., Kattawar, G., Morel, A., Reinsersman, P., Stanmes, K. & Stavin, R. (1993). Comparison of numerical models for computing underwater light fields. *Applied Optics*, **32**, 7484–504.

Molina, L. & Molina, M. (1986). Absolute absorption cross sections of ozone in the 185- to 350-nm wavelelngth range. *Journal of Geophysical Research*, **91**, 14501–8.

Molina, M. & Rowland, F. S. (1974). Stratospheric sink for chlorofluoromethanes: chlorine atom-catalysed destruction of ozone. *Nature*, **249**, 810–12.

Moran, M. A. & Zepp, R. G. (1997). Role of photoreactions in the formation of biologically labile compounds from dissolved organic matter. *Limnology and Oceanography*, **42**, 1307–16.

Mühler, R., Crutzen, P. J., Grooß, J.-U., Brühl, C., Russell, J. M., Gernandt, H., McKenna, D. S. & Tuck, A. F. (1997). Severe chemical ozone loss in the Arctic during the winter of 1995–96. *Nature*, **389**, 709–12.

Neumann, G. & Pierson, W. (1966). *Principles of Physical Oceanography*. Prentice-Hall, Englewood Cliffs, NJ.

Pawson, S. & Naujokat, B. (1997). Trends in daily wintertime temperatures in the northern stratosphere. *Geophysical Research Letters*, **24**, 575–8.

Preisendorf, R. (1976). *Hydrological Optics*. US Dept of Commerce NOAA, Honolulu.

Preisendorf, R. & Mobley, C. (1986). Albedos and glitter patterns of the wind-roughened sea surface. *Journal of Physical Oceanography*, **16**, 1293–316.

Pyle, J. (1997). Global ozone depletion: observations and theory. In *Plants and UV-B*, Society of Experimental Biology Seminar Series, ed. P. Lumsden, pp. 3–12. Cambridge University Press, Cambridge.

Renger, G., Volker, M., Eckert, H. J., Fromme, R., Hohm-Veit, S. & Graber, P. (1989). On the mechanism of photosystem II deterioration by UV-B irradiation. *Photochemistry and Photobiology*, **49**, 97–105.

Roy, C. R., Gies, H. P., Tomlinson, D. W. & Lugg, D. L. (1994). Effects of ozone depletion on the ultraviolet radiation environment at the Australian stations in Antarctica. In *Ultraviolet Radiation in Antarctica: Measurements and Biological Effects*, Antarctic Research Series, ed. C. S. Weiler & P. A. Penhale, pp. 1–15. American Geophysical Union, Washington, DC.

Salawitch, R. J. (1998). A greenhouse warming connection. *Nature*, **392**, 551–2.

Scully, N. M. & Lean, D. R. S. (1994). The attenuation of ultraviolet radiation in temperate lakes. *Ergebnisse der Limnologie (Archiv für Hydrobiologie. Beihefte)*, **43**, 135–44.

Shepherd, T. (1997). The influence of dynamical processes on ozone abundance. In *Ozone Science: A Canadian Perspective on the Changing Ozone Layer*, ed. D. Wardle, J. Kerr, C. McElroy & D. Francis, pp. 41–56. Environment Canada, Toronto.

Shindell, D. T., Rind, D. & Lonergan, P. (1998). Increased polar stratospheric ozone losses and delated eventual recovery owing to increasing greenhouse-gas concentrations. *Nature*, **392**, 589–92.

Smith, R. C. & Baker, K. S. (1979). Penetration of UV-B and biologically effective dose-rates in natural waters. *Photochemistry and Photobiology*, **29**, 311–23.

Smith, R. C. & Baker, K. K. (1981). Optical properties of the clearest natural waters. *Applied Optics*, **20**, 177–84.

Smith, R. C., Prézelin, B. B., Baker, K. S., Bidigare, R. R., Boucher, N. P., Coley, T., Karentz, D., MacIntyre, S., Matlick, H. A., Menzies, D., Ondrusek, M., Wan, Z. & Waters, K. J. (1992). Ozone depletion: ultraviolet radiation and phytoplankton biology in Antarctic waters. *Science*, **255**, 952–8.

Solomon, S., Portmann, R., Garcia, R., Thompson, L., Poole, L. & McCormick, M. (1996). The role of aerosol variations in anthropogenic ozone depletion in the northern hemisphere. *Journal of Geophysical Research*, **101**, 6713–27.

Stolarski, R. S. (1988). The Antarctic ozone hole. *Scientific American*, **258**, 30–6.

Stolarski, R., Bojkov, R., Bishop, L., Zerefos, C., Staehelin, J. & Zawodny, J. (1992). Measured trends in stratospheric ozone. *Science*, **256**, 342–9.

Tsay, S.-C. & Stamnes, K. (1992). Ultraviolet radiation in the Arctic: the impact of potential ozone depletions and cloud effects. *Journal of Geophysical Research*, **97**, 7829–40.

UNEP (United Nations Environmental Programme) (1991). *Environmental Effects of Ozone Depletion*: 1991 *Update*. UNEP, Nairobi.

Valentine, R. L. & Zepp, R. G. (1993). Formation of carbon monoxide from the photodegradation of terrestrial organic carbon in natural waters. *Environmental Science and Technology*, **27**, 409–12.

Vassiliev, I. R., Prasil, O., Wyman, K. D., Kolber, Z., Hanson, A. K., Prentice, J. E. & Falkowski, P. G. (1994). Inhibition of PS II photochemistry by PAR and UV radiation in natural phytoplankton communities. *Photosynthesis Research*, **42**, 51–64.

Vernet, M., Brody, E., Holm-Hansen, O. & Mitchell, B. (1994). The response of Antarctic phytoplankton to ultraviolet light: absorption, photosynthesis, and taxonomic composition. In *Ultraviolet Radiation in Antarctica*: *Measurements and Biological Effects*, Antartic Research Series, ed. C. S. Weiler & P. A. Penhale, pp. 143–58. American Geophysical Union, Washington, DC.

Vincent, W. F. & Roy, S. (1993). Solar ultraviolet-B radiation and aquatic primary production: damage, protection, and recovery. *Environmental Review*, **1**, 1–12.

Wardle, D. I., Kerr, J. B., McElroy, C. T. & Francis, D. R. (ed.) (1997). *Ozone Science*: *A Canadian Perspective on the Changing Ozone Layer*. Environment Canada, Toronto.

Weiler, C. S. & Penhale, P. (ed.) (1994). *Ultraviolet Radiation in Antarctica*: *Measurements and Biological Effects*. American Geophysical Union, Washington, DC.

Wells, M., Mayer, L., Donard, O., de Souza Sierra, M. & Ackleson, S. (1991). The photolysis of colloidal iron in the oceans. *Nature*, **353**, 248–50.

Williamson, C. E., Stemberger, R. S., Morris, D. P., Frost, T. M. & Paulsen, S. G. (1996). Ultraviolet radiation in North American lakes: attenuation estimates from DOC measurements and implications for plankton communities. *Limnology and Oceanography*, **41**, 1024–34.

WMO (World Meteorological Organization) (1995). *Scientific Assessment of Ozone Depletion*: 1994, *WMO Global Ozone Research and Monitoring Project*. Geneva, WMO.

Worrest, R., Dyke, H. V. & Thomson, B. (1978). Impact of enhanced simulated solar ultraviolet radiation upon a marine community. *Photochemistry and Photobiology*, **27**, 471–8.

Worrest, R., Wolniakowski, K., Scott, J., Brooker, D. & Dyke, H. V. (1981). Sensitivity of marine phyplankton to UV-B radiation: impact upon a model ecosystem. *Photochemistry and Photobiology*, **33**, 223–7.

Zepp, R., Callaghan, T. & Erickson, D. (1995). Effects of increased solar ultraviolet radiation on biogeochemical cycles. *Ambio*, **24**, 181–7.

2

○ ○ ○ ○ ○ ○ ○ ○ ○ ○ ○ ○ ○ ○ ○ ○ ○ ○ ○ ○

UV physics and optics

Susana B. Díaz*, John H. Morrow and Charles R. Booth

The discovery and persistence of regions of ozone depletion over large areas of the earth's surface (the 'ozone hole') has fuelled a significant increase in interest about UV radiation (UVR). This interest ranges from investigations focused on atmospheric interactions and the physics of light transmission (WMO, 1988, 1989, 1991, 1995, 1999; UNEP, 1992) to photobiological effects and agricultural, economic and social impacts (UNEP, 1991, 1994, 1998). The purpose of this chapter is to present some background information about the solar spectrum and introduce a few of the fundamental physical concepts related to UV in the atmosphere (Section 2.1) and underwater (Section 2.2).

Unless other sources are mentioned, the modelled irradiances at the earth's surface used in Section 2.1 were calculated with a two-stream radiative transfer model (Frederick & Lubin, 1988), the values of measured irradiance were provided by the National Science Foundation (NSF) Radiation Monitoring Program (Booth, 1990–1997), and the values of total column ozone by Nimbus 7/Meteor 3, Total Ozone Mapping Spectrometer (TOMS) data (McPeters, 1994).

2.1 Atmospheric UV radiation

Extraterrestrial solar radiation encompasses a wide region of the electromagnetic spectrum and exhibits a distribution similar to the emission of a black body at a temperature of 5800 K (Koller, 1965; Madronich, 1993). Fortunately, for the purposes of this chapter, only a portion of the solar spectrum needs to be considered. The visible portion of the total solar radiation spectrum covers a region of the spectrum from 400 to 700 nm. Wavelengths from 10 to 400 nm correspond to the UV region and wavelengths higher than 700 nm correspond to the infrared region. The UV portion of the spectrum is typically divided into sub-bands, with three regions having significance for ground measurements: UV-C, UV-B and UV-A. UV-C extends from 200 to 280 nm, but since oxygen and ozone absorb it in the upper atmosphere, measurable solar UV-C does not reach the earth's surface. UV-B is usually defined as the region from 280 to 320 nm, although the International Radiation Protection Association defines it as 280 to 315 nm. This band is strongly absorbed by atmospheric ozone, particularly in the range 280 to 310 nm, and transmission of these wavelengths through the earth's atmosphere is very low (Frederick & Lubin, 1994). UV-A typically extends from 320 to 400 nm. Since this type of radiation is not strongly attenuated by the earth's atmosphere, the flux reaching the ground in this region can be high. In general, infrared radiation accounts for approximately 52.8% of total extraterrestrial radiation, visible radiation for 38.9%, UV-A for 6.3% and UV-B for 1.5%. The percentage of UV-B at the earth's surface diminishes considerably, the proportion varying with time of day, season, and geographic, atmospheric and ground conditions.

Light may be considered as an electromagnetic wave or as a particle. When considered a wave, light is expressed in radiometric units. Radiant energy (E_r) is expressed in joules while the radiant power or flux (Φ_r) is expressed in watts. The relationship between these two terms is:

$$\Phi_r = dE_r/dt$$

where t is time. Described as a wave, the energy of a photon, ε, in joules, is a function of its wavelength:

$$\varepsilon = hc/\lambda$$

where h is Planck constant (6.63×10^{-34} J s), c is the speed of light (3×10^8 m s^{-1}), and λ is wavelength (m). The particle nature of a radiation flux is reinforced when the beam is described as a number of quanta or photons, and the literature often refers to the einstein, E, defined as a mole of quanta. Quanta s^{-1} and E s^{-1} are directly interchangeable, since 6.022×10^{23} quanta s^{-1} is equal to 1 E s^{-1}. Conversion of flux in quantum units (Φ in quanta s^{-1}) to radiometric units (Φ_r in watts) is wavelength dependent, since the energy of a photon is determined by its wavelength:

$$\Phi_r = (1.99\Phi/\lambda) \times 10^{-25}$$

It is possible to convert from quantum units to radiometric units *only* if the complete spectral distribution is known. Measurements with units such as lumens, lux, phots, stibs, nits, footcandles, candelas and lamberts all refer to a wavelength weighting factor that assumes that the measuring device responds like the visual response of the 'Standard Observer'. When expressed in radiometric terms, this means: (a) a wavelength of maximum response at 550 nm and (b) a sharply sloping spectral response (0.41% response at 700 nm relative to the peak, 0.12% at 400 nm).

2.1.1 Direct and diffuse UV radiation

Earth is surrounded by the atmosphere, a gaseous matrix composed of nitrogen (about 78%), oxygen (about 21%), and small amounts of other compounds, as shown in Table 2.1 (Madronich, 1993). For purposes of discussion, the atmosphere is divided into: (a) a surface layer called the troposphere, which extends up to 8 to 18 km, depending on latitude; (b) a middle region referred to as the stratosphere, extending up to around 50 km, in which the temperature increases primarily from absorption of sunlight by ozone; and (c) the mesosphere and other upper atmospheric layers lying above 80 km.

Upon entering the atmosphere, the solar flux may travel unaffected to the ground or be attenuated by the atmosphere and its constituents by the processes of absorption or scattering. The degree of impact of these processes on the flux is a function of the materials present in the

Table 2.1. Composition of the atmosphere

Molecule	Fraction
Nitrogen	78.084 ± 0.004%
Oxygen	20.946 ± 0.06%
Water vapour	0–2%
Carbon dioxide	353 p.p.m.
Argon	9340 ± 10 p.p.b.
Neon	18.18 ± 0.04 p.p.b.
Helium	5.239 ± 0.002 p.p.b.
Krypton	1.14 ± 0.01 p.p.b.
Xenon	0.086 ± 0.001 p.p.b.
Hydrogen	515 p.p.b.
Nitrous oxide	310 p.p.b.
Carbon monoxide	50–100 p.p.b.
Methane	1.72 p.p.b.
Ozone	10 p.p.b.–10 p.p.m.
Chlorofluorocarbons	0.85 p.p.b.

p.p.b., parts per billion; p.p.m., parts per million.
From Madronich, 1993.

atmosphere (e.g. gases, aerosols, clouds), the angle of incidence and wavelength. The majority of the fraction that is taken up by absorption is converted to chemical energy or heat. The fraction of the flux that is scattered in a different direction, but without a change in the wavelength (elastic scattering), may either continue to propagate through the atmosphere or be scattered again (so-called multiple scattering). Thus, attenuation may generally be described as a combination of absorption and scattering processes, with absorption resulting in a reduction in the flux, and scattering resulting in changes in the direction of propagation. That portion of the flux that is scattered is often referred to as the diffuse component, while the portion that continues in the direction of the beam is the direct component. Of course, the wavelength dependence of absorption and scattering processes affects both the absolute quantity and the spectral distribution of light reaching the earth's surface. For a small solar zenith angle (SZA), the direct and diffuse components of the UV-B radiation on the ground have similar values, but as the angle increases, the direct component decreases faster than the diffuse, especially at shorter wavelengths, amounting to a difference of orders of magnitude for large SZA.

Absorption is most generally described by the Lambert–Beer law, which states that the flux of photons of wavelength λ, travelling through a uniformly distributed material of concentration N, will be reduced exponentially as a function of the molecular absorption coefficient, α, the vertical thickness, z, and the angle of incidence, θ:

$$\frac{I(\lambda,z)}{I(\lambda,0)} = e^{-\alpha(\lambda)\,Nz/\cos(\theta)} \tag{2.1}$$

Wavelength dependence of the molecular absorption coefficient is a characteristic of the specific atmospheric component, and laboratory studies have determined absorption coefficients for a wide variety of gases and conditions. Some components (e.g. ozone) show a sharp change in the absorption coefficient with wavelength, which results in strong variations in irradiances on the ground between 320 and 280 nm (see Figure 2.1).

Likewise, scattering processes that change the direction of propagation of a photon but not the wavelength (elastic scattering) may be described similarly to Equation (2.1). For the purposes of describing the radiation field, the probability of scattering by atmospheric gases is small, and molecular scattering by atmospheric gases falls into the domain of Rayleigh scattering, which is an inverse function of the fourth power of the wavelength (Koller, 1965). In contrast, the angular distribution of photons scattered from the direct beam by larger particles (e.g. aerosols or clouds) is a complex function of particle size and wavelength. This process is best described by Mie scattering theory (van de Hulst, 1981).

2.1.2   Natural variability

UVR at ground level has a natural variation. Among the factors influencing this variation are geometric factors (time of day, latitude, altitude, season, and earth–sun distance), atmospheric factors (ozone concentration, clouds, and other gases in the atmosphere such as NO_2 and other pollutants), and the earth's albedo (which causes solar radiation to bounce between the surface and the atmosphere). Figure 2.1 depicts a typical solar spectrum taken at local apparent noon in San Diego, California. The most obvious features of a solar spectrum are the rapidly decreasing flux with decreasing wavelength in the UV and numerous Fraunhofer lines resulting from atomic absorption by material in the outer layers of the sun. The flux in the UV portion of the spectrum is a very small percentage of the total UVR and visible spectrum – 0.7% in this example. Large variations in UVR reaching the earth's surface are observed, however, and most may be accounted for by the geometry of the earth–sun system. Two factors – the distance between the earth and the

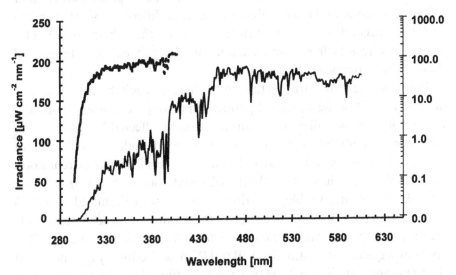

Figure 2.1.   Irradiance plotted on a log scale (left) and linear scale (right) for local apparent noon in March 1997, at San Diego, California (data courtesy NSF UV Spectroradiometer Monitoring Network). The log scale emphasises the dynamic range of the flux between the UV and visible regions of the spectrum. The generalised structure of the spectrum, including numerous Fraunhofer lines, is evident when the same spectrum is plotted normally. (Unless otherwise indicated, Figures 2.1 to 2.19 were all generated by Biospherical Instruments and CADIC.)

sun and the SZA – are the main geometric factors that govern changes in the intensity of solar radiation with latitude and season.

The earth's orbit around the sun is not a perfect circle; it is rather elliptical, with the sun at one focus. The distance between the earth and the sun varies from perihelion (3 January) to aphelion (5 July) by about 3.5%. As a result, extraterrestrial solar radiation during the summer solstice in the Southern Hemisphere is about 7% or $100\,W\,m^{-2}$ (Frölich, 1987) larger than at the summer solstice in the northern hemisphere. Likewise, the position of the sun in the sky has a significant impact on the measured flux. To position the sun, two parameters are used: SZA and azimuth angle. The SZA is the angle formed between the local vertical (zenith) and the centre of the solar disc relative to a position on the surface of the earth. The azimuth angle gives, in a horizontal plane, the position of the sun relative to the geographic north. Changes in the incident flux resulting from latitudinal position, season and time of day are all generalised results from changes in the SZA. Figure 2.2a illustrates the pattern of SZA during daylight hours for summer solstice at differing latitudes in the southern hemisphere, while Figure 2.2b shows the UV-B irradiances at the earth's surface for the same latitudes. As illustrated, there are asymmetries in the irradiance patterns of the northern and southern hemispheres, with higher values in summer and lower values in winter (a difference of between 7% and 10% depending on the latitude) in the southern hemisphere. This asymmetry results from both changes in the sun–earth distance and differences in ozone concentration.

Photons must pass through the atmosphere to reach the earth's surface and, as a result of the increased pathlength through the atmosphere and its constituents, not all wavelengths are equally affected by SZA. UV-B, which is strongly absorbed by ozone, varies by almost an order of magnitude between equatorial and polar regions at local apparent noon as a result of the increase in both SZA and total column ozone with latitude. UV-A and visible radiation show a less pronounced variation with SZA. In the UV-A, the variation of the maximum values with latitude is a factor of about 4, and in the visible range the variation is even smaller. As a consequence, the ratio of UV-B to UV-A is also highly dependent on SZA (Figure 2.3). Figure 2.4 shows the variation of irradiance with SZA and latitude for wavelengths in the UV-B (305 nm) and UV-A (340 nm) range. UV-B is attenuated more rapidly towards the polar regions due to decreasing solar elevation and therefore the longer path that the irradiance must take through the atmosphere. The 340 nm (UV-A) component is much less absorbed by ozone, and therefore drops less rapidly.

Another important factor in determining integrated irradiances is day length. Figure 2.5 shows the variation of day length with latitude for winter and summer. Part of the radiation that reaches the earth's surface results from reflections on the ground and any material covering it. The ratio of the reflected energy to the incident energy is called albedo. The fraction of radiation that is reflected varies with the type of reflecting surface, the wavelength, and the angle of incidence. For example, in the UV-B, snow has an albedo of 0.2 to 1, grass between 0.01 and 0.03, desert

(a)

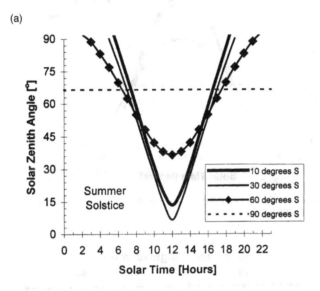

(b)

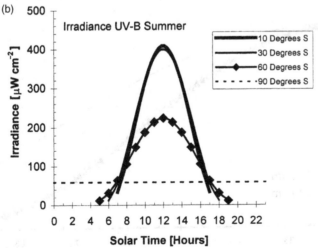

Figure 2.2. Pattern of solar zenith angle (SZA) for summer solstice at varying southern latitudes (a) and resulting UV-B irradiances (b). The smallest SZA results in the highest irradiances.

sand 0.04, and white cement 0.17 (Kondratyev, 1969; Doda & Green, 1980, 1981; Dickerson, Stedman & Delany, 1982; Blumthaler & Ambach, 1988; Madronich, 1993).

Clouds and other aerosols may be responsible for large changes in the total flux of UV (290–400 nm). Clouds usually attenuate UV, but broken

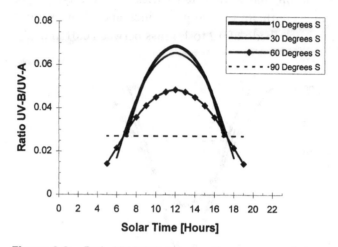

Figure 2.3.   Ratio UV-B/UV-A during the summer solstice at varying southern latitudes.

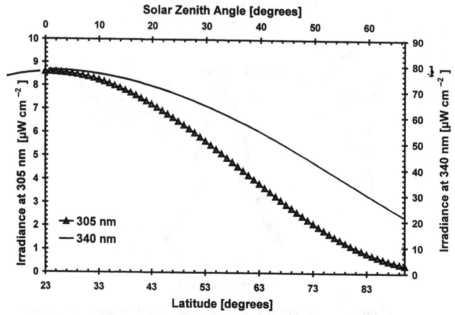

Figure 2.4.   Variation of irradiances at 305 and 340 nm as function of latitude and SZA.

cloud conditions may briefly produce an increase of up to 27% above clear sky values (Estupinan *et al.*, 1996). The composition of clouds varies with altitude. In the lower and middle troposphere they are usually composed of liquid water, while in the upper troposphere and stratosphere they are composed typically of ice crystals. Although cloud optical properties and their impact on the radiation budget of the earth is a complex and dynamic field, simple relationships have been developed to calculate the attenuation due to clouds as a function of fractional cloud cover (Cutchis, 1980; Josefsson, 1986; Ilyas, 1987; Madronich, 1993). Although, for practical purposes, the effect of clouds on irradiances was considered to be the same in the UV-B and UV-A, recent studies have shown spectral differences (Seckmeyer, Erb & Albold, 1996; Frederick & Erlick, 1997).

2.1.3 Atmospheric UV instrumentation

The rapidly diminishing flux with decreasing wavelength in the UV, particularly in the UV-B, places special emphasis on instrumentation, and an instrument that works well in the visible may produce a serious bias when extended into the UV. A full discussion of available designs and pitfalls is beyond the scope of this chapter, but a good introduction can be found in Josefsson (1993) and Webb (1998) is also a good source. In general, however, it is important to understand how wavelength accuracy, calibration, stray light, sampling rate, and other instrument-specific characteristics affect the measurement. Given a 1.1 nm bandpass instrument, a 0.1 nm bias in the reported wavelength results in a difference in the measured irradiance of $\sim 50\%$ at 295 nm, $\sim 8\%$ at 300 nm and $\sim 4\%$ at 305 nm. Although significant improvements have been made in UV

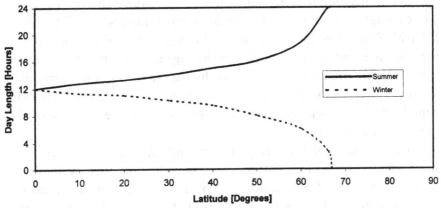

Figure 2.5. Day length hours for winter and summer as function of latitude.

technology, it is still difficult to obtain an absolute calibration accuracy better than 3% to 5% (Blumthaler, 1993; WMO, 1998). The dynamic range between the flux in the UV-B and the visible (Figure 2.1) may easily exceed the stray light specification of an instrument, causing an overestimate of the flux at shorter wavelengths. When measuring the impact of UVR, this overestimate would in turn lead one to *underestimate* the actual impact of a given UV flux. Clouds and long sampling intervals, particularly at large SZAs, can cause subtle or even not-so-subtle distortions in the spectral distribution of the measurement. For the most sensitive work, design and operational details such as temperature coefficients in the optical elements, cosine collector biases and instrument linearity must be well understood.

Depending on the research question, there is a wide variety of instruments available. These include broadband and multiple-filter radiometers, pyranometers, scanning spectroradiometers, actinometers and even biological dosimeters (DeLuisi & Harris, 1983; Quintern *et al.*, 1992; Kerr & McElroy, 1993; Booth *et al.*, 1994). Here, we discuss two of the most common types of instrument: filter radiometers and scanning spectroradiometers. The simplest filter radiometers consist of a diffuser, a UV-sensitive detector, and perhaps an additional filter. The convolution of the response of the detector with the spectral transmittance of the filter and diffuser results in a measurement of the desired part of the spectrum, excluding most of the flux from outside the region of interest. Application of an appropriate calibration relates the signal from the sensor to a known UV flux. In these designs, filters may be described as wideband (e.g. UV-B) or narrowband (e.g. 2 nm). Moderate and narrowband responsivities are often described by the width of the bandpass at one-half the maximum filter transmission or full-width-half-max (FWHM). Typical spectral resolution in filter radiometers ranges from 2 to 10 nm FWHM. Specialised filters may be used to apply a weighting function to the response and thereby emphasise biologically effective wavelengths (e.g. erythemal weighting function for human sunburn). Furthermore, a number of different sensors covering different wavelength ranges may be combined in a single instrument to sample multiple regions of the spectrum (Figure 2.6).

An assortment of filter radiometers and accessories (e.g. rotating shadowband, specialised filters, radiance telescopes) is available to support a variety of applications, such as global, direct or diffuse sky measurements. If the irradiance flux is changing rapidly, such as with changing cloud conditions, filter radiometers have the advantage of fast sampling, with data available on the order of milliseconds per sample. Moderately large bandwidths allow sufficient flux into the detector to support additional blocking filters and improved stray light rejection. Unique detector technologies, such as UV-sensitive films, are also available. UV-sensitive films convert the flux of absorbed UV into visible wavelengths, which are subsequently detected and related to the UV flux impinging upon the film (DeLuisi & Harris, 1983).

Broadband filter instruments are difficult to compare, since they show differences in the weighting function from one manufacturer to the other, and even from a single manufacturer, as well as differences in the corrections in the cosine response function. Then the correction factor shows a strong dependence on ozone concentration and SZA (Leszczynsky et al., 1997).

In general, finer spectral resolution can be obtained from spectrograph and monochromator-based scanning spectroradiometer systems, ranging from 8 nm wide, or wider in some single monochromator designs, to subnanometre resolution in advanced scanning systems. A spectrograph uses a dispersive element to separate wavelengths, which are projected onto a multiwavelength detector, usually a diode array or charge-coupled device (CCD). In contrast, scanning spectroradiometers rely on a moveable monochromator for wavelength selection, projecting a narrow band of the spectrum on a detector such as a photomultiplier tube (Figure 2.7). To increase wavelength selection and reduce stray light, a double mono-chromator may be used, with the first element acting as a disperser,

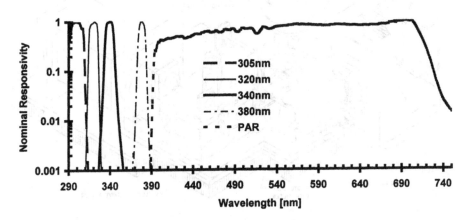

Figure 2.6. Nominal spectral response of the Biospherical Instruments Inc. GUV/PUV-500 series filter radiometers. This is a semi-log plot where each channel has been normalised to 1 at the peak. PAR, photosynthetically available radiation.

projecting a spectrum onto the second element, which acts further to select a narrow wavelength interval.

Comparisons of spectral instruments have been performed periodically (Koskela, 1994; Seckmeyer *et al.*, 1995; Disterhoft *et al.*, 1997), and have shown that, for most instruments, the variation relative to the reference spectra was approximately 5%.

Recently, an algorithm was developed to infer UVR at the earth's surface using satellite measurements. These estimates are obtained from: ozone amounts, cloud and ground reflectivities, aerosols amounts, extraterrestrial solar flux, surface pressure, temperature profiles, and absorption and scattering coefficients (Cebula *et al.*, 1996; Grant, Heisler & Gao, 1996; Herman *et al.*, 1996; Meerkoetter, Wissinger & Seckmeyer, 1997; Krotkov *et al.*, 1998). They allow a worldwide coverage with a single calibrated instrument, which is especially favourable when one is trying to determine trends. Their spatial resolution, periodicity of measurement (one per day) and inability to manage certain type of aerosols and clouds, on the other hand, limit their use in some other studies.

2.1.4   Ozone

Ozone (O_3) is a blue-coloured gas. Although measurable quantities of ozone are present in the troposphere, about 90% of the total ozone in the

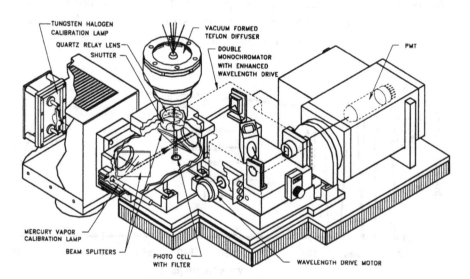

Figure 2.7. Main internal structures of an SUV-100 scanning spectroradiometer. The system, based on a scanning monochromator with photomultiplier tube (PMT) detector, is the heart of the US National Science Foundation UV Spectroradiometer Monitoring Network.

atmosphere is located in the stratosphere. Despite its toxicity, the presence of ozone in the stratosphere has had an important impact on the evolution of life on earth because ozone strongly absorbs solar radiation below 295 nm, radiation that is harmful to living systems. The total amount of ozone in the atmosphere is extremely small, averaging near 0.3 parts per million (Rowland, 1990).

Ozone is constantly being created and destroyed in the atmosphere. Its concentration varies with altitude in the stratosphere, reaching a maximum near the middle of the atmospheric layer. It is created by an interaction between oxygen and sunlight, and is destroyed, normally, through a series of processes that involve sunlight, oxygen, nitrogen and hydrogen.

Ozone was discovered in 1839 by Schöenbein, and was established as a natural component of the atmosphere in 1850. Through experiments in the laboratory and with sunlight, Hartley showed in 1880 that ozone strongly absorbs UV-B radiation. In 1913 it was established that most of this gas (now known to be about 90%) lay in the stratosphere.

An instrument to measure total column ozone was developed by Dobson in 1920. It uses pairs of wavelengths in the UV, which present differences in the ozone absorption coefficient. In 1934 balloon ozone-soundings showed that the maximum ozone concentration is found at a height of 20–30 km. This ozone distribution is explained by Chapman's Photochemical Theory, developed in 1930.

Molecular and atomic oxygen are necessary to form ozone. Atomic oxygen, obtained by photodissociation of molecular oxygen, is maximum near 95 km, and decreases towards the earth's surface as solar radiation, necessary for photodissociation, attenuates. Recombination of atomic oxygen, on the other hand, increases toward the earth's surface, as a result of a larger atmospheric density, converting more atomic oxygen into molecular oxygen. As a consequence, molecular oxygen is more abundant near the earth's surface. There is, therefore, an optimum height for ozone formation (Chapman's Photochemical Theory) (Goody & Walker, 1982). Total column ozone is usually expressed in Dobson units (DU). One hundred DU are equivalent to an ozone layer 1 mm thick at 0 °C and 1 atm pressure.

2.1.4.1 Natural variability

Total column ozone demonstrates large geographic and temporal variations. Under normal conditions, total column ozone varies between 230 and 500 DU, with a worldwide average of aprroximately 300 DU. Total column ozone also varies with latitude. Although most ozone is formed near the equator, total column ozone is lower (250–300 DU) in tropical regions than at higher latitudes as a result of Brewer–Dobson circulation. This behaviour is observed in the northern hemisphere for all latitudes, and in the southern hemisphere for latitudes up to around 60°. Stratospheric east–west winds distribute ozone relatively evenly around the earth (Goody & Walker, 1982; Madronich, 1993), although longitudinal incongruities also exist as a result of the monsoon circulation, for example (Wardle *et al.*, 1997; WMO, 1989).

In the absence of the occurrence of an ozone hole, seasonal variation in total column ozone reaches a maximum in spring and a minimum in autumn, with a more pronounced variation near the poles than near the tropics. On average, natural total column ozone in the southern hemisphere is about 15% smaller than in the northern hemisphere. Figure 2.8 shows seasonal variations in total column ozone for several 2° wide latitudinal bands based on TOMS data (McPeters, 1994) for the year 1980. There are no values for the polar regions during winter, since the TOMS sensor requires sunlight for measurement. The latitudinal and seasonal variations and the asymmetry between both hemispheres, especially for polar regions, are illustrated.

Large day-to-day variations and a decrease in values from morning to

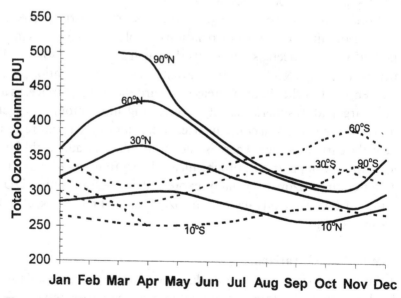

Figure 2.8.   Ozone concentration in latitudinal bands, 1980. (TOMS data, Courtesy of NASA/GSFC, Code 916.)

afternoon are also observed (Koller, 1965). The 11 year solar cycle also produces changes in ozone concentrations of 1% to 2% (WMO, 1995). Total column ozone shows other temporal variations, related to the 27 day solar rotation period and the quasibiennial oscillation (QBO).

2.1.4.2 Ozone depletion

There is evidence that a reduction in stratospheric ozone concentration has been occurring since the mid-1970s, both as a small, worldwide depletion and as a severe seasonal depletion localised over Antarctica. Ozone abundance over the period 1994–7 suffered a year-round average depletion for mid-latitudes (25° to 60°) of around 4% for the northern hemisphere and 4% (satellite) or 5% (ground) for the Southern Hemisphere, when compared with 1979 (WMO, 1999). Depletion was stronger in winter/spring and at higher latitudes (Niu *et al.*, 1992; Harris *et al.*, 1997). Trend acceleration was larger in middle and high latitudes for the period 1979–97 than over the period 1964–97. Non-statistically significant depletion is observed at tropical regions (WMO, 1999).

There is also a strong localised depletion, commonly known as the ozone hole, centred over Antarctica and surrounding regions. It appears in late winter or early spring and persists until late spring, when ozone values rise to almost normal levels. This phenomenon results from the destruction of a high percentage of the regional stratospheric ozone (more than 50%) that occurs when three factors coincide: the presence of halocarbons in the stratosphere, high altitude ice crystals clouds (polar stratospheric clouds), and isolation by the Antarctic polar vortex. Of these, the presence of halocarbons has been found to have a strong anthropogenic component.

Halocarbons are compounds formed by halogens (fluorine, chlorine, bromine, iodine and astatine). From the standpoint of ozone destruction, the most important halocarbons are chlorofluorocarbons (CFCs). These compounds have many applications: aerosol propellants, refrigerants, foam-blowing agents, solvents and fire extinguishers. The first CFCs were synthesised in 1928 and production increased sharply in the middle of the twentieth century, slowing only recently due to concern over ozone depletion (Rowland, 1990).

Methyl bromide is another important ozone-depleting halocarbon, generated by natural and anthropogenic sources (oceans, automobile emissions, fumigation and biomass burning). Although the oceans were thought for many years to be a source of methyl bromide (Singh, Salas & Stiles, 1983; Khalil, Rasmussen & Gunawardena, 1993; Singh & Kanakidou,

1993; WMO, 1995), recent studies indicate that they are, in fact, both a sink and a source, but on the whole act as a sink. Terrestrial systems must therefore provide the source necessary to balance the methyl bromide budget (WMO, 1999). Asymmetric atmospheric distributions of the compound are observed between northern and southern hemispheres, even when the anthropogenic component is removed. The northern hemisphere shows a mean surface concentration that is 1.3 times larger than that of the southern hemisphere (WMO, 1999).

Halocarbons are chemically stable, which means that once they are released into the atmosphere they do not dissociate in the troposphere and are not affected by rain or the chemical reactions that clean most other substances from the troposphere. The compounds reach the mid-stratosphere, where short wavelength solar radiation is strong enough to cause the photodissociation of the CFCs, leaving free chlorine (and other halogens) in the stratosphere. It is this reactive chlorine that is implicated in the destruction of ozone (Rowland, 1990).

The Antarctic polar vortex forms a natural barrier in the stratosphere at a latitude about 70°, which isolates polar air from the rest of the world during winter and part of the spring. In the presence of sunlight during Antarctic spring, chemical reactions occurring on the surface of ice crystals in stratospheric clouds, involving free chlorines from the CFCs, promote the destruction of molecular ozone. It is the combination of halogens, the polar vortex and ice crystals that results in this seasonal depletion of stratospheric ozone called the ozone hole.

Free halogens are most active in ozone destruction during spring. Then, as summer approaches, temperatures increase, polar stratospheric clouds begin to vaporise and the Antarctic vortex becomes weaker, allowing ozone-rich air from other latitudes to reach the polar region. Finally, the vortex breaks up and, in a normal interchange of atmospheric gases, the ozone concentration returns to more normal levels. Figure 2.9 shows the evolution of the ozone hole from September to December.

The possibility of human disruption of the ozone layer has been a topic of research since the early 1970s. In 1971, concern arose about the possible effects of supersonic flights on the ozone layer (Crutzen, 1971; Johnston, 1971). Later, in 1974, F. S. Rowland and M. Molina who, together with P. Crutzen, received the Nobel Prize for Chemistry 1995 for their studies in this field, identified the potential effect of chlorofluorocarbons (CFCs) on stratospheric ozone (Molina & Rowland, 1974). In 1985, J. Farman and co-workers of the British Antarctic Survey published an article in which they reported a strong decline in ozone during the spring over Halley Bay (Farman, Gardiner & Shanklin, 1985), which was later confirmed by other groups and satellite measurements.

The first measure to limit CFCs was the Montreal Protocol, which was signed in 1987. It proposed: (a) cutting the consumption and production of CFC-11, -12, -113, -114 and -115 to 1986 levels by 1990; (b) reducing consumption and production by 20% by 1994; (c) reducing production and consumption by an additional 30% in 1999; and (d) reducing the production and consumption of halons 1301, 1211, and 2402 to 1986 levels after 1990 (Brühl & Crutzen, 1990).

Later scientific evaluations of the evolution of the ozone layer, made by modelling and monitoring, showed that the measures proposed in the protocol were not strong enough. As a result, the protocol has been amended several times – London (1990), Copenhagen (1992), Vienna (1995) and Montreal (1997). As a result of the long lifetime of a CFC compound, it was not possible to observe an immediate improvement in the ozone layer as a result of the Montreal Protocol and its amendments.

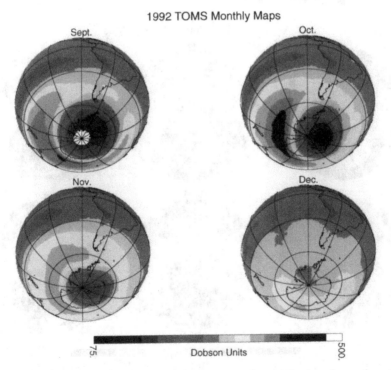

Figure 2.9. Monthly total column ozone for southern polar regions for September, October, November and December 1992. (Courtesy of NASA/GSFC, Code 916.)

The extent and depth of the ozone hole has been increasing in the last decades, as shown in Figure 2.10.

Models predicted that maximum concentration of tropospheric halogens attributable to halocarbons would occur by 1994 (WMO, 1995). Studies based on the analysis of air sampled worldwide confirmed that the tropospheric chlorine peak occurred in 1994 and, although the concentration of bromide from halons was still increasing by mid-1995, the sum of both was decreasing by that time (Montzka *et al.*, 1996). Taking into account that it requires about four years for these compounds to travel from the troposphere and mix in the stratosphere, the above-mentioned study conclude that the abundance of halogen compounds would peak in the stratosphere between 1997 and 1999.

The last WMO Scientific Assessment (WMO, 1999) has concluded that: (a) during the last years the monthly total column ozone in September and October inside the ozone hole has continued at a level around 50% lower than the pre-ozone hole values; (b) the area of the hole showed a maximum in 1996; and (c) the breakdown of the polar vortex, as

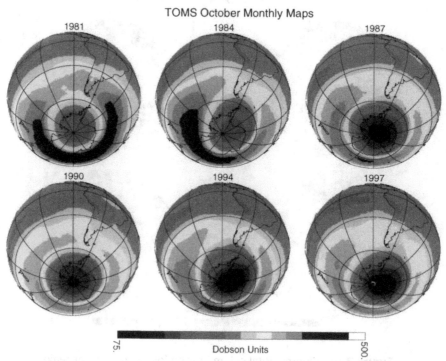

Figure 2.10. Evolution of the 'ozone hole' from 1981 to 1997. (Courtesy of NASA/GSFC, Code 916.)

observed at Syowa Station, has been occurring about a month later than in pre-ozone-hole years (except for 1997).

2.1.4.3 Impact of ozone depletion in surface irradiances

With all other parameters remaining constant, ozone depletion will lead to an increase in the UV-B radiation at the earth's surface. This increase will show a spectral dependence, since the spectral absorption coefficient for ozone increases with decreasing wavelength, reaching a maximum at 260 nm (Figure 2.11). This inverse relationship between ozone and UV-B radiation has been well established through radiative transfer models, and good-quality spectral measurements have confirmed this result experimentally (Kerr & McElroy, 1993; Díaz et al., 1994; Madronich et al., 1995; Varotsos & Kondratyev, 1995; Bodhaine et al., 1996). Many models that derive UV radiation from the total ozone column and vice versa have been developed (Green, Cross & Smith, 1980; Frederick & Lubin, 1988; Stamnes et al., 1988; Madronich, 1992; Smith, Wan & Baker, 1992; Dahlback, 1996), although they differ in algorithm complexity and accuracy of results.

Ozone depletion and SZA combine to produce large changes in incident irradiance at short wavelengths. For instance, any given reduced ozone concentration when combined with larger SZA (a longer pathlength

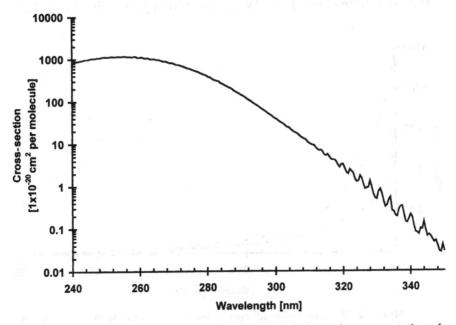

Figure 2.11. Semi-log plot of the modelled absorption cross-section of ozone at 298 K.

through atmospheric constituents) produces larger changes in the relative variation of irradiance (Díaz, Booth & Mestechkina, 1996). For example, 30% depletion during the summer, at latitude 30° S produces a change in the irradiance at 295 nm of near 600%, while at 60° S the change is about 1500% because of the SZA effect (Figure 2.12). For large ozone amounts, SZA larger than 50°, and wavelengths smaller than 300 nm, however, a decrease is observed instead of a continuing increase. This is because, under these conditions, the direct component of the radiation is strongly attenuated, so the radiation reaching the earth's surface is mostly diffuse; less radiation is arriving at the earth's surface than expected (Fioletov, Kerr & Wardle, 1997).

Conditions leading to ozone depletion do not always produce the largest irradiance values (Díaz et al., 1994). Figure 2.13 contrasts the effect of season using modelled irradiances at 305 nm for total column ozone of 200, 300 and 400 DU, and for latitude 30° and 60° S. For a relatively low ozone concentration of 200 DU at 60° irradiance values at 305 nm are always less than at 30° and a total column ozone of 300 DU, regardless of season. At 60°, however, an ozone concentration of 200 DU in mid-September in the southern hemisphere results in an irradiance at 305 nm which is similar to that from 300 DU in mid-October, or 400 DU in mid-November.

The calculation of trends in UV radiation from ground-based spectral

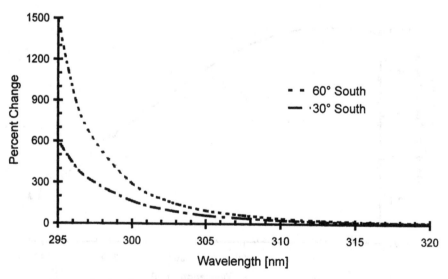

Figure 2.12. Spectral irradiance increase for 30% ozone depletion during the summer solar noon at two different latitudes in the Southern Hemisphere.

measurements has been difficult because of (a) the relatively short series of time (around nine years), (b) the calibration error (5%) is of the same magnitude as the UV trends that may be expected at mid-latitudes based on observed ozone trends, and (c) variability of atmospheric factors such as clouds and aerosols. Nevertheless some trends in UV-B have been well established, based on a few good-quality ground measurements (Wardle *et al.*, 1997; Zerefos *et al.*, 1997; Booth *et al.*, 1998).

As explained before, an algorithm was developed to calculate UV radiation from satellite TOMS data. Although they have smaller spatial resolution, they are a good option for calculating trends, since the time

Table 2.2. Decadal changes in zonally averaged UV irradiances weighted by erythemal action spectrum, derived from TOMS satellite data (1979–92)

Latitude	January	April	July	October	Annual $\pm 2\sigma$
50° to 65° N	6	4	2	4	3.7 ± 3
35° to 50° N	3	3	2	2	3 ± 2.8
30° N to 30° S	0	0	0	0	0 ± 2
35° to 50° S	4	2	2	6	3.6 ± 2
50° to 65° S	4	5	8	14	9 ± 6

σ, standard deviation.
From WMO, 1999.

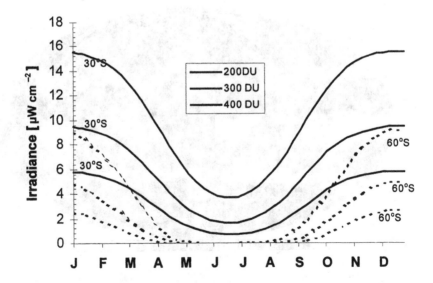

Figure 2.13. Modelled noon irradiance at 305 nm for latitude 30° and 60° S, and total column ozone of 200, 300 and 400 DU.

series are longer and global data are available. Decadal increases in UV radiation, weighted by the erythemal action spectrum, for the period 1979–92 estimated from TOMS satellite data are shown in Table 2.2 (WMO, 1999). Worldwide values of ozone TOMS- and UV TOMS-derived irradiances for October 1979 and 1992 are shown in Figures 2.14 and 2.15.

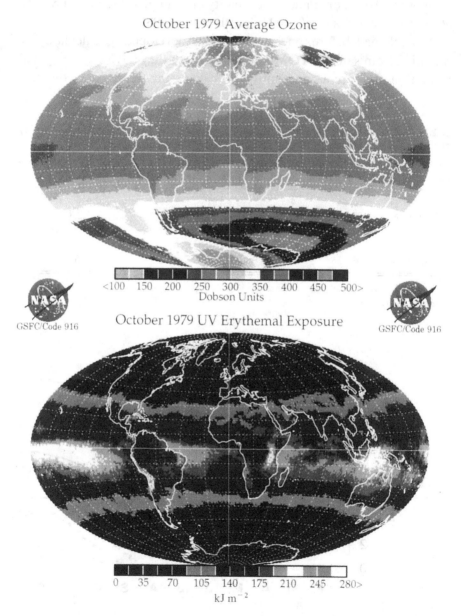

Figure 2.14. (a) Ozone and (b) UV radiation for 1979 from TOMS data. (Courtesy of NASA/GSFC, Code 916.)

Note the differences in both parameters, especially at high latitudes in the southern hemisphere.

2.1.5   Artificial sources of UV

Not all wavelengths contribute equally to produce a given effect on biological systems. The relative effectiveness of different wavelengths

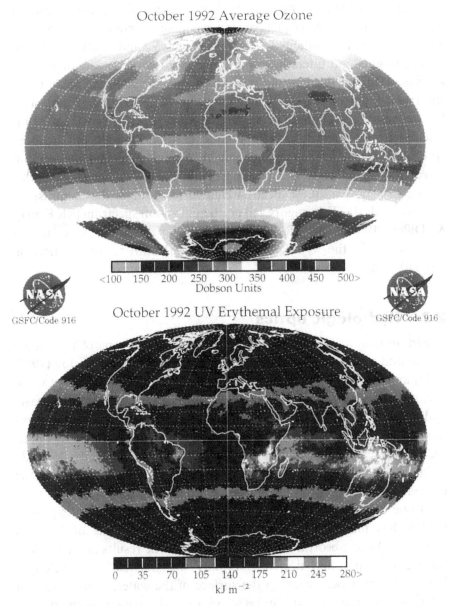

Figure 2.15.   (a) Ozone and (b) UV radiation for 1992 from TOMS data. (Courtesy of NASA/GSFC, Code 916.)

acting to produce a specified response is called an 'action spectrum'. The effective irradiance level required to produce a specified effect is obtained by weighting the spectral irradiance by the action spectrum and integrating over the range of wavelengths where both functions have a different value from zero (cf. Chapter 3).

Historically, researchers investigating the effects of UVR frequently conducted experiments by placing UV-blocking filters over samples exposed to sunlight. Increasing concern over ozone depletion has produced a need to simulate anticipated increases in UV, and combinations of artificial lights and filters have been employed with varying degrees of success. Artificial sources may not accurately simulate the solar UVR spectrum. Figure 2.16a shows the spectrum of QTH (quartz, tungsten, halogen) and xenon lamps in comparison with a modelled solar spectrum calculated for latitude 30° S at the summer solstice. The values of the three spectra are such that the same value is obtained at 305 nm. However, the three curves show large differences, particularly at the shorter wavelengths. Figure 2.16b presents these spectra weighted according to the Commission Internatonale de l'Éclairage (CIE) erythemal action spectrum (McKinlay & Diffey, 1987). The effective irradiance for the solar spectrum is $24.5 \, \mu\text{W cm}^{-2}$, the QTH lamp is $131.5 \, \mu\text{W cm}^{-2}$, and xenon is $161.3 \, \mu\text{W cm}^{-2}$.

2.2 Hydrologic optics

The transmission of UVR through aquatic media is the topic of hydrologic optics, and a variety of excellent books including those of Jerlov (1968, 1976) and Mobley (1994), who build upon Preisendorfer's earlier work (Preisendorfer, 1976) are currently available. Kirk (1994a) combines basic hydrologic optics with strong insight into related problems in photosynthesis. The transmission of solar radiation into natural waters has been the topic of several reviews (Smith & Tyler, 1976; Smith & Baker, 1979a,b; Baker & Smith, 1982; Zafiriou *et al.*, 1984; Kirk, 1994b), some with the emphasis on the UV and UV effects on aquatic systems.

Ozone depletion and the resulting potential for increased exposure to UV has fuelled increasing interest in measurements of UV. As discussed above, absorption by atmospheric ozone results in an incident surface spectrum that decreases rapidly with decreasing wavelength below 340 nm. Upon reaching the surface of the water, transmission of UVR through the air–water interface varies with time owing to changes in the incident irradiance angular distribution and the effect of surface waves (Dera & Stramski, 1986; Stramski, Booth & Mitchell, 1992), which both modulate the intensity at a particular depth and the amount of light reflected back into the air. After entering water, the UV region of the spectrum is reduced rapidly because of the combined influences of water and particulate absorption and scattering (Figure 2.17). Thus, the flux from wavelengths below 280 nm (perhaps even 290 nm) is negligible and mostly insignificant in a practical sense. In the water column, the flux in the UV becomes a vanishingly small

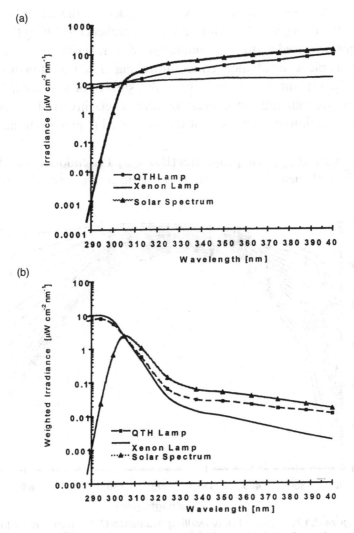

Figure 2.16. Sun and lamp spectra (a), and results from the convolution of the spectra with the CIE erythemal action spectrum (b). QTH, quartz, tungsten, halogen.

signal that must be measured in the presence of a much larger visible component.

2.2.1   Underwater light and UV penetration

The geometry of the light field has important implications depending on the research question. Radiance, $L(\theta, \phi, \lambda)$ is the radiant flux at wavelength λ emanating from the volume described by nadir angle θ and azimuthal angle ϕ. The most commonly reported radiance value is upwelling (nadir) radiance, $L_u(\lambda)$. Scalar irradiance, $I_o(\lambda)$, is the radiance distribution incident on one point integrated over all angles. In the water column scalar irradiance is generally of interest to researchers modelling the light field or those involved in studies of photosynthesis and primary production. In contrast, plane irradiance, either upwelling ($I_u(\lambda)$) or downwelling ($I_d(\lambda)$), is the radiant flux incident on a flat surface of unit area. Plane irradiances are generally of interest to researchers involved in remote sensing applications, as well as to others interested in models of the light field.

Termed inherent optical properties (IOPs) by Prisendorfer (1961), the absorption coefficient $a(\lambda)$, the scattering coefficient, $b(\lambda)$, and the volume

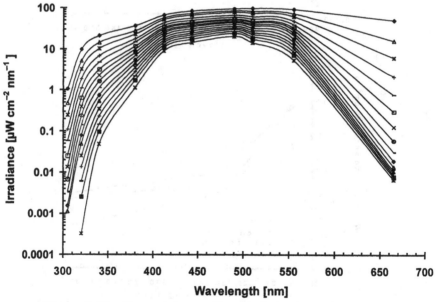

Figure 2.17. Spectral downwelling irradiance ($E_d(\lambda)$) from the surface to 32 m measured off the coast of San Diego, California, USA, using filter radiometers. Note that measurable penetration at 305 nm ends at approximately 22 m depth.

scattering function $B(\theta)$ can be used to describe the optical properties of water. The spectral absorption coefficient is used to describe the loss of photons by water, dissolved substances such as organic material and suspended particles. The scattering coefficient describes how much the original light path is changed by water and its constituents, and the volume scattering function quantifies the angular distribution of this change. IOPs are properties of the medium and, although not affected by the light field itself, are spectrally dependent.

Preisendorfer also recognised apparent optical properties (AOPs), which depend upon both the water and its constituents and upon the geometrical distribution of the light field. This geometrical distribution is influenced by local environmental conditions such as solar angle and clouds and by the properties of the water surface (waves). AOPs include scalar and vector irradiance, attenuation coefficients for various measures of irradiance, reflectance, and functions that describe the geometrical state of the light field (e.g. the average cosine). Theoretically the AOPs of the water can be calculated directly from a complete description of the IOPs and the radiance distribution below the water surface. Likewise, knowledge of the IOPs can lead to calculation of AOPs. However, practical measurements support estimates of these types with varying degrees of success (O'Mongain et al., 1997).

One of the most common descriptors of the penetration of light in water is the diffuse attenuation coefficient, K or K_d in the case of downwelling irradiance, I_d:

$$K_d(\lambda, z) = \frac{-1}{I_d(z, \lambda)} \left[\frac{dI_d(z, \lambda)}{dz} \right] \qquad (2.2)$$

Smith & Baker (1979a) observed that K is generally determined by the inherent optical properties and may be relatively insensitive to changes in the light field, and they termed it a 'quasi-inherent' property. Gordon (1989) presented a comprehensive review suggesting a method of compensating for the effects of changes in the above surface radiance distribution. This trend has led some to assume that such parameters as $K(\text{PAR})$, $K(\text{UVR})$, and to a lesser extent, $K(\text{UV-B})$ also possess this quasi-inherent property. It is important to note, however, that the diffuse attenuation coefficient is not constant with wavelength (Figure 2.18), and spectrally integrated attenuation coefficients, such as $K(\text{PAR})$, $K(\text{UVR})$ and $K(\text{UV-B})$ may be poor descriptors of the water column, particularly in turbid waters.

In all but the most scattering waters, absorption controls the penetration of UVR. The spectral absorption of a diverse sampling of marine waters was characterised by Bricaud, Morel & Prieur (1981) over the 200 to 700 nm range. Scully & Lean (1994) described the relationship of the attenuation of UV in lakes between chlorophyll, dissolved organic carbon (DOC), and particulate organic carbon (POC), and found that the attenuation coefficients of UV-B and UV-A could be predicted using empirically derived equations. Of 20 sites, all but one exhibited K values in the UV-B greater than $1.0 \, \mathrm{m}^{-1}$.

2.2.2   Models

Models describing the penetration of light through water can be divided into two types: numerical and empirical. Numerical models use various forms of the radiative transfer equation as applied to a particular situation. Several examples using numerical models were presented by Mobley *et al.* (1993), who concluded that the numerical models were more 'accurate' than the instruments currently available for validation. However, such models require detailed knowledge of the inherent optical properties (absorption and scattering coefficients, and phase functions) to be used as inputs, which are very difficult to obtain in the field. Empirical models that

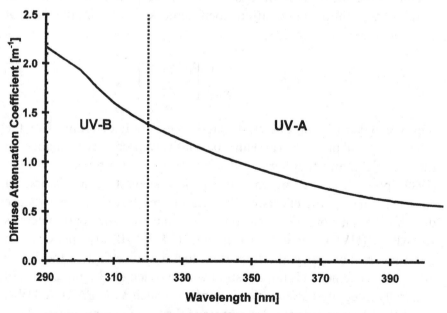

Figure 2.18. Spectral diffuse absorption coefficient (Equation 2.2) modified from a model after Smith and Baker (1981).

predict the $K(UV)$ have been devised by Smith & Baker (1981) to predict attenuation and irradiance penetration from spectral estimates of water absorption, DOC load and chlorophyll concentration. Empirical models are effective at predicting the average optical parameters, since their formulation is based on statistical averages. Both types of model require the incident sky irradiance in order to set the magnitude of the underwater light field.

2.2.3 Underwater UV instrumentation

It is surprisingly difficult to obtain accurate and meaningful measurements of UV in water. The combined influence of water absorption and attenuation of flux by dissolved and suspended materials reduces the flux quickly, especially in the shorter UV wavelengths. The result is a highly unbalanced spectrum (Figure 2.17), with a very wide dynamic range spectrally and little resemblance to the spectral distribution commonly used for calibration standards.

Instruments used to measure UVR in water vary significantly in their design, operation, features and weaknesses. The general evolution of new devices is occurring slowly. A wide variety of radiometer designs have been used in aquatic environments including electrically or mechanically scanned spectroradiometers, filter radiometers, and film (Fleischman, 1989) or biological actinometers (Karentz & Lutze, 1990; Yoshida & Regan, 1997). Much early work used instruments weighting the irradiance spectrum to resemble a biological function such as human erythema, or to report irradiance in a spectral region ('UV-B'). Calkins (1975) and Smith & Calkins (1976) discussed the use of broadband sensors ('R-B meters') in marine applications, including the possible errors in the measurement. In the late 1970s, Smith & Baker (1979b, 1981) measured the spectral penetration of UV (280–340 nm) using an instrument equipped with a double monochromator and an UV-sensitive photomultiplier tube (PMT). The LI-1800UW (LI-COR) submersible instrument uses a holographic single grating monochromator, order-sorting filter wheel and silicon photodetector (Behrenfeld *et al.*, 1993; Kirk *et al.*, 1994; Gulko, 1995). Still another approach to in-water measurements has been taken by Optronic Laboratories in the OL-740 and 750 series of scanning spectroradiometers (Kirk *et al.*, 1994; Piazena & Häder, 1997). In these instruments, the flux is coupled into a scanning double monochromator through a fibre optic bundle attached to a submersible collector. Detailed spectral information concerning irradiance profiles in the Southern Ocean was obtained in support of Icecolors '90 (Prézelin, Boucher & Smith, 1994) and other programmes using the Light and Ultraviolet Submersible Spectroradiometer (LUVSS) system developed by Smith *et al.* (1992). The LUVSS divides the spectrum into two regions (250–380 nm at 0.2 nm resolution and 350–700 nm at 0.8 nm resolution) to address the imbalance in the spectrum between the visible and UV. Specialised submersible instruments such as the PUV-500 (Biospherical Instruments) are also available based on filter-photodetector technology, in which different regions of the spectrum are isolated using specialised filters coupled to individual photodetectors (Kirk *et al.*, 1994). Although the number of channels or spectral resolution of these systems may be less than spectrograph-based instruments, filter radiometers may be built with extremely high out-of-band blocking and are rugged and stable (Booth, Morrow & Neuschuler, 1992).

2.2.4   Measurements

Smith and co-workers using the LUVSS reported that 'UV penetrates to ecologically significant depths' – 'in excess of 60–70 meters' (Smith *et al.*, 1992), while Piazena & Häder (1994) reported similar levels in less than 1 m in coastal lagoons. Figure 2.19 presents measurements of UVR from the literature to examine the range of variability that has been reported. Included are data from temperate lakes measured by Scully & Lean

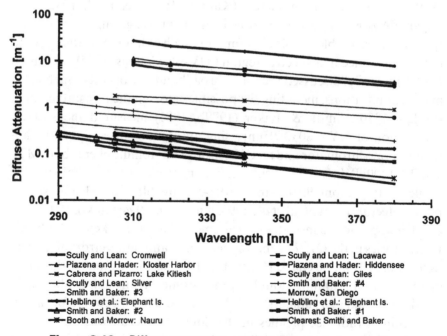

Figure 2.19. Diffuse attenuation coefficient for various natural waters.

(1994), an Antarctic lake (Cabrera & Pizarro, 1994), turbid coastal waters (Piazena & Häder, 1994), clear coastal waters (J. H. Morrow unpublished data), Antarctic waters (Helbling, Villafañe & Holm-Hansen, 1994), and clear oceanic waters (Booth *et al.*, 1992). As a boundary, the 'clearest' natural water as defined by Smith & Baker (1979a) was included, along with their four examples of oceanic water (numbers 1 to 4).

It is evident that UVR penetration in natural waters varies over a wide range and there will probably be a strong relationship between the mixing dynamics of the system under investigation and the penetration of UV to significant depth. The values for attenuation coefficients in the UV-B range from around $20\,\mathrm{m}^{-1}$ to less than $0.2\,\mathrm{m}^{-1}$ for the clearest ocean waters. Likewise, for these examples, the 10% light level for UV-B ranges more than two orders of magnitude between temperate lakes and the clearest oceanic waters.

Acknowledgements

The United States National Science Foundation's (NSF) Polar Programs Ultraviolet (UV) Monitoring Network was established in 1988 under the guidance of Dr Peter Wilkniss, former Director of the Office of Polar Programs. The network is operated and maintained by Biospherical Instruments Inc. under contract to NSF via Antarctic Support Associates (ASA). We would especially like to thank Dr Polly A. Penhale, the NSF network program director, as well as the ASA personnel who have supported project operation including D. C. Sheperd, Dr S. Kottmeier and R. Koger. We also thank the Centro Austral de Investigaciones Científicas (CADIC) and the director Dr Eduardo Olivero for their support, and Dr R. McPeters and Dr J. Herman of NASA Goddard Space Flight Center (GSFC) for providing TOMS data. Lastly, we would like to acknowledge Ms Joan Schwenker for her contributions to the final manuscript.

References

Baker, K. S. & Smith, R. C. (1982). Spectral irradiance penetration in natural waters. *The Role of Solar Ultraviolet Radiation in Marine Ecosystems*, ed. J. Calkins, pp. 233–46. Plenum Press.
Behrenfeld, M., Hardy, J., Gucinski, H., Hanneman, A., Lee, H. & Wones, A. (1993). Effects of ultraviolet-B radiation on primary production along latitudinal transects in the South Pacific Ocean. *Marine Environmental Research*, **35**, 349–63.
Blumthaler, M. (1993). *Solar UV Measurements. UV-B Radiation and Ozone Depletion*, ed. M. Tevini. Lewis Publishers, Boca Raton, FL.

Blumthaler, M. & Ambach, W. (1988). Solar UV-B albedo of various surfaces. *Photochemistry and Photobiology*, **48**, 85–8.

Bodhaine, B. A., McKenzie, R. L., Johnston, P. V., Hoffmann, D. J., Dutton, E. G., Schnell, R. C., Barnes, J. E., Ryan, S. C. & Kotkamp, M. (1996). New ultraviolet spectroradiometer measurements at Mauna Loa Observatory. *Geophysical Research Letters*, **23**, 2121–4.

Booth, C. R. (1990–1997). *NSF UV Radiation Monitoring Network*, vols. 1–6. Biospherical Instruments Inc., San Diego, CA.

Booth, C. R., Ehramjian, J. E., Mestechkina, T., Cabasug, L. W., Robertson, J. S. & Tusson, J. R. (1998). *NSF Polar Programs UV Spectroradiometer Network 1995–1997 Operations Report*. Biospherical Instruments Inc., San Diego, CA.

Booth, C. R., Lucas, T. B., Morrow, J. H., Weiler, C. S. & Penhale, P. A. (1994). The United States National Science Foundation's Polar Network for monitoring ultraviolet radiation. In *Ultraviolet Radiation in Antarctica Measurements and Biological Effects*, Antarctic Research Series, ed. C. S. Weiler & P. A. Penhale, pp. 17–37. American Geophysical Union, Washington, DC.

Booth, C. R., Morrow, J. H. & Neuschuler, D. A. (1992). A new profiling spectroradiometer optimized for use in the ultraviolet. *SPIE Ocean Optics XI*, **1750**, 354–65.

Bricaud, A., Morel, A. & Prieur, L. (1981). Absorption by dissolved organic matter of the sea (yellow substance) in the UV and visible domains. *Limnology and Oceanography*, **26**, 43–53.

Brühl, C. & Crutzen, P. (1990). Ozone and climate changes in the light of the Montreal Protocol: a model study. *Ambio*, **19**, 293–300.

Cabrera, S. & Pizarro, G. (1994). Changes in chlorophyll *a* concentration, copepod abundance and UV and PAR penetration in the water column during the ozone depletion in Antarctic Lake Kitiesh, 1992. *Ergebnisse de Limnologie (Archiv für Hydrobiologie. Beihefte)*, **43**, 71–99.

Calkins, J. (1975). Measurements of the penetration of solar UV-B into various natural waters. Impacts of climatic change on the biosphere. *Climatic Impact Assessment Program (CIAP) Monographs*, **5**, 267–96.

Cebula, R. P., Thuillier, G. O., VanHoosier, M. E., Hilsenrath, E., Herse, M., Brueckner, G. E. & Simon, P. C. (1996). Observations of the solar irradiance in the 200–350 nm interval during the ATLAS I mission: A comparison among three sets of measurement SSBUV, SOLSPEC and SUSIM. *Geophysical Research Letters*, **23**, 2289–92.

Crutzen, P. J. (1971). Ozone production rates in an oxygen, hydrogen, nitrogen-oxide atmosphere. *Journal of Geophysical Research*, **76**, 7311–27.

Cutchis, P. (1980). A formula for comparing actual damaging ultraviolet (DUV) radiation doses at tropical and mid-latitude sites. *Federal Aviation Administration Report* FAA-EE 80-21. US Department of Transportation, Washington, DC.

Dahlback, A. (1996). Measurements of biologically effective UV doses, total ozone abundances, and cloud effects with multichannel, moderate bandwidth filter instrument. *Applied Optics*, **35**, 6514–21.

DeLuisi, J. J. & Harris, J. M. (1983). A determination of the absolute radiant energy of a Robertson–Berger meter sunburn unit. *Atmospheric Environment*, **17**, 751–8.

Dera, J. & Stramski, D. (1986). Maximum effects of sunlight focusing under a wind-disturbed sea surface. *Oceanologia*, **23**, 15–42.

Díaz, S. B., Booth, C. R., Lucas, T. B. & Smolskaia, I. (1994). Effects of ozone depletion on irradiances and biological doses over Ushuaia. *Ergebnisse de Limnologie (Archiv für Hydrobiologie. Beihefte)*, **43**, 115–22.

Díaz, S. B., Booth, C. R. & Mestechkina, T. (1996). UV radiation variations over Ushuaia. *Proceedings of the First SPARC General Assembly*, Melbourne Australia, II, pp. 533–6.

Dickerson, R. R., Stedman, D. H. & Delany, A. C. (1982). Direct measurement of ozone and nitrogen dioxide photolysis rates in the troposphere. *Journal of Geophysical Research*, **87**, 4933–46.

Disterhoft, P., Early, E. A., Thompson, A., DeLuisi, J., Mestechkina, Y., Sun, T., Wardle, J., Rives, J., Lucas, T., Neale, P. & Kerr, J. (1997). The 1994 North American interagency intercomparison of ultraviolet monitoring spectroradiometers. *Journal of Research of the National Institute of Standards and Technology*, **102**, 279–322.

Doda, D.D. & Green, A. E. S. (1980). Surface reflectance measurements in the UV from an airborne platform, part I. *Applied Optics*, **19**, 2140–5.

Doda, D. D. & Green, A. E. S. (1981). Surface reflectance measurements in the UV from an airborne platform, part II. *Applied Optics*, **20**, 636–42.

Estupinan, J. G., Raman, S., Crescenti, G. H., Streitcher, J. J. & Barnard, W. F. (1996). The effects of clouds and haze on UV-B radiation. *Journal of Geophysical Research*, **101**, 807–16.

Farman, J. C., Gardiner, B. G. & Shanklin, J. D. (1985). Large losses of total ozone in Antarctica reveals seasonal ClO_x/NO_x interaction. *Nature*, **315**, 207–10.

Fioletov, V., Kerr, J. B. & Wardle, D. J. (1997). The relationship between total ozone and spectral UV irradiance from Brewer spectrophotometer observations and its use for derivation of total ozone from UV measurements. *Geophysical Research Letters*, **24**, 2705–8.

Fleischman, E. M. (1989). The measurement and penetration of ultraviolet radiation into tropical marine water. *Limnology and Oceanography*, **34**, 1623–9.

Frederick, J. E. & Erlick, C. (1997). The attenuation of sunlight by high latitude clouds. Spectral dependence and its physical mechanisms. *Journal of Atmospheric Science*, **64**, 2813–19.

Frederick, J. E. & Lubin, D. (1988). The budget of biologically active ultraviolet radiation in the earth-atmosphere system. *Journal of Geophysical Research*, **93**(D4), 3825–32.

Frederick, J. E. & Lubin, D. (1994). Solar ultraviolet irradiance at Palmer Station, Antarctica. In *Ultraviolet Radiation in Antarctica Measurements and Biological Effects*, Antarctic Research Series, ed. C. S. Weiler & P. A. Penhale, pp. 43–52. American Geophysical Union, Washington, DC.

Frölich, C. (1987). Variability of the solar "constant" on time scales of minutes to years. *Journal of Geophysical Research*, **92**(D1), 796–800.

Goody, R. M. & Walker, J. C. G. (1982). *Atmospheres*. Prentice Hall Inc., Englewood Cliffs, NJ.

Gordon, H. R. (1989). Can the Lambert–Beer law be applied to the diffuse attenuation coefficient of ocean water? *Limnology and Oceanography*, **34**, 1389–409.

Grant, R. H., Heisler, G. M. & Gao, W. (1996). Clear sky radiance distribution in ultraviolet wavelength bands. *Theoretical and Applied Climatology*, **371**, 1–13.

Green, A. E. S., Cross, K. R. & Smith, L. A. (1980). Improved analytic characterization of ultraviolet skylight. *Photochemistry and Photobiology*, **53**, 717–25.

Gulko, D. (1995). The ultraviolet radiation environment of Kaneohe Bay, Oahu. In *Ultraviolet Radiation and Coral Reefs*, ed. D. Gulko, & P. L. Jokiel, pp. 25–35. HIMB Technical Report no. 41, UNIHI-Sea Grant-CR-95-03.

Harris, N. P. R., Ancellet, G., Bishop, L., Hoffman, D. J., Kerr, J. B., McPeters, R. D., Prendez, M., Randel, W. J., Staehelin, J., Subbaraya, B. H., Volz-Thomas, A., Zawoday, J. & Zerefos, C. S. (1997). Trends in stratospheric and free tropospheric ozone. *Journal of Geophysical Research*, **102**(D1), 1571–90.

Helbling, E. W., Villafañe, V. & Holm-Hansen, O. (1994). Effects of ultraviolet radiation on antarctic marine phytoplankton photosynthesis with particular attention to the influence of mixing. In *Ultraviolet Radiation in Antarctica: Measurements and Biological Effects*, Antarctic Research Series, ed C. S. Weiler & P. A. Penhale, pp. 207–26. American Geophysical Union, Washington, DC.

Herman, J. R., Bhartin, P. K., Kiemke, J., Ahmad, Z. & Larko, D. (1996). UV-B increases (1979–1992) from decreases in total ozone. *Geophysical Research Letters*, **23**, 2117–120.

Ilyas, M. (1987). Effect of cloudiness on solar ultraviolet radiation reaching the surface. *Atmospheric Environment*, **21**, 1483–4.

Jerlov, N. G. (1968). *Optical Oceanography*. Elsevier, New York.

Jerlov, N. G. (1976). *Marine Optics*. Elsevier, New York.

Johnston, H. S. (1971). Reduction of stratospheric ozone by nitrogen oxide catalysts from supersonic transport exhaust. *Science*, **173**, 517–22.

Josefsson, W. A. P. (1986). Solar ultraviolet radiation in Sweden. *National Institute of Radiation Protection in Stockholm*, SMHI Report-53, Norrköping, Sweden.

Josefsson, W. A. P. (1993). Monitoring ultraviolet radiation. In *Enironmental UV Photobiology*, ed. A. R. Young, L. O. Björn, J. Moan & W. Nultsch. Plenum Press, New York.

Karentz, D. & Lutze, L. H. (1990). Evaluation of biologically harmful ultraviolet radiation in Antarctica with a biological dosimeter designed for aquatic environments. *Limnology and Oceanography*, **35**, 549–61.

Kerr, J. B. & McElroy, C. T. (1993). Evidence for large upward trends of ultraviolet-B radiation linked to ozone depletion. *Science*, **262**, 1032–4.

Khalil, M. A. K., Rasmussen, R. A. & Gunawardena, R. (1993). Atmospheric methyl bromide: trends and global mass balance. *Journal of Geophysical Research*, **98**, 2887–96.

Kirk, J. T. O. (1994a). *Light and Photosynthesis in Aquatic Ecosystems*, 2nd edn. Cambridge University Press, Cambridge.

Kirk, J. T. O. (1994b). Optics of UV-B radiation in natural waters. *Ergebnisse de Limnologie* (*Archiv für Hydrobiologie. Beihefte*), **43**, 1–16.

Kirk, J. T. O., Hargreaves, B. R., Morris, D. P., Coffin, R. B., David, B., Frederickson, D., Karentz, D., Lean, D. R. S., Lesser, M. P., Madronich, S., Morrow, J. H., Nelson, N. B. & Scully, N. M. (1994). Measurements of UV-B radiation in two freshwater lakes: an instrument intercomparison. *Ergebnisse de Limnologie* (*Archiv für Hydrobiologie. Beihefte*), **43**, 71–99.

Koller, L. R. (1965). *Solar Radiation. Ultraviolet Radiation*. 2nd edn. John Wiley & Sons Inc. New York.

Kondratyev, K. Y. (1969). *Radiation in the Atmosphere*. Academic Press, New York.

Koskela, T. (1994). *The Nordic Intercomparison of Ultraviolet and Total Ozone Instruments at Izaña from 24 October to 5 November 1993*. Final Report. Finnish Meteorological Institute, Helsinki.

Krotkov, N. A., Bhartia, P. K., Herman, J. R., Fioletov, V. & Kerr, J. (1998). Satellite estimation of spectral surface UV irradiance in the presence of tropospheric aerosols. 1. Cloud-free case. *Journal of Geophysical Research*, **103**, 8779–93.

Leszczynsky K., Jokela, K., Yliantila, L., Visuri, R. & Blumthaler, M. (1997). *Report of the WMO/Stuk Intercomparison of Erythemally-weighted Solar UV Radiometers* (*Spring/Summer 1995, Helsinki, Finland*). World Meteorological Organization, Global Atmospheric Watch, Report no. 112, p. 90.

Madronich, S. (1992). Implications of recent total atmospheric ozone measurements for biologically active ultraviolet radiation reaching the earth's surface. *Geophysical Research Letters*, **19**, 37–40.

Madronich, S. (1993). *The Atmosphere and UV-B Radiation at Ground Level, Environmental UV Photobiology*, ed. A. R. Young, L. O. Björn, J. Moan & W. Nultsch, pp. 1–39. Plenum Press, New York.

Madronich, S., Mc Kenzie, R. L., Caldwell, M., Björn, L. O. (1995). Changes in ultraviolet radiation reaching the earth's surface. *Ambio*, **24**, 143–152.

McKinlay, A. F. & Diffey, B. L. (1987). *A Reference Action Spectrum for Ultra-violet Induced Erythema in Human Skin, Human Exposure to Ultraviolet Radiation: Risks and Regulations*, ed. W. R. Passchler & B. M. F. Bosnajanovic, pp. 83–7. Elsevier, Amsterdam.

McPeters, R. B. (1994). *Nimbus 7/Meteor 3, TOMS Data*. NASA/Goddard Space Flight Centre, Houston.

Meerkoetter, R., Wisinger, B. & Seckmeyer, G. (1997). Surface UV from ERS 2/GOME and NOAA/AVHRR data: a case study. *Journal of Geophysical Research*, **24**, 1939–42.

Mobley, C. D. (1994). *Light and Water: Radiative Transfer in Natural Waters*. Academic Press, San Diego, CA.

Mobley, C. D., Gentili, B., Gordon, H. R., Jin, Z., Kattawar, G., Morel, A., Reinersman, P., Stanmes, K. & Stavn, R. H. (1993). Comparison of numerical models for computing underwater light fields. *Applied Optics*, **32**, 7484–504.

Molina, M. J. & Rowland, F. S. (1974). Stratospheric sink of chlorofluormethanes: chlorine atom-catalysed destruction of ozone. *Nature*, **249**, 810–12.

Montzka, S. A., Butler, J. H., Myers, R. C., Thompson, T. M., Swanson, T. H., Clarke, A. D., Lock, L. T. & Elkins, J. W. (1996). Decline in the tropospheric abundance of halogen from halocarbons: implications for stratospheric ozone depletion. *Science*, **232**, 1318–22.

Niu, X., Frederick, J. E., Stern, M. L. & Tiao, C. (1992). Trends in column ozone based on TOMS data: dependence on month, latitude and longitude. *Journal of Geophysical Research*, **97**(D13), 14661–9.

O'Mongain, E., Buckton, D., Green, S., Bree, M., Moore, K., Doerffer, R., Danaher, S., Hakvoort, H., Kennedy, J., Fischer, J., Fell, F., Papantoniou, D. & McGarrigle, M. (1997). Spectral absorption coefficient measured *in situ* in the North Sea with a marine radiometric spectrometer system. *Applied Optics*, **36**, 5162–7.

Piazena, H. & Häder, D.-P. (1994). Penetration of solar UV irradiation in coastal lagoons of the Southern Baltic Sea and its effect on phytoplankton communities. *Photochemistry and Photobiology*, **60**, 463–9.

Piazena, H. & Häder, D.-P. (1997). Penetration of solar UV and PAR into different waters of the Baltic Sea and remote sensing of phytoplankton. In *The Effects of Ozone Depletion on Aquatic Ecosystems*, ed. D.-P. Häder, pp. 45–96. Academic Press, New York.

Preisendorfer, R. W. (1961). Generalized invariant imbedding relation. *Proceedings of the National Academy of Sciences, USA*, **47**, 591–4.

Preisendorfer, R. W. (1976). *Hydrological Optics*. 6 vols. US Department of Commerce, NOAA, Honolulu, HA.

Prézelin, B. B., Boucher, N. P. & Smith, R. C. (1994). Marine primary production under the influence of the Antarctic ozone hole: Icecolors 90. Ultraviolet radiation in Antarctica: measurements and biological effects. *Ergebnisse de Limnologie (Archiv für Hydrobiologie. Beihefte)*, **43**, 71–99.

Quintern, L. E., Horneck, G., Eschweiler, U. & Bücker, H. (1992). A biofilm used as ultraviolet dosimeter. *Photochemistry and Photobiology*, **55**, 389–95.

Rowland, F. S. (1990). Stratospheric ozone depletion by chlorofluorocarbons. *Ambio*, **19**, 281–92.

Scully, N. M. & Lean, D. R. S. (1994). The attenuation of ultraviolet radiation in temperate lakes. In: *Ergebnisse de Limnologie* (*Archiv für Hydrobiologie. Beihefte*), **43**, 135–44.

Seckmeyer, G., Erb, R. & Albold, A. (1996). Transmittance of a cloud is wavelength-dependent in the UV range. *Geophysical Research Letters*, **23**, 2753–5.

Seckmeyer, G., Mayer, B., Bernhard, G., McKenzie, R. L., Johnston, P. V., Kotkamps, M., Booth, C. R., Lucas, T., Mestechkina, T., Roy, C. R., Gies, H. P. & Tomlinson, D. (1995). Geographical differences in the UV measured by intercomparerd spectroradiometers. *Geophysical Research Letters*, **22**, 1889–92.

Singh, H. B. & Kanakidou, M. (1993). An investigation of the atmospheric sources and sinks of methyl bromide. *Geophysical Research Letters*, **20**, 133–6.

Singh, H. B., Salas, L. J. & Stiles, R. E. (1983). Methyl halides in and over the eastern Pacific (406y N–32° N). *Journal of Geophysical Research*, **88**, 3684–90.

Smith, R. C. & Baker, K. S. (1979a). Optical classification of natural waters. *Limnology and Oceanography*, **23**, 260–7.

Smith, R. C. & Baker, K. S. (1979b). Penetration of UV-B and biologically effective dose-rates in natural waters. *Photochemistry and Photobiology*, **29**, 311–23.

Smith, R. C. & Baker, K. S. (1981). Optical properties of the clearest natural waters (200–800 nm). *Applied Optics*, **20**, 177–84.

Smith, R. C. & Calkins, J. (1976). The use of the Roberson Meter to measure the penetration of solar middle-ultraviolet radiation (UV-B) into natural waters. *Limnology and Oceanography*, **21**, 746–9.

Smith, R. C. & Tyler, J. E. (1976). Transmission of solar radiation into natural waters. In *Photochemistry and Photobiology Review*, ed. R. C. Smith, pp. 117–55. Plenum Press, New York.

Smith, R. C., Prézelin, B. B., Baker, K. S., Bidegare, R. R., Boucher, N. P., Colery, T., Karentz, D., MacIntyre, S., Matlick, H. A., Menzies, D., Ondrusek, M., Wan, Z. & Waters, K. J. (1992). Ultraviolet radiation and phytoplankton biology in Antarctic waters. *Science*, **255**, 952–9.

Smith, R. C., Wan, Z. & Baker, K. S. (1992). Ozone depletion in Antarctica: modelling its effect on solar UV irradiance under clear sky conditions. *Journal of Geophysical Research*, **97**(C5), 7383–97.

Stamnes, K., Tsay, S. C., Wiscombe, W. & Jayaweera, K. (1988). Numerically stable algorithm for discrete-ordinate-method radiative transfer in multiple scattering and emitting layered media. *Applied Optics*, **27**, 2502–9.

Stramski D., Booth, C. R. & Mitchell, B. G. (1992). Estimation of downward irradiance attenuation from a single moored instrument. *Deep-Sea Research*, **39**, 567–84.

UNEP (1991). *Environmental Effects of Ozone Depletion: 1991*. United Nations Environmental Program, Nairobi.

UNEP (1992). *Methyl Bromide: Its Atmospheric Science, Technology and Economics* (*Assessment Supplement*). United Nations Environmental Program, Nairobi.

UNEP (1994). *Environmental Effects of Ozone Depletion: 1994*. United Nations Environmental Program, Nairobi.

UNEP (1998). *Environmental Effects of Ozone Depletion: 1998*. United Nations Environmental Program, Nairobi.

van de Hulst, H. C. (1981). *Light Scattering by Small Particles*. Dover Publications Inc., New York.

Varotsos, C. & Kondratyev, K. Y. (1995). On the relationship between total ozone and solar ultraviolet radiation at St. Petersburg. Russia. *Geophysical Research Letters*, **22**, 3481–4.

Wardle, D. L., Kerr, J. B., McElroy, C. T. & Francis, D. R. (1997). *Ozone Science: Canadian Perspective on the Changing Ozone Layer*. Environmental Canada, Toronto.

Webb, A. R. (1998). *UV-B Instrumentation and Application*. Gordon and Breach, Reading.

WMO (1988). *International Ozone Trends Panel Report: 1988*. World Meteorological Organization, Global Ozone Research and Monitoring Project, Report no. 18.

WMO (1989). *Scientific Assessment of Ozone Depletion: 1989*. United Nations Environment Program, World Meteorological Organization, Global Ozone Research and Monitoring Project, Report no. 20.

WMO (1991). *Scientific Assessment of Ozone Depletion: 1991*. United Nations Environment Program, World Meteorological Organization, Global Ozone Research and Monitoring Project, Report no 25.

WMO (1995). *Scientific Assessment of Ozone Depletion: 1994*. United Nations Environment Program, World Meteorological Organization, Global Ozone Research and Monitoring Project, Report no. 37.

WMO (1999). *Scientific Assessment of Ozone Depletion: 1998*. United Nations Environmental Program, World Meteorological Organization.

Yoshida, H. & Regan, J. D. (1997). UVB DNA dosimeters analyzed by polymerase chain reactions. *Photochemistry and Photobiology*, **66**, 82–8.

Zafiriou O. C., Joussot-Dubien, J., Zeep, R. G. & Zika, R. G. (1984). Photochemistry of natural waters. *Environmental Science and Technology*, **18**, 358A–371A.

Zerefos, C. S., Balis, D. S., Bais, A. F., Gillotay, D., Simon, P. C., Mayer, B. & Seckmayer, G. (1997). Variability of UV-B at four stations in Europe. *Geophysical Research Letters*, **24**, 1363–6.

3

○ ○ ○ ○ ○ ○ ○ ○ ○ ○ ○ ○ ○ ○ ○ ○ ○ ○ ○ ○

Spectral weighting functions for quantifying effects of UV radiation in marine ecosystems

Patrick J. Neale*

3.1 Introduction

Most studies of the effects of UV radiation (radiation with wavelength, λ, < 400 nm) on biological and chemical processes include some component that distinguishes how response varies with wavelength. How fine the spectral distinctions are made depends on the hypothesis to be tested. The simplest hypothesis is whether UV has any effects at all, which is tested using a presence/absence experiment. For example, the effects of treatment exposure to UV and visible radiation ($400 < \lambda < 700$ nm) is compared to exposure to visible radiation alone. If there is a difference, then a more complex hypothesis can be tested, e.g. 'Do effects differ between the three arbitrary divisions of the UV waveband: UV-C ($\lambda = 200$–280 nm), UV-B (280–320 nm)[1], and UV-A (320–400 nm)?' Radiation in the UV-C band is emitted by the sun but does not penetrate the earth's atmosphere. The UV-B band is the shortest wavelength UV in solar irradiance incident to the earth's surface, generally having strong biological effects per unit energy. The energy in the UV-B is small compared to the longer wavelength UV-A band, which has generally weaker negative effects per unit energy. This three-part distinction indicates whether solar UV radiation is capable of inducing effects. Other classifications of UV have been used that even further subdivide each band (Moeller, 1994).

At the other end of the spectrum (so to speak) are investigations that seek the finest resolution possible, i.e. a continuous measure of UV effects as a function of wavelength. A general term for such a description is a 'spectral weighting function' (SWF). It is any function that quantifies the

---

[1] The Commission Internationale de l'Éclairage (CIE) defines UV-B as 280–315 nm and UV-A as 315–400 nm. Here, the alternative boundary of 320 nm, which has been the convention in ecological studies, is used. With respect to biological processes, there are no consistent discontinuities in response at either 315 nm or 320 nm.

effectiveness (or 'weight') of UV at causing some effect in relation to wavelength. Two specific types of SWFs are action spectra and biological weighting functions. Action spectra are based on responses to narrowband (monochromatic) irradiance and are defined for both biological and chemical effects. Biological weighting functions are determined under broadband (polychromatic) irradiance and reflect the simultaneous (and sometimes competing) effects of multiple wavelength-dependent processes as they occur in Nature. The differences between these two types of SWFs are discussed in detail below.

A mathematically continuous weighting function is a useful theoretical concept. However, it is unattainable in practice. Both optical resolution and the sheer effort involved limits how narrowly the spectral window (bandwidth) can be defined around a measurement. In experimental exposures, it is difficult to obtain better than 2 nm bandwidth in the UV and in most instances the bandwidth will be considerably wider. Indeed, the concept of bandwidth unites all descriptions of the wavelength dependence of responses to UV, no matter how coarse or fine the distinctions are made. Bandwidth has a practical side to it, too. For example, an SWF with a wide bandwidth (e.g. all UV) is inappropriate for use in predicting effects of UV at a narrower bandwidth (e.g. short wavelength UV-B only).

In this chapter, the SWF is used as a unifying concept for all spectral descriptions used in UV studies. SWFs are examined from the standpoint of theory, experimental determination and analytical application. This introduction to SWFs, along with process-specific discussions in subsequent chapters, will lay the basis for answering questions about the effects of UV on processes in marine ecosystems, e.g. 'Is this process affected by UV in the marine environment? What is the mechanism of the effect? Will changes associated with ozone depletion (which mainly affect UV-B, see Chapter 2), have an effect on this process?' The discussion starts with the basic principles of action spectra, show how these principles are modified for environmentally relevant biological weighting functions and finally considers the use of weighting functions to predict responses of processes in marine ecosystems.

3.2   Action spectra

The term action spectrum is relatively recent; some history of the term is given by Coohill (1991). However, the concept is probably as old as the discovery that sunlight is a composite of radiation of many colours (wavelengths). This discovery stimulated a question: 'Do these component wavelengths of light have differential importance in inducing the effects of solar exposure?' An early example of an attempt to derive an action spectrum was the experiment in which an algal filament was illuminated with a solar spectrum from a prism (Engelmann, 1882). Observation was then made of the chemoattraction of bacteria to oxygen emitted by parts of the filament under different colours. Using the relative strength of attraction to different colours, i.e. the spectral response, Engelmann inferred the spectral characteristics of the photoactive agent (chromophore) involved. The qualitative action spectra defined by Engelmann and others in the nineteenth century were instrumental in the identification of chlorophyll as the chromophore involved in oxygenic photosynthesis (Coohill, 1991). Later, more quantitative work on the killing effects of UV illumination on bacteria revealed the spectral characteristics of the 'genetic material', later confirmed to be DNA (Gates, 1930).

The basic approach is to measure independently the effects of narrowband (monochromatic) radiation at many wavelengths and plot the results as a function of wavelength. The quantitative result is a set of weighting coefficients, $\varepsilon(\lambda)$, that measure the strength of UV action at each wavelength. These coefficients can be given as absolute weights, or as dimensionless ratios of action at each wavelength relative to that at a standard wavelength (e.g. 300 nm). The units of absolute weights are generally 'effect per unit exposure'. Specific units will depend on the units of the process measured and UV exposure, some commonly used units for measurement of spectral UV are $J\,m^{-2}\,nm^{-1}$ and $\mu mol\,photons\,m^{-2}\,nm^{-1}$. For example, Jones & Kok (1966, Table II) measured a specific rate of photoinhibition of chloroplast activity in 328 nm light of $0.0067\,s^{-1}$ at an incident irradiance (15 nm half-bandwidth) of $162\,\mu mol\,photons\,m^{-2}\,s^{-1}$. From this can be calculated a weighting coefficient of 4×10^{-5} ($= 0.0067/162$) in units of reciprocal $\mu mol\,photons\,m^{-2}$. If desired, a weighting coefficient is easily converted to energy units by using the ratio of the Planck constant in appropriate units and wavelength. In the case of the units already mentioned, multiplication by the ratio of $119.67/\lambda$ converts $\varepsilon(\lambda)$ in quantum units to $\varepsilon(\lambda)$ in energy units. In the Jones & Kok example, the weight at 328 nm becomes *circa* 1.5×10^{-5} in units of reciprocal $J\,m^{-2}$.

The precise determination of an action spectrum is a demanding exercise. Coohill (1991) discussed six criteria that are required for analytically useful action spectra. The criteria include transparency of the sample at the wavelengths of interest, minimisation of scattering, and reciprocity – the equivalence of intensity and time in determining total exposure. The criteria are all directed towards being able to quantify exactly the energy incident to, and absorbed by, a chromophore.

The requirements for determining precise action spectra can usually be satisfied for investigations of marine photochemical processes (see Chapter 4), but it is much more difficult to define action spectra for biological processes. Most cellular structures and molecules absorb and scatter light, and this can have significant effects on intracellular light fields even for marine micro-organisms (Morel, 1990). It can be difficult to know how much light reaches an intracellular chromophore. Take, for example, UV-induced cross-linking of intercellular DNA (see Chapter 6). The action spectrum of the effect has been determined for DNA molecules (Setlow, 1974), yet it is difficult to determine cellular DNA damage *a priori* because of uncertainty as to how much UV incident on the cell is transmitted to the DNA. However, intracellular optical effects can be empirically estimated (Quaite, Sutherland & Sutherland, 1992).

For UV effects on many physiological processes, the chromophore(s) are either uncertain or unknown. Chromophore(s) may be polypeptide or ligand molecules that are bound to a large complex *in vivo*. The absorption spectrum of the chromophore in the bound state can be quite different from the isolated molecule *in vitro*, as is well known for pigment molecules like chlorophyll (Goedheer, 1996). For example, the electron donor 'Z' within the photosystem II (PS II) reaction centre complex (a specific amino acid, a tyrosine, near the reaction centre) has been suggested to be an important target in UV effects on PS II (Greenberg *et al.*, 1989; Renger *et al.*, 1989). Tryptophan is another aromatic amino acid that could be a chromophore for UV damage to specific peptides (see Section 6.2.2).

Despite the lack of analytical action spectra for marine biological processes, it is still instructive to examine the typical features of biological action spectra. Two types of effect of UV that have been intensively studied are damage to the 'light reactions' of photosynthesis and damage to DNA. Example action spectra for these effects (Figure 3.1) reveal characteristics that are seen for almost all weighting functions for UV responses. Firstly, weights increase (effects become stronger) as wavelength decreases. For the same amount of energy, UV-B is more harmful than UV-A and within the UV-B band the shorter wavelengths are more harmful than the longer ones. Secondly, though weight generally increases at shorter wavelengths, the rate of increase varies between processes. For the spectra in Figure 3.1, weight increases much more rapidly at shorter wavelengths for the spectrum of UV-induced dimerisation of adjacent pyrimidine bases in DNA as opposed to the spectrum for inhibition of the chloroplast light reactions. This spectral slope is important because the greater is the relative increase in weight at shorter wavelengths, the more a process could be affected by ozone depletion, which preferentially enhances the shortest wavelengths of UV-B (see Chapter 2, also see Section 3.8). Finally, absolute weights can differ markedly for different types of damage, some processes having a greater overall sensitivity to UV than others do.

The action spectra in Figure 3.1 were not derived using marine organisms and so caution should be exercised in using them to interpret results of marine studies. Nevertheless, they are 'benchmark' spectra that are used for comparison with newly derived SWFs (Section 3.6). These two spectra typify processes that are strongly UV-B dependent (induction

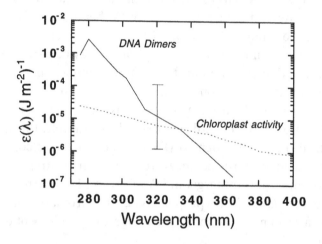

Figure 3.1. Examples of action spectra. These spectra show absolute weights ($\varepsilon(\lambda)$, reciprocal J m^{-2}) for UV damage to cellular DNA by formation of cyclobutane dimers (solid line) and inhibition of chloroplast electron transport (dashed line). Each spectrum has been scaled to 'lethality' (specific effect per unit exposure in J m^{-2}). The DNA spectrum is based on the Quaite et al. (1992) spectrum for DNA dimer induction in alfalfa seedlings, which was scaled to lethality based on the average dimer load for 1/e survival in Antarctic diatoms in the absence of photorepair (Karentz et al., 1991). The chloroplast spectrum is based on inhibition of electron transport in isolated spinach chloroplasts (Jones & Kok, 1966) and was converted to specific inhibition per unit exposure using the UV optical density (nominally, 0.3) and exposure cross-section (10^{-2}) stated by Jones & Kok (1966). The vertical bar indicates the likely variation in these spectra between organisms due to variations in the intracellular transparency to UV and, for DNA, lethality of dimers (Karentz et al., 1991). The analysis gives an overall indication of inherent sensitivity of biological targets to UV exposure in the absence of repair. In particular, UV damage through formation of cyclobutane dimers appears to be more lethal than inhibition of chloroplast electron transport in the UV-B, whereas the opposite is true in the UV-A.

of DNA dimers) versus processes for which both UV-B and UV-A are important (inhibition of chloroplast activity).

The effects of UV on the photosynthetic apparatus have been defined from exposure of isolated photosynthetic membranes (Jones & Kok, 1966; Bornman, Björn & Åkerlund, 1984; Renger *et al.*, 1989). These studies all point to PS II as being the most sensitive part of the photosynthetic electron transport chain. However, these spectra lack any prominent spectral peak (they are 'flat'), leading to some uncertainty as to what is the precise site of UV damage to PS II (Vass, 1997). Even in these simplified systems, the complex mixture of UV-absorbing molecules and structures can mask chromophore identity. An elegant study of UV effects on the photosynthetic apparatus was performed by Greenberg *et al.* (1989) using an aquatic plant (*Spirodela oligorhiza*) grown in intermittent light. Under these conditions, the pigment content of the tissue is very low, and the plants (which are small to begin with) had limited absorption of UV. The investigators then measured the rate of degradation of one polypeptide (D1) of the PS II reaction centre as an indicator of light-dependent damage. They found that the UV spectral response for degradation closely resembled the absorption by one electron transport intermediate, plastosemiquinone (Figure 3.2). The results implicated plastosemiquinone

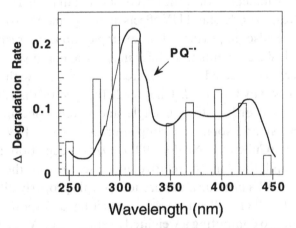

Figure 3.2. Application of an action spectrum to diagnose the mechanism of UV effects. The bars show the degradation rate of the photosystem (PS) II polypeptide D1 in the aquatic vascular plant *Spirodela oligorhiza* in response to equal incident quanta of monochromatic UV (6 μmol m^{-2}s^{-1}). Shown is the enhancement in the rate of degradation (h^{-1}) in intermittent light grown (low pigment) plants over the rate in normal plants. The action spectrum is compared to the absorption spectrum of the plastosemiquinone anion radical (PQ$^-$), an intermediate in PS II electron transport. The results suggest that PQ$^-$ is a photosensitiser for increased turnover of D1 in UV, see text for details. (Redrawn from Greenberg *et al.* (1989), with permission.)

as one of several candidate moieties that mediate UV damage effects on photosynthesis (for more information, see Chapter 6).

Action spectra are also useful tools for diagnosing screening effects. Comparative studies can reveal how absorption and scattering within an organism reduces UV impacts. Comparison of the quantum yield of DNA dimer formation in naked DNA (Setlow, 1974) and the reduced yield in alfalfa seedlings (Quaite *et al.*, 1992) suggested tissue screening of short wavelength UV. A similar conceptual approach was taken by Neale, Banaszak & Jarriel (1998a) to reveal the screening effects of mycosporine-like amino acids (see Chapter 7) for UV inhibition of photosynthesis in the dinoflagellate *Gymnodinium sanguineum*. However, Neale *et al.* did not use the standard action spectrum approach, instead they inferred weighting functions from responses to a combination of UV and photosynthetically available radiation (PAR) illumination. This brings up the question of what are the alternatives to measuring UV responses under precisely controlled, but highly artificial, monochromatic illumination. This is the subject of the next section.

3.3   Biological weighting functions

Identification of sensitising chromophores and detection of the screening effects are one part of understanding the effects of UV radiation. However, if the main objective is to understand UV effects in the marine environment, other questions are also important. For example, how do responses compare between UV treatments that differ in spectral composition and how might a process be affected by UV increases associated with ozone depletion? As discussed by Díaz *et al.* (Chapter 2), the spectral distribution of solar UV differs according to sun angle and ozone concentration (among other factors), and solar and lamp sources can also have large spectral differences (Cullen & Neale, 1997). Ozone depletion affects primarily UV-B, and the largest relative increases are in the short wavelength UV-B. SWFs provide an approach to scale properly different exposure spectra for their 'effectiveness' in eliciting a response. The concept is analogous to computing a weighted average of a UV spectrum, in that the weighting coefficients emphasise the short wavelengths relative to the long wavelengths. The approach is to calculate a single measure of effectiveness, E^*, by summing the product of spectral exposure, $E(\lambda)$, and a scaling coefficient, $\varepsilon(\lambda)$, over a series of narrow wavelength bands ($\Delta\lambda$):

$$E^* = \sum_{\lambda = 280\,\text{nm}}^{400\,\text{nm}} \varepsilon(\lambda)E(\lambda)\Delta\lambda \qquad (3.1)$$

If the $\varepsilon(\lambda)$ are chosen correctly, then an effect of UV exposure will be the same function of E^* independent of the spectral composition. The units of ε and E^* depend on how the effects and the UV are measured, as has already been mentioned in Section 3.2, so specific units are discussed in connection with specific examples in Section 3.5. In general, if the ε are in absolute units, then E^* will be a non-dimensional measure of UV damage. In contrast, if the ε are in relative units, the weighted UV will have the same units as spectrally integrated E, e.g. mWm^{-2}, and is termed biologically effective UV, E_{Beff}.

An action spectrum would seem to be an obvious choice for defining the $\varepsilon(\lambda)$ in Equation (3.1). However, using action spectra as measured by monochromatic radiation implicitly assumes that each waveband ($\Delta\lambda$) contributes *independently* to the overall effect, i.e. that there is no interaction between different wavebands. For photochemical effects, this is probably not a bad assumption; UV photochemistry in the marine environment seems to be predicted well from action spectra defined from monochromatic exposures (see Chapter 4).

If an action spectrum is to be used in Equation (3.1), another consideration is whether there is adequate coverage of the complete UV spectrum. There may be limited coverage of the UV-A band by analytical action spectra that were measured to define a chromophore with fine structure in the UV-B. However, good UV-A coverage is needed to predict responses to the high solar output in the UV-A (Coohill, 1991). Predicting response to UV-A is actually part of predicting the effects of ozone depletion, since the relative effect of an increase in UV-B depends on how much effects are also induced by UV-A (Madronich, 1992).

In contrast to chemical processes, the effects of UV on biological processes *in vivo* usually involve interactions between spectral regions. Illumination with UV-A and blue light induces photorepair of DNA (Hanawalt & Setlow, 1975; Friedberg, 1985), so net DNA damage by UV-B in intact organisms depends on whether the UV-B illumination is accompanied by exposure to a longer wavelength background (Hanawalt & Setlow, 1975). Similarly, repair of at least some components of the photosynthetic apparatus is dependent on whether photosynthesis is active, i.e. on the presence of PAR (Neale, 1987). Indeed, Jones & Kok (1966) recognised that repair processes confound studies of UV effects on PS II and chose to work with isolated membranes in which repair is inactive. Given that processes in living organisms are the net result of competing photodependent mechanisms that are all active during polychromatic treatment (Smith *et al.*, 1992; Prézelin, Boucher & Smith,

1994; Quesada, Mouget & Vincent, 1995), beneficial effects can outweigh damage over parts of the UV spectrum (e.g. Neale, Lesser & Cullen, 1994; see also Prézelin *et al.*, 1994; Boucher & Prézelin, 1996). Thus, monochromatic action spectra may not accurately describe the interplay between damage and repair processes in living organisms.

To be more representative of biological responses to environmental UV, measurements are made using treatments composed of a range of wavelengths, i.e. with polychromatic sources. One approach is to add monochromatic UV to a broadband background, which shows how much each wavelength enhances a baseline response. This approach retains the precision of an action spectrum, but is still quite unlike natural illumination (Coohill, 1991). A more realistic polychromatic approach is to generate a set of spectra using cut-off filters, i.e. filters that pass longer wavelength light starting at successively shorter cut-off wavelengths. The trade-off for greater realism is that effects can no longer be precisely attributed to specific wavelengths. Instead, the weights are composites of the effects at that wavelength and the interactive effects of other wavelengths (Coohill, 1991)[2]. These empirical weighting functions are referred to here as 'biological weighting functions' (BWFs), though they have also been referred to as polychromatic action spectra (Coohill, 1991) or just considered to be another type of action spectrum (Caldwell *et al.*, 1986).

3.4   Relating response to exposure

A crucial point that we now consider is how to go from properly weighted exposure to a predicted biological response. Biological responses reflect the net result of damage and processes that counteract that damage (see Chapters 6 and 7). Counteracting processes include repair or protection, which can be immediate and ongoing. Other counteracting processes may be induced sometime after exposure begins (acclimation) or after exposure ends (recovery). Thus the response to UV exposure will depend on the time scale over which the prediction is to be made. In this section, the basic concept of the exposure response curve (ERC; Coohill, 1994) is introduced. This provides a tool for developing the most appropriate response model for a given exposure time scale.

---

[2] In theory, interactive effects could be directly, and precisely, accounted for by adding coefficients for cross-products of irradiance at different wavelengths to Equation (3.1). But in most cases the specification of the relevant coefficients of the cross-products would be an arduous, if not impossible, experimental objective.

3.4.1 Kinetics of UV damage

The result of weighting irradiance according to Equation (3.1) is a single measure of the biological effectiveness, namely effective irradiance (E^* or E_{Beff}). A simple way to view the biological response to this irradiance is to use an analogy with a one-component photochemical reaction, in which a 'reactant' (the chromophore) results in a 'product' (an inactivated target) at a rate, k, dependent on weighted irradiance ($k \propto E^*$). Such a reaction is described by first-order (exponential) kinetics (see Chapter 4). Specifically, the quantity of a target, P, varies with time according to the model:

$$\frac{\partial P}{\partial t} = -kP \qquad (3.2)$$

where k is now the rate of inactivation of P. This equation states that effects are proportional to weighted irradiance as long as the incremental response is small and there are no counteracting processes to reactivate P. This would be a good approximation to Nature if the pool size of susceptible targets (P) is large compared to the rate of inactivation ($\partial P/\partial t$). An example could be DNA. Under environmental conditions only a small proportion of potential sites are damaged by UV, e.g. dimerisation of adjacent pyrimidine bases. It has, therefore, been assumed that variation in cancers induced by UV, thought to arise from DNA damage, is directly proportional to variation in weighted irradiance (Madronich, 1992).

A linear relationship, an approximate solution of Equation (3.2), is the simplest form of an ERC. But the assumptions needed to apply a linear approximation are not met for many biological processes. For example, rates of photosynthesis much less than 50% of a PAR-only control are measured for samples incubated several hours at surface irradiance (e.g. Helbling, Villafañe & Holm-Hansen, 1994; Prézelin et al., 1994; Vernet et al., 1994). In this case, the effect will no longer be proportional to exposure and the effects must be integrated over a declining pool size. For the simple damage model (Equation (3.2)), the time course of effects will be described by an exponential 'survival curve' (Harm, 1980),

$$P(t) = P_0 e^{-H^*}$$

$$H^* = \int k \partial t = \int E^* \partial t \qquad (3.3)$$

where $P(t)$ is the amount of target, initially at level P_0, after a period, t, during which the integrated weighted exposure is H^*. For simplicity, k is scaled directly with E^*.[3] Biological processes would be expected to obey this type of ERC if no significant renewal of targets occurs during the exposure period. Examples include acute sunburn (erythema) and inhibition of photosynthesis in deeply mixed assemblages of Antarctic phytoplankton (Neale, Cullen & Davis, 1998b).

An important property of both Equations (3.2) and (3.3) is that the total amount of damage (decrease in P) is a function of cumulative exposure (H^*). Thus reciprocity holds: the overall effect will not depend on the time or intensity of exposure as long as the total amount of exposure is the same. This behaviour is sometimes called 'dose-dependence'. Comparing Equations (3.2) and (3.3), it is also seen that reciprocity does not necessarily imply that effects are directly (linearly) proportional to exposure. Processes following Equation (3.3) obey reciprocity (intensity and time are interchangeable); nevertheless the ERC is non-linear.

3.4.2   Kinetics of UV damage and recovery

In other biological processes, renewal of targets (by repair, reactivation or resynthesis) is ongoing during exposure and therefore reciprocity would not apply. However, there are few direct tests of reciprocity for UV effects in marine ecosystems. Reciprocity for the inhibition of photosynthesis by UV-B radiation has been tested in the laboratory (Cullen & Lesser, 1991). Cultures of a marine diatom growing at 20 °C were exposed to different irradiances of supplementary UV-B for periods of 15 min to 4 h. The rate of photosynthesis declined in response to UV-B, and within about 30 min reached a rate that was maintained for the remainder of the experiment. The time course was consistent with an ERC in which repair (r) is proportional to the pool size of inactivated targets (Lesser et al., 1994):

$$\frac{\partial P}{\partial t} = -kP + r(P_0 - P) \tag{3.4}$$

A steady state ($\partial P/\partial t = 0$) is eventually reached that is a function of the ratio of damage and repair rates:

$$\frac{P}{P_0} = \frac{r}{(r+k)} = \frac{1}{\left(1 + \dfrac{k}{r}\right)} \tag{3.5}$$

The hyperbolic function specified by Equation (3.5) described steady-state

---

[3] Biologically effective irradiance, E_{Beff}, could also be used, but an additional parameter has to be introduced to scale E_{Beff} to k.

rates of photosynthesis in the laboratory culture under UV + PAR exposure consistent with a dynamic balance between damage and repair (Lesser, Cullen & Neale, 1994). The steady-state rate of photosynthesis depends on k, which in turn depends on UV exposure rate (weighted irradiance). Thus, reciprocity fails and inhibition cannot be described solely as a function of cumulative exposure. In contrast to the ERCs in Equations (3.2) and (3.3), the steady-state rate (Equation (3.5)) shows 'dose rate' dependence.

The ERCs that depend solely on the cumulative exposure (Equation (3.3)) or an exposure rate (Equation (3.5)) are special cases of a general kinetic model of UV damage (Neale, 1987; Lesser *et al.*, 1994). More generally, repair could be significant during exposure but not fast enough to attain steady state over the time scale of interest (Figure 3.3). In this situation, the kinetics of the transition become important. Some example cases are presented here. For dynamics following Equation (3.4), i.e. damage and repair proportional to the number of targets that are

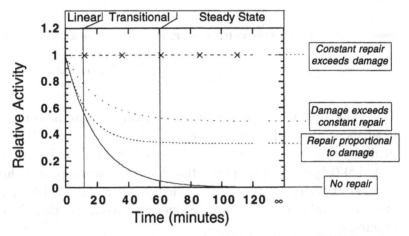

Figure 3.3. Theoretical time courses of UV damage (e.g. inhibition of photosynthesis) to illustrate how the presence and regulation of repair affects UV responses. There is one curve for each of the kinetic cases presented in Section 3.4. Solid line: case of no repair, activity decreases to zero (Equation (3.3)). Small dashes: repair is active and proportional to number of damaged sites (Equation (3.6)), activity decreases to a steady state. Large dashes: repair is active and constant, damage exceeds the repair rate, and activity decreases to a steady state (Equation (3.7)). Dashes with x symbols: repair is active, constant, and exceeds damage rate – no net decrease in activity. In each case, the equation was evaluated with a k of 0.05 min⁻¹ and r of 0.025 min⁻¹, except the upper curve, which is Equation (3.7) with $k < r$. Labels above the plot indicate time scale dependence: for short exposures (<15 min in this case) response is approximately linear (Equation (3.2)); for long exposures (>60 min) steady-state levels apply (e.g. Equation (3.5)), for intermediate time-scales (15–60 min), the transitional (non-linear) kinetics apply (Equation (3.6)).

inactivated, and assuming a fixed weighted irradiance, the time course of decrease in targets follows:

$$\frac{P(t)}{P_0} = \left(\frac{r}{k+r}\right) + \left(\frac{k}{k+r}\right)e^{-(k+r)t} \tag{3.6}$$

This solution has a combination of exposure and irradiance dependence that is shown graphically in Figure 3.3. This equation simplifies to the cumulative exposure model when r is zero (cf. Equation (3.3) and remembering that $kt = H^*$ if irradiance is constant). On the other hand, if r is greater than zero, Equation (3.6) simplifies to the steady-state model for large t (i.e. $t > [1/(k+r)]$).

As another example, consider the case where repair is constant at fixed irradiance, irrespective of the pool size of inactivated targets. This would apply to repair that is light regulated through a photoreceptor, i.e. to UV-A or PAR. In this case, the time course of activity would be in the following form:

$$\frac{P(t)}{P_0} = \text{for } k \leq r, 1$$

$$\text{for } k > r, \frac{r}{k} + \frac{(1-r)}{k}e^{-kt} \tag{3.7}$$

This model is notable in that there is an explicit threshold in effective irradiance below which no net effects would be observed, a distinct qualitative difference from the repair model described in Equations (3.4) and (3.5) (Figure 3.3). Response thresholds do not appear to hold for photosynthetic response to UV (Cullen, Neale & Lesser, 1992), but are more frequently observed for organism survival (see Chapters 7 to 10 for examples).

3.5   Marine biological weighting functions

The specific application of the principles of BWFs and exposure response curves are now discussed in the context of BWFs for marine processes. A number of BWFs have been defined, but almost all of them concern a single process, marine phytoplankton photosynthesis. The only other published marine BWF is for UV effects on the survival of fish larvae (Hunter, Taylor & Moser, 1979). Hopefully, BWFs for more processes will become available in the near future. For example, efforts are underway to define BWFs for UV effects on bacterial assimilation and growth (P. J. Neale & W. H. Jeffrey, unpublished data). Even for phytoplankton, an understanding of the spectral effects on growth (versus photosynthesis) is lacking (Jokiel & York, 1984). Presently, we have very little information on the spectral sensitivity of other marine organisms that have high potential for exposure to UV, including zooplankton and larvae of various invertebrates, together with coral.

Phytoplankton photosynthesis has been a focus for marine UV research, since phytoplankton production is a major component of the global carbon cycle and, as photoautotrophs, they need solar exposure for growth (see Chapter 9). Consequently, over the last two decades several research groups have investigated the BWF for inhibition of photosynthesis by UV. Efforts to define BWFs for UV effects on phytoplankton photosynthesis in the Southern Ocean have been particularly intense, since marked enhancement of incident UV-B occurs during the spring period of intense ozone depletion (the ozone hole) and Antarctic food chains depend on phytoplankton productivity (Smith & Nelson, 1986).

A selection of BWFs for UV inhibition of photosynthesis in Antarctic, temperate and tropical marine environments, and a temperate estuary, is presented in Figure 3.4. The comparison shows a wealth of variability in the overall sensitivity to UV and comparative effects of different parts of the UV band. A complication is that inhibition of photosynthesis has been related to both weighted exposure ($J m^{-2}$) and weighted irradiance ($mW m^{-2}$). Apparently reciprocity (Section 3.4) applies in some instances but not in others. In either case, sensitivity (effect per unit UV) apparently varies by several orders of magnitude between environments. Within oceanic regions, sensitivity of phytoplankton assemblages can vary by more than a factor of 3, e.g. Figure 3.4b, Weddell–Scotia Confluence (WSC). There is also considerable variation in spectral shape (relative effect as a function of wavelength), though spectral shape seems less variable within environments as opposed to between environments. The slope of BWFs for UV inhibition is less variable within the UV-B compared to the UV-A. Indeed, Behrenfeld et al. (1993) proposed that a single log-linear slope of − 0.134 per nanometer could be used to express the spectral shape of the BWF in the UV-B. The proposed slope does approximate the shape of other BWFs, e.g. the < 300 nm portion of the WSC BWF (Figure 3.3b), but slopes are variable and generally lower than − 0.134 per nm in the longer wavelength UV-B.

The BWFs display the most variability in the UV-A region, both within and between environments. UV-A weights can vary from being positive and only one to two orders of magnitude less than UV-B weights, to being near zero or negative. Among other factors, the high variability in UV-A weights probably reflects the multiplicity of effects that UV-A can have,

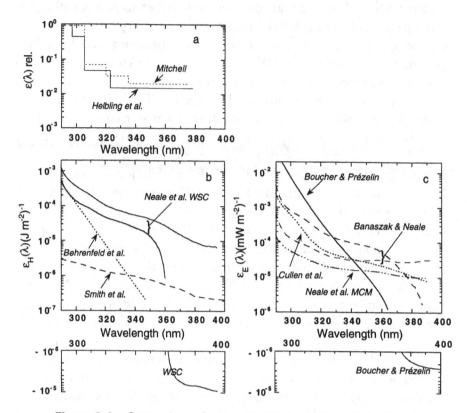

Figure 3.4. Comparison of selected BWFs for UV inhibition of marine photosynthesis. Annotations and arrows indicate source of each BWF. Panels (b) and (c) include lower sections below the wavelength legend, which show negative weights. (a) BWFs estimated by simple difference. A single weight ($\varepsilon(\lambda)$, dimensionless relative to maximum weight) is plotted for each wavelength band, based on nominal cut-off wavelengths of treatment spectra. Shown are spectra for coastal assemblages of Antarctic phytoplankton from Helbling *et al.* (1992) and Mitchell (1990) as described by Lubin *et al.* (1992). (b) BWFs for weighting cumulative exposure ($\varepsilon_H(\lambda)$, reciprocal J m^{-2}) estimated by template fitting. Two types of BWFs are shown. (1) General BWFs for phytoplankton from a range of environments as presented by Behrenfeld *et al.* (1993) and Smith *et al.* (1980). The BWFs shown are the product of the published $\varepsilon(\lambda)$ (non-dimensional) and a linear dose–response coefficient (reciprocal J m^{-2}) for the specific decrease in photosynthesis (P) due to UV [$= (P_{control} - P_{UV\,exposed})/P_{control}$] per unit of biologically effective exposure (H_{Beff}, J m^{-2}). (2) BWFs for two stations in the Weddell–Scotia Confluence (WSC) having assemblages inferred to be acclimated to high light (lower line) and low light (upper line) during austral spring 1993 (Neale *et al.*, 1998b). The lower panel shows negative $\varepsilon_H(\lambda)$ for the high light WSC assemblage. (c) BWFs for weighted irradiance ($\varepsilon_E(\lambda)$, reciprocal mW m^{-2}) estimated by template fitting for (1) the effect of average daily irradiance on a coastal Antarctic assemblage near Palmer Station (Boucher & Prézelin, 1996), (2) the effect of 1 h exposures on a diatom (*Phaeodactylum* sp.) in culture (Cullen *et al.*, 1992), natural assemblages from the Chesapeake Bay with high (upper line) and low (lower line) sensitivity to UV (A. T. Banaszak & P. J. Neale, unpublished data) and 30-min exposures of a natural culture of diatoms from McMurdo Station (MCM, Neale *et al.*, 1994). The lower panel shows the negative $\varepsilon_E(\lambda)$ of the Boucher & Prézelin (1996) BWF. Further details on the comparison are given in the text.

inducing both damaging (see Chapter 6) and restoring (see Chapter 7) processes. Furthermore, screening compounds, such as mycosporine-like amino acids and scytonemin, usually have the strongest absorption in the UV-A region (see Chapter 7). In two of the example BWFs, weights are negative in the longer UV-A wavelengths (Figure 3.4b,c). Overall, the variability in UV-A weights relative to UV-B weights has a strong influence on the sensitivity of photosynthesis to ozone depletion (Section 3.8).

3.6   Biological weighting function methodology

While much of the variation in BWFs shown in Figure 3.4 probably does relate to real differences in sensitivity and spectral response, it is also true that diverse experimental and analytical approaches have been used to estimate the BWFs. In this section, different approaches will be discussed with a view towards understanding how approach affects estimation of the BWF.

3.6.1   Experimental considerations

The basic approach for obtaining environmentally relevant BWFs is to measure responses to polychromatic irradiance, which are obtained using cut-off filters (Figure 3.5). The particular method used to measure UV effects will depend on the process; relevant discussion can be found in Chapters 5 and 6. However, a necessary step in designing any experimental protocol is the choice of UV source and the number of filters to be used. Sources that have been used include filtered incident solar irradiance alone, solar irradiance supplemented by UV lamps, and solar simulator (xenon arc) lamps. Solar simulators can approximate the spectral distribution of solar irradiance, but there is no lamp/filter combination that exactly reproduces solar irradiance. Spectral features of solar irradiance that are difficult to simulate are (a) the sharp drop in energy with wavelength in the UV-B (see Chapter 2), and (b) the high ratio of UV-A and PAR to UV-B. Furthermore, no artificial source can simulate the changes of spectral irradiance with depth in surface waters containing various constituents (Kirk *et al.*, 1994). Fluorescent UV lamps, which have wide commercial availability due to their use as tanning lamps, have much higher amounts of short wavelength versus long wavelength UV-B than are found in solar irradiance even after filtering through cellulose acetate sheets to remove the UV-C component (Cullen & Neale, 1997). Since the BWF weight at any wavelength implicitly includes interactions with other wavelengths, there can be some question about the capability of a BWF that was estimated using artificial sources to predict responses to solar irradiance. On the other hand, use of solar irradiance alone provides little reliable variation in the shortest UV-B wavelengths (< 305 nm, cf. Figure 3.5b), though the rotation of the ozone hole provides a natural source of spectral variation in Antarctica (Smith *et al.*, 1992). Variability in cloud cover also complicates experiments with natural irradiance. Questions about the predictive power of a BWF measured with any source can be addressed by comparing BWF-based predictions to observations of UV effects using an independent set of response – irradiance measurements (Lesser, Neale & Cullen, 1996).

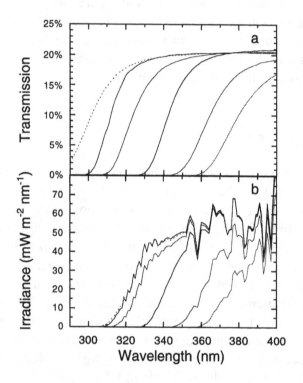

Figure 3.5. Use of cut-off filters to generate polychromatic irradiance regimes. (a) Spectral transmission curves for a set of example treatments that were obtained by combinations of nylon window screen, UV transparent acrylic sheet (UVT Plexiglas™), and Schott longpass filters (from left to right) WG305 (dotted line), WG320, WG335, WG345, WG360 and WG385. (b) Actual treatment irradiance obtained by applying transmission curves in (a) to solar spectral irradiance in this example for the late summer at 74° N. This illustrates the interaction of the fine structure of solar irradiance with filter transmission. In particular, the short wavelength cut-off of solar irradiance in the WG305 treatment (dotted line) occurs at much longer wavelength than the spectral transmission cut-off for the filter in (a), due to the very low amount of solar energy at the short wavelengths. (A. T. Banaszak, P. J. Neale, K. H. Dunton & H. L. Miller, unpublished data.)

Materials used for longpass cut-off filters include Schott filter glass, various types of other glasses and plastics. Examples of UV treatment spectra obtained by longpass filters are shown in Figure 3.5b (see also Caldwell *et al.*, 1986; Cullen *et al.*, 1992). Schott filter glass is the best choice from the standpoint of durability and optics; however, it is expensive and some filter types are no longer manufactured. Other materials used include acrylic sheet (Plexiglas™), Mylar, and polycarbonate. It is well to keep in mind that cut-off properties vary between manufacturers (Karentz *et al.*, 1994) and can change during exposure. Choice of the number of filters used is a trade-off: resolution will improve as the number of filters is increased but there are practical limits to the number of spectral treatments (for further discussion of resolution, see Section 3.6.2). Typically, filter cut-offs are spaced at 10 to 20 nm intervals. Finally, determining a BWF, and its subsequent application in any other context, depends directly on measuring spectral irradiance at adequate resolution. Minimum resolution for spectral irradiance should be at least 2 nm in the UV-B, this will ensure that the BWF will have the maximum resolution possible for the spectral treatments used. Irradiance spectra defined by filtering either solar or lamp sources with cut-off filters can have complex features (Figure 3.5b) and it is important to take this into account during the data analysis (Neale *et al.*, 1994).

3.6.2 Analytical considerations

Calculation of an action spectrum is unambiguous. The effect is related to exposure at each wavelength through an exposure response curve (Sections 3.3 and 3.4). Calculation of a BWF is more complex. As for an action spectrum, one starts with a set of responses and their associated treatment spectra (e.g. Figure 3.5). There are three strategies that can be used to derive a spectrum from such a data set: (a) simple difference, (b) adjusting an assumed function, (c) component analysis.

In general, there is an increasingly greater effect as successively shorter wavebands of UV are transmitted to the treatment. A simple analysis is to order the results by increasing UV exposure and to compute the difference in effect between successive treatments. Then the biological weight for each waveband is estimated as the differential effect divided by the difference in energy between the treatments that include and exclude that waveband. However, as Rundel (1983) pointed out, the estimated weight may be inaccurate if the actual response changes rapidly over the wavelengths for which the treatments differ. This could be the case if the differential bandwidth exceeds 5 nm in the UV-B or 10–15 nm in the

UV-A (cf. Figure 3.4). Moreover, there is no objective method to determine the effective centre wavelength for an estimated weight without already knowing a high resolution spectrum (Rundel, 1983).

The advantage of the difference method is its simplicity, and it was used to estimate useful, albeit crude, BWFs for the effect of UV on photosynthesis by Antarctic phytoplankton. Smith *et al.* (1992) measured *in situ* photosynthesis under PAR, PAR + UV-A and PAR + UV-A + UV-B for phytoplankton from the Bellingshausen Sea. The simple difference method was used to calculate what is, in effect, a two-point BWF for UV-B and UV-A wavebands. Subsequently, a greater number of spectral treatments was used in exposures of coastal Antarctic assemblages of phytoplankton (Mitchell, 1990; Helbling *et al.*, 1992; Lubin *et al.*, 1992) and enabled calculation of four relative weights (Figure 3.4a). Despite the limitations of the difference approach, these initial results were useful in obtaining the first estimates of the relative increase of biologically damaging irradiance associated with Antarctic ozone depletion.

The only way to improve upon the coarse resolution of the difference method is to make some assumption about the shape of the BWF. The overall concept is to start with a spectral 'template', some function that represents the general features of a BWF, e.g. that weights increase as wavelength decreases. This template is then adjusted so as to mimic best the observed responses to the spectral treatments. There are several variations on this approach, which mainly differ in how modification of the template is constrained. An approach that tightly constrains the BWF shape is to assume that the BWF is similar to an already known action spectrum. A good example is the pioneering study of Smith *et al.* (1980) on the spectral sensitivity of marine photosynthesis to UV. The relative change in photosynthesis in each treatment was compared to irradiance weighted by a DNA damage spectrum (Setlow, 1974) and the spectrum for inhibition of chloroplast electron transport (Jones & Kok, 1966; cf. Figure 3.1). Inhibition was more consistently predicted by the latter spectrum. The authors did formulate a dose–response relationship based on the chloroplast spectrum (Figure 3.4b) but were cautious in assigning general significance to this result: they did not conclude that the Jones & Kok spectrum predicts effects of UV on marine photosynthesis, rather that the DNA spectrum was not an acceptable predictor. In retrospect, it does appear that constraining the shape of the BWF probably led to a substantial underestimation of the weights in the UV-B (Figure 3.4). Hunter *et al.* (1979) related weighted exposure to the survival of anchovy and mackerel eggs using several UV-B action spectra and found that survival was best predicted when UV exposure was weighted by the Setlow DNA action spectrum. However, the data were later reanalysed by Smith & Baker (1982), who found that a shallower spectral slope gave a better fit. Buma, Engelen & Gieskes (1997) found that inhibition of diatom growth by UV was in agreement with the predictions of the Setlow spectrum.

Constraining the BWF to the same shape as a pre-existing action spectrum is probably only necessary if a limited number of spectral treatments are used (three in the case of Smith *et al.*, 1980). If a larger number of treatments is available, the constraints can be relaxed. A general property of most action spectra is an approximate log-linear increase in weight as wavelength decreases (Rundel, 1983; Behrenfeld *et al.*, 1993). Thus, a simple, but general, template for a BWF is an exponential function, i.e. $\varepsilon(\lambda) \propto e^{-a\lambda}$, with the characteristic log-linear slope (a) to be determined for each particular BWF. The slope giving the best fit to observed responses given corresponding spectral irradiance can be readily estimated using non-linear regression (Marquardt, 1963). Behrenfeld *et al.* (1993) used this approach to estimate a generalised BWF for UV-B effects on photosynthesis that extended to 340 nm (Figure 3.4b). However, a single log-linear slope is not apparent in other BWFs for inhibition of photosynthesis, even when just considering weights for wavelengths < 340 nm (Figure 3.4). If a single log-linear slope is assumed, even though multiple slopes are present, then the fitted slope will correspond to the spectral region that contributed most to the differences between experimental treatments (Cullen & Neale, 1997; Neale *et al.* 1998b). Indeed, the Behrenfeld *et al.* (1993) analysis (Figure 3.4b) was relatively insensitive to the effects of longer wavelength UV-B, since it emphasised explaining the relative decrease between treatments that differed primarily in irradiance at wavelengths shorter than 310 nm.

Rundel (1983) recognised that BWFs could have a more complex shape than a single log-linear slope and advocated using a more generalised exponential function, e.g. that the natural log of the BWF is a polynomial:

$$\varepsilon(\lambda) = e^{-(a_0 + a_1\lambda + a_2\lambda^2 + \ldots)} \tag{3.8}$$

Many BWFs can be approximated by such a general exponential function (Figure 3.1; Boucher & Prézelin, 1996). Again, the parameters for the BWF polynomial (a_i) can be estimated by non-linear regression (a complete analytical protocol is given by Cullen & Neale, 1997). The constraints on the template can be relaxed even further by introducing thresholds, offsets (which allow negative weights), etc. (Rundel, 1983), but these complexities place additional statistical demands on the base data set. The principle of Occam's razor always applies: the model should be chosen to explain the most variation possible with the least number of parameters. The regression results can also be used to estimate the uncertainty in the BWF, based on the technique of propagation of errors (Bevington, 1969).

A third method of estimating BWFs places a minimum of constraints on the template. In this technique (Cullen et al., 1992; Neale et al., 1994; Cullen & Neale, 1997), UV spectral irradiance in each treatment, normalised to PAR, is analysed by principal component analysis (PCA) to generate up to four principal components (essentially, statistically independent shapes defined by weights for each wavelength), which account for nearly 100% of the variance of the treatment spectra relative to the mean spectrum. Component scores (the relative contribution, c_i, of the $i = 1$ to 4 principal components to a given UV spectrum) are derived for each of the normalised treatment spectra. The estimation of the BWF then proceeds by non-linear regression as for the Rundel method, except that the estimated coefficients represent the contribution of each spectral component to weighted irradiance. For the case of the response of photosynthesis to UV irradiance, coefficients m_i (reciprocal $W\,m^{-2}$) are estimated such that

$$E^* = E_{PAR}\left(m_0 + \sum_{i=1}^{i=4} m_i c_i\right)$$

where E_{PAR} is PAR in $W\,m^{-2}$, m_0 is the coefficient for any PAR effects that are not part of the spectral treatments, plus the weight of the mean spectrum, and m_i are the contributions of the spectral components. Again, only as many components are incorporated into the final estimate as can be justified on the basis of the variance explained. Usually it is necessary to include no more than two or, at most, three spectral components. Finally, once the m_i are estimated by regression, the estimates of $\varepsilon(\lambda)$ and their respective statistical uncertainties can be calculated via the original spectral components. Further details of the method are given by Cullen et al. (1992) and Neale et al. (1994), and a step by step protocol is provided by Cullen & Neale (1997). Examples of spectra calculated using the PCA method are shown in Figure 3.4b (Neale et al., WSC) and 3.4c (Banaszak & Neale, Neale et al., MCM). The BWFs explained >90% of the variance in photosynthesis of 72 spectral treatments (eight cut-off filters and nine intensities), and the weighting functions show features more complex than simple exponential slopes. For simplicity, these graphs do not show the associated confidence intervals. As a general guide, the pairs of BWFs shown for the same environment are statistically different ($p < 0.05$, t-test), though differences much smaller than those shown would not be statistically significant.

3.7   Comprehensive models of UV effects

The previous sections have shown that BWFs and exposure response curves are closely linked. In this section, approaches that combine BWFs and exposure response curves into a single comprehensive model of UV response are considered. As indicated above, the structure of the model will depend on what aspects of the process are to be predicted. If repair is active and the objective is to model the steady-state response during UV and PAR exposure, the correct approach is to weight irradiance ($E(\lambda)$, $mW\,m^{-2}\,nm^{-1}$, Figure 3.4c) and predict response as a hyperbolic function of weighted irradiance. This approach was taken to develop an integrated model of photosynthetic response to UV and PAR, i.e. the BWF-PI (BWF-photoinhibition) model, with weightings, $\varepsilon(\lambda)$, having units of reciprocal $mW\,m^{-2}$ (Cullen et al., 1992). In this model, the achieved rate of photosynthesis (P^B, $g\,C\,g\,Chl^{-1}\,h^{-1}$) is the product of a PAR-dependent optimal rate in the absence of photoinhibition (P^B_{opt}) and the ERC for UV and PAR inhibition:

$$P^B = \frac{P^B_{opt}}{1 + E^*_{inh}}$$

$$E^*_{inh} = \sum_{\lambda=280\,nm}^{\lambda=700\,nm} \varepsilon(\lambda)E(\lambda)\Delta\lambda \qquad (3.9)$$

In theory, weighted irradiance for inhibition (E^*_{inh}, dimensionless) is evaluated for the full UV and PAR spectrum, since it is known that both UV and supraoptimal intensities of PAR can be inhibitory (Neale, 1987). If the spectral variation of $\varepsilon(\lambda)$ within PAR is not being investigated, then a single average weighting coefficient is defined for the whole PAR spectrum (ε_{PAR}). This BWF-PI model was modified for the case when repair is not active (Equation (3.3) applies) and the objective was to predict average photosynthetic rate over a specific exposure period (Neale et al., 1998b). In this case, we weight radiant exposure ($H(\lambda)$, $J\,m^{-2}\,nm^{-1}$) to compute weighted exposure for inhibition, H^*_{inh}, using an equation similar to that for E^*_{inh}:

$$H^*_{\text{inh}} = \sum_{\lambda = 280 \text{ nm}}^{\lambda = 700 \text{ nm}} \varepsilon_{\text{H}}(\lambda)H(\lambda)\Delta\lambda \tag{3.10}$$

where $H(\lambda) = \int E(\lambda)dt$. The weightings, ε_{H} (reciprocal J m^{-2}, cf. Figure 3b), define the effects of radiant exposure in contrast to irradiance. The average P over the period of cumulative exposure is obtained by integrating the ERC (Equation (3.3)):

$$P^B_{\text{avg}} = P^B_{\text{opt}} \frac{(1 - e^{-H^*_{\text{inh}}})}{H^*_{\text{inh}}} \tag{3.11}$$

This second model is termed the BWF$_{\text{H}}$-PI model because of the dependence on H, compared to the earlier BWF$_{\text{E}}$-PI model. Thus for the case of photosynthesis models, the type of BWF defined depends strongly on which ERC is applied. To define accurately both of these functions, observations will usually be needed of the variation in UV effects in relation to both varying spectral composition (for the BWF) and exposure times (for the ERC). Measurements of photosynthesis during time courses and under 72 polychromatic UV-PAR treatments defined the BWF$_{\text{E}}$-PI model for cultures of diatoms and dinoflagellates at 20 °C (Cullen et al., 1992). More recently, time course and spectral response data have been measured for phytoplankton assemblages in a temperate estuary (A. T. Banaszak & P. J. Neale, unpublished data). As in the previous culture measurements, photosynthesis by these assemblages attained a steady-state level during UV + PAR exposure, and thus photosynthesis was modelled using the BWF$_{\text{E}}$-PI approach. In sharp contrast, recovery processes made a negligible contribution to the response of phytoplankton sampled from the open waters of the WSC in the Southern Ocean (Neale et al., 1998b). Photosynthesis continuously declined during time courses of UV exposure and rates remained low even after return to a benign irradiance regime. Consequently, the BWF$_{\text{H}}$-PI model was developed in which the cumulative effect of inhibition is described as the integral of a semi-log survival curve.

Time course observations were critical to justifying the use of the BWF$_{\text{H}}$ model in the WSC. Both the BWF$_{\text{E}}$ or BWF$_{\text{H}}$ models were statistically equivalent in accounting for photosynthesis over a single incubation period (1 h). Each model predicts that photosynthesis over a 1 h incubation will exhibit a quasihyperbolic decrease with increasing exposure to inhibiting irradiance. Thus, measurements over a single incubation time are not sufficient to unambiguously define the exposure response curve.

In the estimation of several of the BWFs in Figure 3.4, simplifying assumptions were made as to the exposure response curve. In cases where reciprocity was assumed to apply, a further assumption was made that inhibition of photosynthesis was linearly related to weighted exposure (Smith *et al.*, 1980; Behrenfeld *et al.*, 1993). When a response is observed to a 'long' incubation period, i.e. sufficient duration that effects become target limited and/or a steady state is reached, a linear analysis will underestimate weights. This is because the ratio of response to exposure over a long incubation will be less than the ratio over a short interval. This analytical factor may contribute, along with between environment differences in sensitivity, to the particularly low weights of the Smith *et al.* (1980) and Behrenfeld *et al.* (1993) BWFs compared with the Neale *et al.* (1998b) BWFs in the same figure.

3.8   Predicting the effects of ozone depletion

As a final point, the implications of these discussions on the estimation and prediction of the effects of increased UV-B that result from decreased atmospheric ozone concentration are briefly considered. Some of the fundamental principles have already been introduced in Chapter 2. For completeness, I restate the approach here and consider how it might be extended. Given an appropriate SWF and calculations of solar spectral irradiance as a function of column ozone, it is possible to estimate a radiation amplification factor (RAF), a coefficient that quantifies the non-linear change in effective irradiance as a function of the change in column ozone (Booth & Madronich, 1994; Smith & Cullen, 1995):

$$\frac{E_1^*}{E_2^*} = \left(\frac{\omega_2}{\omega_1}\right)^{-RAF} \tag{3.12}$$

where E_1^* and E_2^* represent biologically effective irradiance at column ozone amounts ω_1 and ω_2, respectively. If biological effects scale linearly with effective irradiance, then the RAF gives a good first-order indication of the sensitivity of a process to ozone-mediated effects of UV. The greater the RAF, the more a process will be affected as ozone decreases. In other cases, caution should be exercised in overinterpreting this first-order indicator. As the discussion in this chapter shows, modelling UV effects frequently involves more than defining weighted exposure. If the biological effect is a non-linear function of E^*, then the ratios of effective irradiance predicted by a RAF will not reflect the actual effects that ozone depletion would have. Conceptually, the more complex relationship between biological effect and column ozone has been described as being governed by a total amplification factor (TAF) that incorporates the non-linear relationship between E^* and biological effect (Smith & Cullen, 1995). Functionally, defining such a relationship almost certainly involves simulation models that integrate responses over time and space (particularly depth) including variations in exposure that occur during vertical mixing (Gieskes & Buma, 1997). It is beyond the scope of this chapter to examine such simulations; however, there are a number of reports that illustrate the techniques involved. Example applications of BWFs to predicting changes in water column productivity are given by Cullen & Neale (1994), Arrigo (1994), and Boucher & Prézelin (1996). More recently, models have incorporated time dependence of UV effects in response to the time variation of exposure, for example plankton undergoing vertical movement in surface mixing layers (Neale, Davis & Cullen, 1998c).

3.9   Summary

This chapter has presented an introduction to the concepts, methodology and application of SWFs, with particular attention to their use in research on the effects of UV on marine ecosystems. Both theory and practice of SWFs depends on the research objective. Definition of an action spectrum is a desirable experimental objective when asking basic questions about the spectral properties of chromophores and intracellular shielding, and relevance to environmental UV is secondary. On the other hand, BWFs are necessary to predict cellular/organismal response to environmental light fields, since BWFs implicitly incorporate multiwavelength interactions that are characteristic of most UV photobiological processes. No matter what type of SWF is used or is to be defined, it must be used in the context of an ERC in order to be an accurate predictive tool. This is probably the most important point that the reader can take to the following chapters of this book: interpreting the environmental significance of UV responses and determining reasonable treatments for future studies require consideration of both spectral (BWF) and temporal (ERC) responses.

For further reading, there are several reviews published in recent years that discuss spectral weighting functions in marine ecosystems and related subjects including in greater detail (Smith & Baker, 1989; Cullen & Neale, 1993, 1994, 1997; Smith & Cullen, 1995). Further inspiration for developing techniques to study spectral sensitivity of UV effects can be found in literature on the health effects of UV (de Gruijl, 1997). For comprehensive discussions of technical issues relating to weighting functions and their application, the benchmark reviews by Caldwell *et al.* (1986) and Coohill (1991) have not lost their relevance.

Acknowledgements

Support from NSF Polar Programmes and Biological Oceanography Programmes, and the Smithsonian Institution scholarly studies programme is gratefully acknowledged. Suggestions from Anastazia Banaszak, John Beardall, John Cullen and Winfried Gieskes are appreciated.

References

Arrigo, K. R. (1994). Impact of ozone depletion on phytoplankton growth in the Southern Ocean: large-scale spatial and temporal variability. *Marine Ecology Progress Series*, **114**, 1–12.

Behrenfeld, M. J., Chapman, J. W., Hardy, J. T. & Lee, H. I. (1993). Is there a common response to ultraviolet-B radiation by marine phytoplankton? *Marine Ecological Progress Series*, **102**, 59–68.

Bevington, P. R. (1969). *Data Reduction and Error Analysis for the Physical Sciences.* McGraw Hill, New York.

Booth, C. R. & Madronich, S. (1994). Radiation amplification factors: improved formulation accounts for large increases in ultraviolet radiation associated with Antarctic ozone depletion. In *Ultraviolet Radiation in Antarctica: Measurements and Biological Effects*, Antartic Research Series, ed. C. S. Weiler & P. A. Penhale, pp. 39–42. American Geophysical Union, Washington, DC.

Bornman, J. F., Björn, L. O. & Åkerlund, H.-E. (1984). Action spectrum for inhibition by ultraviolet radiation of photosystem II activity in spinach thylakoids. *Photobiochemistry and Photobiophysics*, **8**, 305–13.

Boucher, N. P. & Prézelin, B. B. (1996). Spectral modeling of UV inhibition of *in situ* Antarctic primary production using a field derived biological weighting function. *Photochemistry and Photobiology*, **64**, 407–18.

Buma, A. G. J., Engelen, A. H. & Gieskes, W. W. C. (1997). Wavelength-dependent induction of thymine dimers and growth rate reduction in the marine diatom *Cyclotella* sp. exposed to ultraviolet radiation. *Marine Ecology Progress Series*, **153**, 91–7.

Caldwell, M. M., Camp, L. B., Warner, C. W. & Flint, S. D. (1986). Action spectra and their key role in assessing biological consequences of solar UV-B radiation change. In *Stratospheric Ozone Reduction, Solar Ultraviolet Radiation and Plant Life*, ed. R. C. Worrest & M. M. Caldwell, pp. 87–111. Springer-Verlag, New York.

Coohill, T. P. (1991). Photobiology school. Action spectra again? *Photochemistry and Photobiology*, **54**, 859–70.

Coohill, T. P. (1994). Exposure response curves, action spectra and amplification factors. In *Stratospheric Ozone Depletion/ UV-B Radiation in the Biosphere*, ed. R. H. Biggs & M. E. B. Joyner, pp. 57–62. Springer-Verlag, Berlin.

Cullen, J. J. & Lesser, M. P. (1991). Inhibition of photosynthesis by ultraviolet radiation as a function of dose and dosage rate: results for a marine diatom. *Marine Biology*, **111**, 183–90.

Cullen, J. J. & Neale, P. J. (1993). Quantifying the effects of ultraviolet radiation on aquatic photosynthesis. In *Photosynthetic Responses to the Environment*, ed. H. Yamamoto & C. M. Smith, pp. 45–60. American Society of Plant Physiologists, Washington, DC.

Cullen, J. J. & Neale, P. J. (1994). Ultraviolet radiation, ozone depletion, and marine photosynthesis. *Photosynthesis Research*, **39**, 303–20.

Cullen, J. J. & Neale, P. J. (1997). Biological weighting functions for describing the effects of ultraviolet radiation on aquatic systems. In *Effects of Ozone Depletion on Aquatic Ecosystems*, ed. D.-P. Häder, pp. 97–118. R. G. Landes, Austin, TX.

Cullen, J. J., Neale, P. J. & Lesser, M. P. (1992). Biological weighting function for the inhibition of phytoplankton photosynthesis by ultraviolet radiation. *Science*, **258**, 646–50.

de Gruijl, F. R. (1997). Health effects from solar UV radiation. *Radiation Protection Dosimetry*, **72**, 177–96.

Engelmann, T. W. (1882). Ueber Sauerstoffausscheidung von Pflanzensellen im Microspectrum. *Botanische Zeitung*, **40**, 419–26.

Friedberg, E. C. (1985). *DNA Repair*. W. H. Freeman, New York.

Gates, F. L. (1930). A study of the bactericidal action of ultraviolet light. III. The absorption of ultraviolet light by bacteria. *Journal of General Physiology*, **14**, 31–42.

Gieskes, W. W. C. & Buma, A. G. J. (1997). UV damage to plant life in a photobiologically dynamic environment: the case of marine phytoplankton. *Plant Ecology*, **128**, 16–25.

Goedheer, J. C. (1966). Visible absorption and fluorescence of chlorophyll and its aggregates in solution. In *The Chlorophylls*, ed. L. P. Vernon & G. R. Seely, pp. 147–84. Academic Press, New York.

Greenberg, B. M., Gaba, V., Canaani, O., Malkin, S., Mattoo, A. K. & Edelman, M. (1989). Separate photosensitizers mediate degradation of the 32-kDa photosystem II reaction centre protein in the visible and UV spectral regions. *Proceedings of the National Academy of Sciences, USA*, **86**, 6617–20.

Hanawalt, P. C. & Setlow, R. B. (1975). *Molecular Mechanisms for Repair of DNA*. Plenum Press, New York.

Harm, W. (1980). *Biological Effects of Ultraviolet Radiation*. Cambridge University Press, Cambridge.

Helbling, E. W., Villafañe, V., Ferrario, M. & Holm-Hansen, O. (1992). Impact of natural ultraviolet radiation on rates of photosynthesis and on specific marine phytoplankton species. *Marine Ecology Progress Series*, **80**, 89–100.

Helbling, E. W., Villafañe, V. & Holm-Hansen, O. (1994). Effects of ultraviolet radiation on Antarctic marine phytoplankton photosynthesis with particular attention to the influence of mixing. In *Ultraviolet Radiation and Biological Research in Antarctica*, Antarctic Research Series, ed. C. S. Weiler & P. A. Penhale, pp. 207–27. American Geophysical Union, Washington, DC.

Hunter, J. R., Taylor, J. H. & Moser, H. G. (1979). Effect of ultraviolet irradiation on eggs and larvae of the northern anchovy, *Engaulis mordax*, during the embryonic stage. *Photochemistry and Photobiology*, **29**, 325–38.

Jokiel, P. L. & York, R. H., Jr (1984). Importance of ultraviolet radiation in photoinhibition of microalgal growth. *Limnology and Oceanography*, **29**, 192–9.

Jones, L. W. & Kok, B. (1966). Photoinhibition of chloroplast reactions. I. Kinetics and action spectra. *Plant Physiology*, **41**, 1037–43.

Karentz, D., Bothwell, M. L., Coffin, R. B., Hanson, A., Herndl, G. J., Kilham, S. S., Lesser,

M. P., Lindell, M., Moeller, R. E., Morris, D. P., Neale, P. J., Sanders, R. W., Weiler, C. S. & Wetzel, R. G. (1994). Impact of UV-B radiation on pelagic freshwater ecosystems: report of working group on bacteria and phytoplankton. *Ergebnisse der Limnologie (Archiv für Hydrobiologie. Beiheft)*, **43**, 31–69.

Karentz, D., Cleaver, J. E. & Mitchell, D. L. (1991). Cell survival characteristics and molecular responses of Antarctic phytoplankton to ultraviolet-B radiation. *Journal of Phycology*, **27**, 326–41.

Kirk, J. T. O., Hargreaves, B. R., Morris, D. P., Coffin, R. B., David, B., Frederickson, D., Karentz, D., Lean, D. R. S., Lesser, M. P., Madronich, S., Morrow, J. H., Nelson, N. B. & Scully, N. M. (1994). Measurements of UV-B radiation in two freshwater lakes: an instrument intercomparison. *Ergebnisse der Limnologie (Archiv für Hydrobiologie. Beiheft)*, **43**, 71–99.

Lesser, M. P., Cullen, J. J. & Neale, P. J. (1994). Carbon uptake in a marine diatom during acute exposure to ultraviolet B radiation: relative importance of damage and repair. *Journal of Phycology*, **30**, 183–92.

Lesser, M. P., Neale, P. J. & Cullen, J. J. (1996). Acclimation of Antarctic phytoplankton to ultraviolet radiation: ultraviolet-absorbing compounds and carbon fixation. *Molecular Marine Biology and Biotechnology*, **5**, 314–25.

Lubin, D., Mitchell, B. G., Frederick, J. E., Alberts, A. D., Booth, C. R., Lucas, T. & Neuschuler, D. (1992). A contribution toward understanding the biospherical significance of Antartic ozone depletion. *Journal of Geophysical Research*, **97**, 7817–28.

Madronich, S. (1992). Implications of recent total atmospheric ozone measurements for biologically active ultraviolet radiation reaching the earth's surface. *Geophysical Research Letters*, **19**, 37–40.

Marquardt, D. W. (1963). An algorithm for least-squares estimation of non-linear parameters. *Journal of the Society of Industrial and Applied Mathematics*, **11**, 431–41.

Mitchell, B. G. (1990). Action spectra of ultraviolet photoinhibition of Antarctic phytoplankton and a model of spectral diffuse attenuation coefficients. In *Response of Marine Phytoplankton to Natural Variations in UV-B Flux*, ed. B. G. Mitchell, O. Holm-Hansen & I. Sobolev, Appendix H. Chemical Manufacturers Association, Washington, DC.

Moeller, R. E. (1994). Contribution of ultraviolet radiation (UV-A, UV-B) to photoinhibition of epilimnetic phytoplankton in lakes of differing UV transparency. *Ergebnisse der Limnologie (Archiv für Hydrobiologie. Beiheft)*, **43**, 157–70.

Morel, A. (1990). Optics of marine particles and marine optics. In *Particle Analysis in Oceanography*, ed. S. Demers, pp. 141–88. Springer-Verlag, Berlin.

Neale, P. J. (1987). Algal photoinhibition and photosynthesis in the aquatic environment. In *Photoinhibition*, ed. D. J. Kyle, C. B. Osmond & C. J. Arntzen, pp. 35–65. Elsevier, Amsterdam.

Neale, P. J., Banaszak, A. T. & Jarriel, C. R. (1998a). Ultraviolet sunscreens in dinoflagellates: mycosporine-like amino acids protect against inhibition of photosynthesis. *Journal of Phycology*, **34**, 928–38.

Neale, P. J., Cullen, J. J. & Davis, R. F. (1998b). Inhibition of marine photosynthesis by ultraviolet radiation: variable sensitivity of phytoplankton in the Weddell–Scotia Sea during the austral spring. *Limnology and Oceanography*, **43**, 433–48.

Neale, P. J., Davis, R. F. & Cullen, J. J. (1998c). Interactive effects of ozone depletion and vertical mixing on photosynthesis of Antarctic phytoplankton. *Nature*, **392**, 585–9.

Neale, P. J., Lesser, M. P. & Cullen, J. J. (1994). Effects of ultraviolet radiation on the photosynthesis of phytoplankton in the vicinity of McMurdo Station (78° S). In

Ultraviolet Radiation in Antarctica: Measurements and Biological Effects, Antartic Research Series, ed. C. S. Weiler & P. A. Penhale, pp. 125–42. American Geophysical Union, Washington, DC.

Prézelin, B. B., Boucher, N. B. & Smith, R. C. (1994). Marine primary production under the influence of the Antarctic ozone hole: Icecolors '90. In *Ultraviolet Radiation in Antartica: Measurements and Biological Effects*, Antartic Research Series, ed. C. S. Weiler & P. A. Penhale, pp. 159–86. American Geophysical Union, Washington, DC.

Quaite, F. E., Sutherland, B. M. & Sutherland, J. C. (1992). Action spectrum for DNA damage in alfalfa lowers predicted impact of ozone depletion. *Nature*, **358**, 576–8.

Quesada, A., Mouget, J.-L. & Vincent, W. F. (1995). Growth of Antarctic cyanobacteria under ultraviolet radiation: UVA counteracts UVB inhibition. *Journal of Phycology*, **31**, 242–8.

Renger, G., Völker, M., Eckert, H. J., Fromme, R., Hohm-Veit, S. & Gräber, P. (1989). On the mechanism of photosystem II deterioration by UV-B irradiation. *Photochemistry and Photobiology*, **49**, 97–105.

Rundel, R. D. (1983). Action spectra and estimation of biologically effective UV radiation. *Physiologia Plantarum*, **58**, 360–6.

Setlow, R. B. (1974). The wavelengths in sunlight effective in producing skin cancer: a theoretical analysis. *Proceedings of the National Academy of Sciences, USA*, **71**, 3363–6.

Smith, R. C. & Baker, K. S. (1982). Assessment of the influence of enhanced UV-B on marine primary productivity. In *The Role of Solar Ultraviolet Radiation in Marine Ecosystems*, ed. J. Calkins, pp. 509–37. Plenum Press, New York.

Smith, R. C. & Baker, K. S. (1989). Stratospheric ozone, middle ultraviolet radiation and phytoplankton productivity. *Oceanography Magazine*, **2**, 4–10.

Smith, R. C., Baker, K. S., Holm-Hansen, O. & Olson, R. S. (1980). Photoinhibition of photosynthesis in natural waters. *Photochemistry and Photobiology*, **31**, 585–92.

Smith, R. C. & Cullen, J. J. (1995). Effects of UV radiation on phytoplankton. *Reviews of Geophysics*, **33**, 1211–23.

Smith, R. C., Prézelin, B. B., Baker, K. S., Bidigare, R. R., Boucher, N. P., Coley, T., Karentz, D., MacIntyre, S., Matlick, H. A., Menzies, D., Ondrusek, M., Wan, Z. & Waters, K. J. (1992). Ozone depletion: ultraviolet radiation and phytoplankton biology in Antarctic waters. *Science*, **255**, 952–9.

Smith, W. O. & Nelson, D. M. (1986). Importance of ice edge phytoplankton production in the Southern Ocean. *Bioscience*, **36**, 251–7.

Vass, I. (1997). Adverse effects of UV-B light on the structure and function of the photosynthetic apparatus. In *Handbook of Photosynthesis*, ed. M. Pessarakli, pp. 931–49. Marcel Dekker, New York.

Vernet, M., Brody, E. A., Holm-Hansen, O. & Mitchell, B. G. (1994). The response of Antarctic phytoplankton to ultraviolet radiation: absorption, photosynthesis, and taxonomic composition. In *Ultraviolet Radiation in Antarctica: Measurements and Biological Effects*, Antartic Research Series, ed. C. S. Weiler & P. A. Penhale, pp. 143–58. American Geophysical Union, Washington, DC.

4

○ ○ ○ ○ ○ ○ ○ ○ ○ ○ ○ ○ ○ ○ ○ ○ ○ ○ ○ ○

Marine photochemistry and its impact on carbon cycling

Kenneth Mopper* and David J. Kieber

4.1 Introduction

Dissolved organic matter (DOM) in seawater is present at concentrations of about 0.5–2 mg l^{-1} or 0.5–2 ppm (MacKinnon, 1981) and represents one of the largest reservoirs of organic carbon on the surface of the earth (Mopper & Degens, 1979; Hedges, 1992). Despite the size of this carbon pool, only about 25% to 50% has been characterised (Zika, 1981; Druffel, Williams & Suzuki, 1989; Benner *et al.*, 1992). In surface waters, the characterised fraction consists mainly of biomolecules that are rapidly turned over by the biota (Lee & Wakeham, 1988; Kirchman *et al.*, 1991; Amon & Benner, 1994). The uncharacterised fraction, >50% of the DOM, consists largely of heteropoly-condensates derived from *in situ* and terrestrial sources (Ehrhardt, 1984; Harvey & Boran; 1985; Brophy & Carlson, 1989; Ishiwatari, 1992; Lee & Wakeham, 1992; Tranvik, 1993; Heissenberger & Herndl, 1994). These substances appear to be biologically refractory, especially in the deep sea (Barber, 1968; Menzel, 1974; Williams & Carlucci, 1976; Carlucci & Williams, 1978), where they have an apparent mean ^{14}C age of about 6000 years (Williams & Druffel, 1987; Bauer, Williams & Druffel, 1992), which is approximately 12-fold greater than the oceanic deep water replacement time (Stuiver, Quay & Ostlund, 1983). Despite this apparent biological inertness, the pool of uncharacterised DOM is the strongest light-absorbing component of seawater, especially in coastal regions (Højerslev, 1982) and, therefore it plays a dominant role in marine photochemical and photophysical processes (Zika, 1981, 1987; Zafiriou, 1983; Zafiriou *et al.*, 1984; Zepp, 1988; Miller, 1994; Zepp, Callaghan & Erickson, 1995; Blough, 1996; Moran & Zepp, 1997), which in turn impact on biogeochemical cycling of elements in the sea (Kieber, McDaniel & Mopper 1989; Kieber, Zhou & Mopper, 1990; Mopper *et al.*, 1991; Miller & Zepp, 1995; Bushaw *et al.*, 1996; Miller & Moran, 1997).

(a)

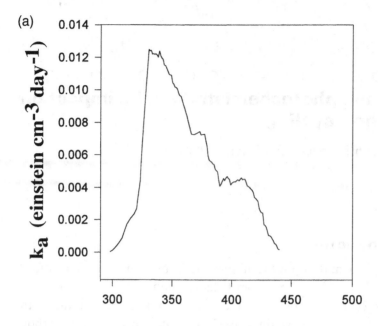

(b)

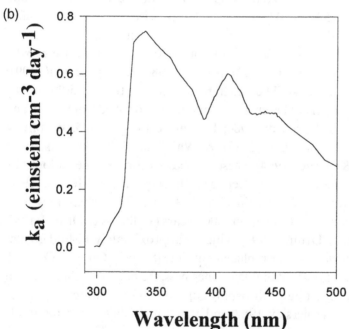

Wavelength (nm)

Figure 4.1.   Specific rate of light absorption of chromophoric DOM (K_a) plotted as a function of wavelength in water from: (a) the Gulf of Mexico and (b) Mobile Bay, Alabama. K_a was determined at each wavelength by multiplying the wavelength-dependent absorbance coefficient by the daily (24 h) average summer irradiance for 30° N latitude in the summer (Leifer, 1988). Absorbance coefficients were calculated from the absorbance spectrum that was determined for each water sample in a 10 cm quartz cell (Hoge *et al.*, 1993).

The uncharacterised fraction of DOM has often been referred to as 'marine humic substances' (Harvey *et al.*, 1983), although it should be pointed out that the operational definitions of these materials as used by soil chemists for humic and fulvic acids are not strictly applicable to seawater. The term 'unknown photoreactive chromophores' (Zafiriou *et al.*, 1984), is probably more appropriate, but, for simplicity, we will continue to equate the uncharacterised fraction of marine DOM with humic substances.

In natural waters, photochemical reactions are initiated by absorption of sunlight mainly by humic substances and, to a lesser extent, inorganic compounds (principally nitrate and nitrite). Published action spectra indicate that only a small portion of the solar spectrum is photochemically active, i.e. mainly wavelengths in the UV and blue regions (Zepp, 1988; Cooper *et al.*, 1989; Kieber *et al.*, 1990, 1996; Moore, Farmer & Zika, 1993). In some strongly absorbing organic-rich waters, such as coastal waters near marsh outflows and perhaps surface microlayer films (Blough, 1996), photoactive wavelengths may extend further out into the visible region of the solar spectrum (Figure 4.1). The energy at the shorter wavelengths (UV and blue regions) is sufficiently high to induce the formation of a variety of highly reactive transient species, such as the hydroxyl radical and carbon-centred radicals, as well as relatively stable species, such as hydrogen peroxide, NO and CO_2. At present, little is known about the photochemical reactions and reactive sites within marine DOM that are responsible for production of reactive transients at the sea surface (Zafiriou, 1977, 1983; Zika, 1981, 1987; Zafiriou *et al.*, 1984; Blough & Zepp, 1990, 1995). This chapter considers the role of DOM-related photochemical reactions in marine processes and, in particular, carbon cycling. Although photochemical reactions also impact on the cycling of sulfur (Zepp *et al.*, 1995; Kieber *et al.*, 1996), nitrogen (Bushaw *et al.*, 1996) and trace metals (Sulzberger, 1990; Palenik, Price & Morel, 1991; Faust, 1994), these topics are not covered in this chapter. Analytical techniques for determining transient species in natural waters are reviewed elsewhere (Zafiriou *et al.*, 1990; Blough & Zepp, 1995). Marine optics and biological effects of photochemical reactions are discussed in Chapters 2 and 5.

4.2 Photochemistry of dissolved organic matter

The overall process of light absorption by DOM and subsequent energy dissipation is described by various models for the photochemical oxidation of organic matter in natural waters (Zepp, 1988, 1991; Hoigné *et al.*, 1989;

Zafiriou *et al.*, 1990). In these models, the absorption of light by DOM initially results in the formation of singlet excited state species (^{1}DOM*) that subsequently decay through a series of photophysical and photochemical pathways, as shown in Figure 4.2. Most of the absorbed energy is dissipated through vibrational and rotational relaxation, internal conversions, collisional deactivation and radiative decay, e.g. fluorescence (Blough & Zepp, 1990). These processes yield heat and light but no net photochemical change. The singlet excited state can also undergo

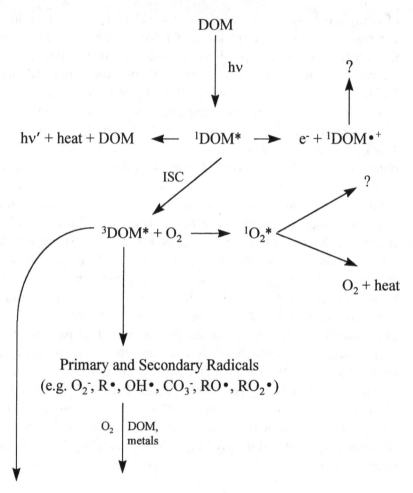

Figure 4.2.   General pathways for the photochemical oxidation of dissolved organic matter (DOM) in natural waters. A detailed discussion of these pathways (and notations) is given in the text.

intersystem crossing (ISC) to the triplet excited state ($^3DOM^*$) (Mill, Hendry & Richardson, 1980; Zepp, Schlotzhauer & Sink, 1985; Faust & Hoigné, 1987). Like the singlet, the longer-lived $^3DOM^*$ may undergo photophysical deactivation, producing heat and light emission (phosphorescence), with no resultant photochemical change.

A small fraction of the singlet and triplet excited states of DOM decays through energy transfer to acceptor molecules and through photochemical reactions in solution. In natural waters, an important decay pathway is energy transfer from $^3DOM^*$ to ground state triplet oxygen ($^3O_2$), which is the dominant acceptor molecule in natural waters (Zepp et al., 1985). The product of this reaction is singlet oxygen ($^1O_2^*$) (Zepp et al., 1977, 1985; Momzikoff, Santus & Giraud, 1983; Haag & Hoigné, 1986). Singlet oxygen decays almost entirely back to the ground state triplet through collisional deactivation with water. Less than 1% of singlet oxygen is expected to react selectively in natural waters with electron-rich centres such as reduced sulfur compounds, e.g. dimethylsulfide (Kieber et al., 1996), although there is very little direct evidence for this type of reaction at ambient levels in the marine environment (Baxter & Carey, 1982; Scully & Hoigné, 1987; Larson, 1995).

Other decay pathways for excited state DOM are molecular rearrangement, fragmentation and/or reaction with metal ions and oxygen (Figure 4.2). These pathways result in the formation of an array of radical (Table 4.1) and non-radical species (*vide infra*). Primary free radicals are formed on the time scale of pico- to nanoseconds (Figure 4.3). However, due to the high concentration of dissolved oxygen in surface waters, primary radicals react with oxygen at diffusion controlled rates on a time scale of microseconds yielding a suite of secondary oxygenated radicals. This oxygen-dependent transition from primary to secondary radicals has been appropriately termed the 'oxygen wall' (Zafiriou et al., 1990), depicted in Figure 4.3.

The radical spectra is composed principally of superoxide ion (O_2^-) (Baxter & Carey, 1982; Petasne & Zika, 1987; Zafiriou & Dister, 1991), hydrated electrons ($e_{(aq)}^-$) (Fischer et al., 1985; Zepp et al., 1987a), carbon-centred radicals ($R\cdot$) (Kieber & Blough, 1990), organoperoxy and alkoxy radicals (Mill, 1980; Mill et al., 1980), and hydroxyl ($OH\cdot$) radicals (Zafiriou, 1974, 1977; Zafiriou & True, 1979; Mill, 1980; Mill et al., 1980; Haag & Hoigné, 1985; Zepp, Hoigné & Bader, 1987; Zhou & Mopper, 1990; Mopper & Zhou, 1990). Of these transient species, the hydroxyl radical is by far the most reactive (Mill, 1980; Zafiriou, 1983; Zafiriou et al., 1984; Mopper & Zhou, 1990). Other reactive species such as the humic acid radical cation, carbonate radical (CO_3^-), and bromine-containing radicals, especially Br_2^-, have also been postulated, but their fates and importance in seawater have not been studied (Zafiriou, 1974; Zafiriou et al., 1984; Mabury & Crosby, 1995). There are numerous comprehensive reviews of the probable sources and reactions of the major reactive transients formed in natural waters (Cooper et al., 1989; Hoigné et al., 1989; Blough & Zepp, 1990, 1995).

The photochemical production rate of radicals varies substantially in seawater depending to a large degree on the absorbance of DOM. In a recent study that considered a broad range of coastal and oligotrophic seawater samples (Zafiriou & Dister, 1991), the radical flux varied over two orders of magnitude from 0.1–$10\,nmol\,l^{-1}\,min^{-1}\,sun^{-1}$ (for the definition of 'sun' unit, see Zafiriou & Dister, 1991). The major transient formed was the superoxide anion (O_2^-), which arises from one electron reduction of dissolved oxygen by DOM via either direct electron transfer or reaction with the hydrated electron (Blough & Zepp, 1995). The superoxide anion and all other radicals produced in natural waters eventually form non-radical (diamagnetic) species through radical–radical reactions. For example, the superoxide anion disproportionates to form

Table 4.1. Photochemically formed transient species in sunlit natural waters

Species	Estimated concentration range (M)	Probable sources
Singlet oxygen ($^1O_2$)	10^{-14}–10^{-13}	DOM energy transfer to triplet oxygen
Hydrated electron ($e_{(aq)}^-$)	10^{-17}–10^{-15}	DOM photolysis
Superoxide anion (O_2^-)	10^{-9}–10^{-8}	e^- transfer to and $e_{(aq)}^-$ reaction with triplet oxygen
Hydrogen peroxide (H_2O_2)	10^{-8}–10^{-7}	Dismutation of O_2^-
Humic cation (humic$^+$)	$\sim 10^{-10}$?	DOM photolysis
Humic triplet (humic*)	$\sim 10^{-10}$?	DOM photolysis
Organoperoxides (ROO')	10^{-14}–10^{-10}	DOM photolysis
Hydroxyl radicals (OH')	10^{-19}–10^{-17}	NO_3^-, DOM photolyses
Dibromide anion radical in SW (Br_2^-)	?	'OH/Br$^-$ reactions
Carbon-centred radicals (RH_2C')	10^{-13}–10^{-11}	DOM photolysis
Carbonate radical (CO_3^-)	$\sim 10^{-14}$?	'OH/HCO$_3^-$ and Br_2^-/HCO$_3^-$ reactions
Cu^+, Mn^{2+}, Fe^{2+}, Cr^{3+}	$< 10^{-12}$	Ligand to metal charge transfer reactions

For references, see the text.
DOM, dissolved organic matter; SW, sea water.

hydrogen peroxide (Petasne & Zika, 1987). For most other radical species, especially organic radicals, the decay/loss pathways in seawater are not known.

In addition to the formation of stable non-radical species through radical decay processes, the decay of excited state DOM can also result in

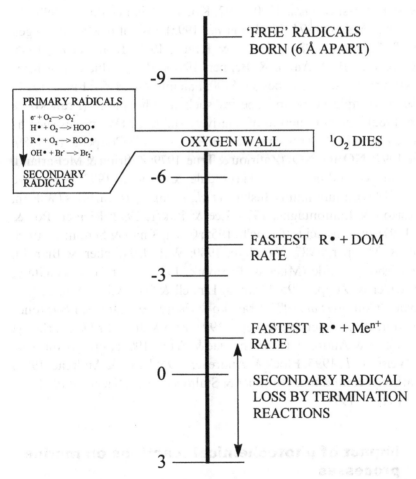

Figure 4.3. Time scales for primary and secondary radical processes in relation to the oxygen wall in sunlit surface waters (for $[O_2] \gg [Me^{+n}]$, [DOM], or [transient]). [DOM] is assumed to be about $50\,\mu\text{mol}\,l^{-1}$, and redox active $[Me^{n+}]$ is assumed to be about $1\,\text{nmol}\,l^{-1}$. ($1\,\text{Å} = 0.1\,\text{nm}$.) (Reprinted from Zafiriou *et al.* (1990), copyright 1990, with permission from Elsevier Science.)

the direct formation of non-radical species, e.g. CO from DOM (Zuo & Jones, 1995). The relative importance of radical versus non-radical processes in the formation of low molecular weight (LMW) carbon products, such as carbonyl compounds, organic acids, and carbon dioxide, is unknown.

The net effects of these photochemical processes and transformations are the oxidation of DOM (Zepp, 1988; Zafiriou *et al.*, 1990), cleavage of humic substances (Geller, 1985, 1986; Lindell, Granéli & Tranvik, 1995), oxidation of reduced sulfur (Pos *et al.*, 1997), photobleaching of absorbance and fluorescence (Stewart & Wetzel, 1981; Kotzias *et al.*, 1987; Kieber *et al.*, 1990; Kouassi & Zika, 1990, 1992; Kouassi, Zika & Plane, 1990; De Haan, 1993; Miller, 1994; Lindell *et al.*, 1995), loss of dissolved oxygen (Miles & Brezonik, 1981; Gieskes & Kraay, 1982; Laane *et al.*, 1985; Lindell & Rai, 1994; Amon & Benner, 1996), photoreduction of trace metals (for reviews, see Sulzberger, 1990; Palenik *et al.*, 1991; Faust, 1994), release of complexed/bound species such as phosphate (Francko & Heath, 1982), and production of relatively stable LMW species such as hydrogen peroxide (for reviews, see Plane *et al.*, 1987; Cooper *et al.*, 1989; Miller, 1994), NO and NO_2 (Zafiriou & True, 1979; Zafiriou & McFarland, 1981), reduced iodine species (Truesdale & Moore, 1992; Moore & Zafiriou, 1994), ammonium (Bushaw *et al.*, 1996), hydrocarbons (Wilson, Swinnerton & Lamontagne, 1970; Lee & Baker, 1992; Riemer, Pos & Zika, 1997), organic acids (Creac'h, 1955; Chen, Khan & Schnitzer, 1978; Kieber & Mopper, 1987; Vaughan, 1989; Wetzel, Hatcher & Bianchi, 1995), carbon dioxide (Miles & Brezonik, 1981; Salonen & Vähätalo, 1994; Miller & Zepp, 1995; Granéli, Lindell & Tranvik, 1996), carbon monoxide (Conrad *et al.*, 1982; Francko & Heath, 1982; Jones, 1991; Jones & Amador, 1993; Valentine & Zepp, 1993; Zuo & Jones, 1995), carbonyl sulfide (Ferek & Andreae, 1984; Andreae & Ferek, 1992; Zepp & Andreae, 1994; Weiss *et al.*, 1995; Flöck & Andreae, 1996; Uher & Andreae, 1997) and carbonyl compounds (Mopper & Stahovec, 1986; Kieber *et al.*, 1990; Mopper *et al.*, 1991).

4.3   Impact of photochemical reactions on marine processes

Photochemically formed reactive transients play important roles in chemical and biological processes in aquatic environments. For example, biologically and chemically recalcitrant components of DOM, e.g. humic substances and organic pollutants, are degraded by photochemical oxidation. A fraction of these degradation products can be further oxidised by biological (bacterial) utilisation. These combined processes lead to enhanced geochemical cycling of organic carbon (Section 4.4), nitrogen and sulfur in the sea and to accelerated degradation of otherwise persistent pollutants, including petroleum and pesticides (Hansen, 1975; Thominette & Verdue, 1984; Lichtenthaler, Haag & Mill, 1989; Amador, Alexander & Zika, 1991; Chen *et al.*, 1994; Burns, Hassett & Rossi, 1997; Guiliano *et al.*, 1997). The photochemical formation of biological substrates may enhance biological productivity in carbon-limited regions of the ocean (see Chapter 5).

Biological productivity can also be enhanced by photoreduction of essential trace metals, principally iron (Miles & Brezonik, 1981; Waite & Morel, 1984; Miller *et al.*, 1995; Voelker, Morel & Sulzberger, 1997), manganese (Sunda, Huntsman & Harvey, 1983; Spokes & Liss, 1995) and copper (Moffett & Zika, 1987, 1988). Iron has been shown to be limiting for growth in large areas of the ocean (Martin & Fitzwater, 1988). Thus, photoreduction of iron colloids may be a critical process affecting productivity in those regions. In contrast to iron, copper has been shown to be toxic to micro-organisms, particularly in some industrialised and populated coastal areas (Sunda & Gillespie, 1979). In those areas, it seems likely that photoreduction of copper will affect its toxicity (Sunda & Gillespie, 1979; Moffett & Zika, 1983, 1987).

UV-B radiation (280–320 nm) is known to cause photo-inhibition of growth, especially in plankton and bacteria (Bailey, Neihof & Tabor, 1983; Smith & Baker, 1989; Cullen & Lesser, 1991; Häder & Worrest, 1991; Karentz, 1991; Paerl, 1991; Smith *et al.*, 1992; Herndl, Müller-Niklas & Frick, 1993; Williamson, 1995; Malloy *et al.*, 1997; Neale, Davis & Cullen, 1998). Photoinhibition is generally thought to be due to intracellular damage from the direct photolysis of biomolecules, especially nucleic acids, and from the reaction of biomolecules with internally produced reactive transients, especially free radicals (see e.g. Buma, Engelen & Gieskes, 1997). However, externally produced free radicals at or near the cell surface may also cause some of the observed photoinhibition (Harm, 1980; Jagger, 1985; Biaglow, 1987; Mee, 1987; Porter & Wujek, 1988; Döhler, 1990). The extent of extracellular damage will depend to a large degree on the photochemistry of the seawater surrounding the cell. A milieu of photochemically generated, reactive transient species are produced in seawater and these species may affect the cell membrane by damaging critical transport sites (Mill, Haag & Karentz, 1990; Karentz, 1991, 1994; Palenik *et al.*, 1991). Depending on the steady-state concentrations of these species and the type of damage that they incur, extracellular free radical attack may be important, particularly for those plankton that have not developed adequate photorepair systems. For example, the highly reactive hydroxyl radical reacts more or less indiscriminately with many biomolecules, resulting in chemical and structural transformations (Jagger, 1985; Biaglow, 1987; Mee, 1987). But, externally produced hydroxyl and other highly reactive free radicals are too short-lived to be damaging to cells, except if they are formed within the hydration sphere near membrane transport sites (Mill *et al.*, 1990). However, reactions of hydroxyl radical with inorganic constituents in seawater produce a cascade of longer lived, reactive secondary radicals that may negatively impact on the cell's immediate environment. In response to this stress, some organisms may excrete protective slimes (Hellebust, 1965; Mague *et al.*, 1980; Zlotnik & Dubinsky, 1989). Thus, energy normally required by organisms for growth and reproduction would be diverted to protect their cell surface enzymes and membrane lipids from extracellular free radical attack (Calkins & Thordardøttir, 1980).

Photochemical reactions have been shown to alter the optical properties of the water column. In particular, photobleaching of absorbance makes the water column more transparent, thereby increasing the depth of the photic zone. This effect is readily seen in lakes (Williamson, 1995; Lindell *et al.*, 1996; Morris & Hargreaves, 1997), but is also observable in certain oceanic environments, e.g. in coastal regions near river outflows (Vodacek *et al.*, 1997) (Figure 4.4a,b) and in the surface waters of the open ocean (Sarpal, Mopper & Kieber, 1995; Nelson, Siegel & Michaels, 1998). Plots of DOM absorbance versus DOC are generally linear, but with a

Figure 4.4. (a) Location of sampling transect from the mouth of Delaware Bay to Sargasso Sea (reprinted with permission from Vodacek *et al.*, 1997. Copyright 1997, the American Society for Limnology and Oceanography, Inc.) (b) Relationship between chromophore-containing DOM (i.e. absorption coefficient at 355 nm and fluorescence at 355/450 nm) and salinity for surface waters along the transect in (a) during August 1993. The triangles with a regression line are from samples from a depth of 20–35 m, which is below the mixed layer (adapted from Vodacek *et al.*, 1997). A fluorescence unit (f.u.) is defined as the emission intensity of a 1 p.p.b. solution of quinine sulfate in 0.05 M H_2SO_4 (Hoge *et al.*, 1993). (c) Relationship between chromophore-containing DOM (i.e. absorption coefficient at 355 nm and fluorescence at 355/450 nm) and dissolved organic carbon (DOC) for waters along the transect in (a) during four seasons in 1993. The crosses with a regression line depict surface samples in November, March and April. The open circles depict samples from below the mixed layer in August, while the solid circles are for surface samples (mixed layer) from August. (Adapted from Vodacek *et al.*, 1997.)

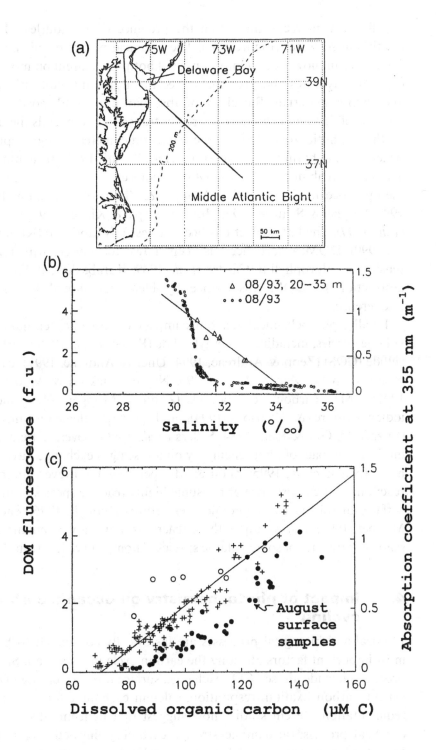

significant x-intercept, indicating the presence of a colourless, DOM fraction in seawater (Figure 4.4c). The chemical nature of this residual fraction is unknown, but it may consist of both transparent biomolecules (e.g. carbohydrates and proteins) of recent origin and photo-bleached, aged humic materials. Bleaching of absorbance has both positive and negative effects on the biota. The greater transparency extends the depth of the euphotic zone, but at the same time increases the depth of penetration of harmful UV radiation (Herndl *et al.*, 1993). Bleaching has also been attributed to the low DOM fluorescence of surface waters in the open ocean (Kramer, 1979; Mopper & Zhou, 1990; Chen & Bada, 1992; Mopper & Schultz, 1993; Mopper, Sarpal & Kieber, 1995; Vodacek *et al.*, 1997). The latter effect has been successfully modelled (Kouassi *et al.*, 1990). DOM absorbance has been shown to interfere with remote sensing of chlorophyll *a* (Carder *et al.*, 1989; Karabashev, 1992). Thus, destruction of DOM absorbance by bleaching should lessen this interference.

Finally, photochemical reactions impact on oceanic ventilation of volatile species, including dimethylsulfide (Kieber *et al.*, 1996), carbonyl sulfide (COS) (Zepp & Andreae, 1994; Uher & Andreae, 1997), carbon monoxide (CO) (Zuo & Jones, 1995; Najjar, Erickson & Madronich, 1995), LMW carbonyl compounds (Zhou & Mopper, 1997), methyl iodide (Moore & Zafiriou, 1994), and perhaps elemental mercury (Fitzgerald, Gill & Kim, 1984; Sellers *et al.*, 1996). Several attempts to model the impact of photochemistry on air–sea gas exchange have been made (Najjar *et al.*, 1995; Zhou & Mopper, 1997); however, unknown reactions that can produce or consume highly reactive species in the sea surface microlayer can confound these efforts (Blough, 1996; Zhou & Mopper, 1997). For example, these reactions may 'cap' or enhance the sea-to-air transfer of some volatile species (Thompson & Zafiriou, 1983).

4.4   Impact of photochemistry on oceanic carbon cycling

Physical and biological processes are historically considered to be the most important factors affecting the carbon cycle in the ocean. Specific processes include air–sea CO_2 exchange, surface mixing, venting of deep waters, carbon fixation, respiration, calcium carbonate formation and sedimentation. Recent studies now suggest that light-initiated (photochemical) processes also impact strongly carbon cycling at the sea surface. In particular, photochemical degradation of DOM may play an important role in oceanic carbon cycling either directly by the photochemical production of volatile carbon species, or indirectly through the production of CO_2 by coupled photochemical/biological oxidation (Figure 4.5) (also see Chapter 5).

Photochemical degradation of marine DOM gives rise to gaseous species, including LMW carbonyl compounds, COS, CO_2 and CO. Recent studies have shown that the photochemical production of the latter two species represents an important remineralisation pathway for carbon in the photic zone (Conrad et al., 1982; Mopper et al., 1991; Valentine & Zepp, 1993; Miller & Zepp, 1995). To illustrate this point, the photochemical flux of CO_2 can be calculated using estimates for the flux of photochemically produced CO, based on the finding that CO_2 photoproduction is approximately 20 times greater than that of CO (Miller & Zepp, 1995). The CO fluxes range from 4–77 Tg-C (teragrams carbon) yr^{-1} in the open ocean to 130–170 Tg-C yr^{-1} in coastal waters (Conrad et al., 1982; Valentine & Zepp, 1993). Using the lower end of the range in CO flux estimates (i.e. 134 Tg-C yr^{-1}), the total oceanic flux of photochemically produced CO_2 is at least 2680 Tg-C yr^{-1}. When CO_2 and CO are considered together, the photochemical carbon flux is conservatively 5% to 10% of the global average for total marine primary production (Martin et al., 1987), and is considerably larger than global estimates for organic carbon burial in marine sediments (e.g. 100 Tg-C yr^{-1};

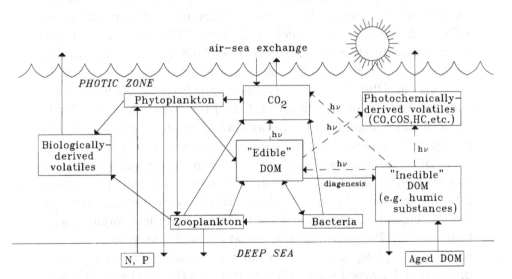

Figure 4.5. Carbon cycling in the upper ocean. The left side depicts the historical view of the carbon cycle, while the right side shows how photochemistry impacts on it.

Hedges, 1992, and references therein). However, because of the scarcity of measurements to date, the quantitative importance of this photochemical remineralisation pathway is not known. In order to evaluate the impact of this process on carbon cycling in the sea, photochemical production rates of CO_2 and CO, as well as their action spectra, need to be measured. These data can then be used to predict how photo-oxidation of DOM will vary in surface waters and at depth, based on measurements of DOM absorbance (or fluorescence) and water column diffuse attenuation coefficients (Hoge, Vodacek & Blough, 1993; Vodecek et al., 1995; Blough, 1996).

In addition to the formation of volatile species, such as CO_2 and CO, photochemical degradation of marine DOM gives rise to biologically labile carbon from biologically refractory carbon (Figure 4.5). In this DOM remineralisation pathway, biologically labile carbon substrates, such as formaldehyde and pyruvate, are formed photochemically from otherwise biologically refractory carbon, such as humic substances (Kieber et al., 1989, 1990; Mopper et al., 1991). Although the photo-production rate of *known* biological substrates from this pathway is much lower than carbon fixation by photosynthesis in the sea (Kieber et al., 1989), photochemical degradation of DOM may still significantly enhance bacterial production in carbon-limited waters (see Chapter 5).

If the *uncharacterised* photo-oxidised DOM products are also considered, in addition to photochemically formed biological substrates, then the photochemical/microbial loop may be substantial (Strome & Miller, 1978; Geller, 1985, 1986; Lindell et al., 1995, 1996). Miller et al. (1994) estimate that 10% to 30% of photo-oxidised DOC products may be taken up by heterotrophs, which is comparable to (or greater than) the photochemical flux of DOC to CO_2 and CO (*vide supra*), and is consistent with the findings of Granéli et al. (1996) for lakes. Thus, this DOM photochemical degradation pathway may significantly speed up carbon cycling and enhance bacterial growth. Photochemical production of bacterial substrates may also largely explain the apparent imbalance between high bacterial carbon utilisation rates and low primary productivity rates, especially in low productivity central gyres (Hansell, Bates & Gundersen, 1995; Carlson, Ducklow & Sleeter, 1996; del Giorgio, Cole & Cimbleris, 1997). Upwelling of deep water is well known to bring 'new' nitrogen and phosphorus into the euphotic zone and fuel new primary production. Upwelled water will also bring DOM into the photic zone (Figure 4.5) (Hansell & Waterhouse, 1997; Bauer et al., 1998). It has been previously shown that deep sea DOM is about 5–10 times more photoreactive than photobleached surface DOM (Mopper & Zhou, 1990; Mopper et al., 1991), and therefore may serve as a source of new carbon for bacteria via photodegradation (Benner & Biddanda, 1998). This hypothesis is supported by the recent finding that about 15% to 20% of carbon assimilated into nucleic acids by bacteria in surface waters at a station in the eastern North Pacific is isotopically old (Cherrier, 1997; Cherrier et al., 1999). Photochemically produced substrates may be more critical (i.e. important) to bacteria in carbon-limited regions, such as central gyres and other oligotrophic waters, compared to productive coastal waters. For open ocean surface waters, bacterial carbon utilisation rates appear to be similar to the production rate of labile substrates from DOM photodegradation, as estimated below.

Several studies indicate that approximately 15% of the total DOC in various water types can be photomineralised to CO_2 and CO, and that a comparable (or greater) percentage of photolysed DOC is taken up and remineralised by heterotrophs (Miller et al., 1994; Miller & Zepp, 1995). The latter result is supported by lake studies (Lindell et al., 1995, 1996; Granéli et al., 1996). For open ocean surface waters, if we assume a bacterial production rate of 0.25 ($\pm$ 0.2) $\mu M C d^{-1}$ (Kirchman, Rich & Barber, 1995; Carlson et al., 1996) and a growth efficiency of 15 ($\pm$ 10)% (Carlson et al., 1996; del Giorgio et al., 1997), the resultant bacterial carbon utilisation rate is about 1.7 ($\pm$ 1.4) $\mu mol C l^{-1} d^{-1}$. Based on the freshwater studies of Miller et al. (1994), the fraction of DOC that potentially becomes available to bacteria via photodegradation can be estimated by assuming that it is approximately equal to the fraction of DOC photomineralised to CO_2. The latter can be estimated assuming that CO_2 photoproduction is about 20 times greater than CO photoproduction (Miller & Zepp, 1995). The latter result was also based mainly on irradiation of terrestrially dominated DOM, but the few marine samples examined (Gulf of Mexico) gave a similar CO_2:CO photoproduction ratio. Assuming the latter can be applied to the open ocean, using a noon-time photoproduction rate of 15 ($\pm$ 4) $nmol CO l^{-1} h^{-1}$ (at about 26° N; Mopper et al., 1991; Zuo & Jones, 1995) and assuming that one day receives 7.6 h noon-time sun (Miller & Zepp, 1995), the photoproduction rate of labile carbon substrates (or photochemically formed CO_2) at the sea surface is estimated to be about 2.3 ($\pm$ 0.7) $\mu mol C l^{-1} d^{-1}$. Comparing the two rates, it is clear that photodegradation of DOC can potentially supply a major percentage of the carbon utilised by bacteria in surface waters in the open ocean (from about 50% to > 100%). The importance of this carbon source will decrease exponentially with depth, depending on the light attenuation rates and action spectra of the photoprocesses (Blough, 1996). When the above calculations are performed for coastal systems, similar percentages are obtained, even though absolute rates are higher. However, it should be pointed out that, although significant rates of DOC photo-oxidation in terrestrial and marine coastal waters have been observed in several studies (e.g. Miller & Zepp, 1995; Granéli et al., 1996; Morris & Hargreaves, 1997; Vodacek et al., 1997), the importance of this source of carbon to bacteria in marine coastal waters may be less important than in carbon-limited open ocean gyres (Hansell et al., 1995; del Giorgio et al., 1997).

The coupled processes of photochemical DOM oxidation and biological uptake of the oxidation products have been hypothesised to play an important role in the geochemical cycling of oceanic DOM and appear to be a major removal pathway for riverine carbon in the sea. Several studies have indicated that the oceanic carbon budget is not in balance with respect to riverine carbon inputs (Mantoura & Woodward, 1983; Williams & Druffel, 1987; Deuser, 1988; Blough, Zafiriou & Bonilla, 1990; Hedges et al., 1992; Opsahl & Benner, 1997; Vodacek et al., 1997). Based on riverine inputs alone, the entire DOM content of the oceans can be accounted for in about 600–2000 years (Deuser, 1988). However, LMW deep sea DOC has an apparent age of > 6000 years and shows little terrestrial character (Meyers-Schulte & Hedges, 1986; Williams & Druffel, 1987; Druffel et al., 1989; Ospahl & Benner, 1997). Thus, the contribution of riverine DOM to the oceanic pool of DOM, especially in the deep sea, appears to be negligible. These findings are puzzling, since riverine DOC shows a more or less conservative behaviour during estuarine mixing, with coagulation and precipitation accounting for only a small fraction of its removal (Sholkovitz, 1976; Mantoura & Woodward 1983; Amon & Benner, 1996; Ospahl & Benner, 1997). Therefore, there must exist major unknown sinks for the rapid removal of riverine DOM from the oceans. Kieber et al. (1990) tested the hypothesis that photochemical chromophoric DOM degradation and oxidation of humic substances is a major removal pathway for carbon. These investigators examined the photochemical production of LMW carbonyl compounds in a wide variety of fresh, estuarine, and sea waters, with and without added humic and fulvic extracts from different terrestrial and marine sources. It was found that photochemical production of carbonyl compounds was closely linked to UV–visible absorbance and photobleaching characteristics of these waters. The photochemical reactivities of different humic extracts, with respect to absorbance-normalised carbonyl photoproduction, were indistinguishable from each other and from the DOM naturally present in the waters within the precision of the experiments. Action spectra for photoproduction of carbonyl compounds were also similar for humic substances from different terrestrial and marine sources (within the precision of the measurements), with the most photoactive wavelengths in the solar spectrum being 300–320 nm. These are also the wavelengths where the greatest DOM photobleaching (i.e. loss of absorbance) was consistently observed. Calculations indicate that the half-life of riverine DOM via photodegradation in the mixed layer is 5–15 years, which is much shorter than the oceanic mixing time. A number of assumptions and simplifications were made in this estimate (Kieber *et al.*, 1990). For example, only a few LMW carbonyl compounds were considered as the photodegradation products of riverine DOM. When additional photo-products, such as CO and CO_2, are considered, the half-life of riverine DOC in the sea is significantly reduced (Miller & Zepp, 1995; Zuo & Jones, 1995; Amon & Benner, 1996). Another uncertainty involves the mixing dynamics. It is not known to what extent optically thick river water needs to spread over and mix into the sea surface to obtain the optically thin conditions necessary for efficient photo-oxidation. The main point is that photochemical oxidation of riverine humic substances at the sea surface, coupled with the biological uptake of the oxidation products, appears to be a major sink for this oceanic carbon input.

A similar study was carried out to examine the role of photo-oxidation on the geochemical cycling of deep sea carbon (Mopper *et al.*, 1991). As discussed above, deep sea DOC is composed mainly of biologically refractory compounds such as humic substances. The average apparent ^{14}C age of this large pool of refractory LMW DOC is > 6000 years in the deep oceans, suggesting that its rate of turnover is slow. In contrast, the much smaller pool of high molecular weight DOC (more than about 1000 daltons) is turned over by microbial activity relatively quickly, e.g. days to months, especially in surface waters (Kirchman *et al.*, 1991; Moran & Buesseler, 1992; Amon & Benner, 1994; Santschi *et al.*, 1995). The pathways and rates responsible for the apparent slow turnover and loss of refractory LMW DOC are unknown. Mopper *et al.* (1991) hypothesised that photochemical processes might be involved. These investigators demonstrated that deep water (below about 1000 m) is more photoreactive with respect to production of LMW carbon species (formaldehyde, acetaldehyde, glyoxylate, pyruvate and CO), by a factor of 10–20, than surface waters. Using ^{14}C-labelled probes, the rate of the back reaction, i.e. the photochemical incorporation (condensation) of LMW organic molecules into DOM in bulk seawater, was found to be negligible (less than 1% of the photochemical production rate). On the basis of these results, the oceanic residence time of biologically refractory, photochemically reactive DOC was estimated to be 500–2000 years, which is several times less than its average apparent ^{14}C age. The injection of 'old' carbon from sediments into the deep sea may explain this discrepancy. Recent studies support the latter hypothesis (Bauer & Druffel, 1998; Bauer *et al.*, 1998).

4.5   Summary

DOM in the sea is one of the major pools of organic carbon on the surface of the earth and it is also one of the strongest light-absorbing components in seawater, especially in the UV region of the solar irradiance spectrum. Most of the light absorbed by DOM is dissipated as heat; however, a small fraction of the absorbed energy results in photochemical reactions with the production of transient species, in particular free radicals. Due to the high reactivity of the latter, photochemistry can strongly impact on a variety of marine processes. These processes include accelerated degradation of recalcitrant materials (e.g. humic substances and organic pollutants), enhancement of biological productivity in carbon-limited regions by the photochemical production of substrates, photoreduction of biologically essential trace metals making them more or less available to organisms, inhibition to growth by internal and external damage to vital biochemical reaction/transport sites, alteration of optical properties of seawater by photobleaching of DOM absorbance and fluorescence, and enhancement of air–sea exchange of gaseous species. Photochemical oxidation/degradation of DOM coupled with biological uptake of photoproducts appears to be a major sink for riverine carbon inputs in the sea. This coupled process also strongly impacts on the geochemical cycling of refractory LMW DOC found in the deep sea.

Acknowledgements

We thank the National Science Foundation Office of Polar Programs (OPP-9527255, K.M. and OPP9610173, D.J.K.) and the Chemical Ocean Program (OCE-9315821 and OCE-9711206, K.M.) for support of this work. We thank R. P. Kiene for donating the Mobile Bay water sample used in Figure 4.1.

References

Amador, J. A., Alexander, M. & Zika, R. G. (1991). Degradation of aromatic compounds bound to humic acid by the combined action of sunlight and micro-organisms. *Environmental Toxicology and Chemistry*, **10**, 475–82.

Amon, R. M. W. & Benner, R. (1994). Rapid cycling of high-molecular weight dissolved organic matter in the ocean. *Nature*, **369**, 549–52.

Amon, R. M. W. & Benner, R. (1996). Photochemical and microbial consumption of dissolved organic carbon and dissolved oxygen in the Amazon River system. *Geochimica et Cosmochimica Acta*, **60**, 1783–92.

Andreae, M. O. & Ferek, R. J. (1992). Photochemical production of carbonyl sulfide in seawater and its emission to the atmosphere. *Global Biogeochemical Cycles*, **6**, 175–83.

Bailey, C. A., Neihof, R. A. & Tabor, P. S. (1983). Inhibitory effect of solar radiation on amino acid uptake in Chesapeake Bay bacteria. *Applied Environmental Microbiology*, **46**, 44–9.

Barber, R. T. (1968). Dissolved organic carbon from deep waters resists microbial oxidation. *Nature*, **220**, 274–5.

Bauer, J. E. & Druffel, E. R. M. (1998). Ocean margins as a significant source of organic matter to the deep open ocean. *Nature*, **392**, 482–5.

Bauer, J. E., Druffel, E. R. M., Wolgast, D. M., Griffin, S. & Masiello, C. A. (1998) Distributions of dissolved organic and inorganic carbon in the eastern north Pacific continental margin. *Deep Sea Research II*, **45**, 689–731.

Bauer, J. E., Williams, P. M. & Druffel, E. R. M. (1992). ^{14}C activity of dissolved organic carbon fractions in the north-central Pacific and Sargasso Sea. *Nature*, **357**, 667–70.

Baxter, R. M. & Carey J. H. (1982). Reactions in singlet oxygen in humic waters. *Freshwater Biology*, **12**, 285–92.

Benner, R., Pakulski, J. D., McCarthy, M., Hedges, J. I. & Hatcher, P. G. (1992). Bulk chemical characteristics of dissolved organic matter in the ocean. *Science*, **255**, 1561–4.

Benner, R. & Biddanda, B. (1998). Photochemical transformations of surface and deep marine dissolved organic matter: effects on bacterial growth. *Limnology and Oceanography*, **43**, 1373–8.

Biaglow, J. E. (1987). The effects of ionizing radiation on mammalian cells. In *Radiation Chemistry: Principles and Applications*, ed. Farhataziz & M. A. J. Rodgers, pp. 527–63. VCH Publishers, New York.

Blough, N. V. (1996). Photochemistry in the sea-surface microlayer. In *The Sea-surface Microlayer and Its Potential: Role in Global Change*, ed. P. S. Liss & R. Duce, pp. 383–424. Sponsored by World Meteorological Organization. Cambridge University Press, Cambridge.

Blough, N. V. & Zepp, R. G. (eds.) (1990). *Effects of Solar Ultraviolet Radiation on Biogeochemical Dynamics in Aquatic Environments*, Woods Hole Oceanographic Institute Technical Report, WHOI 90-09.

Blough, N. V., Zafiriou, O. C. & Bonilla J. (1990). Optical absorption spectra of waters from the Orinoco river outflow: terrestrial input of colored organic matter to the Caribbean. *Journal of Geophysical Research*, **98**, 2271–8.

Blough, N. V. & Zepp, R. G. (1995). Reactive oxygen species in natural waters. In *Active Oxygen in Chemistry*, ed. C. S. Foote, J. S. Pacentine, A. Greenberg & J. F. Liebman, pp. 280–333. Chapman & Hall, New York.

Brophy, J. E. & Carlson, D. J. (1989). Production of biologically refractory dissolved organic carbon by natural seawater microbial populations. *Deep Sea Research*, **36**, 497–507.

Buma, A. G. J., Engelen, A. H. & Gieskes, W. W. C. (1997). Wavelength-dependent induction of thymine dimers and growth rate reduction in the marine diatom *Cyclotella* sp. exposed to ultraviolet radiation. *Marine Ecology Progress Series*, **153**, 91–7.

Burns, S. E., Hassett, J. P. & Rossi, M. V. (1997). Mechanistic implications of the intrahumic dechlorination of mirex. *Environmental Science and Technology*, **31**, 1365–71.

Bushaw, K. L., Zepp, R. G., Tarr, M. A., Schultz-Jander, D., Bourbonniere, R. A., Hodson, R. E., Miller, W. L., Bronk, D. A. & Moran, M. A. (1996). Photochemical release of biologically available nitrogen from aquatic dissolved organic matter. *Nature*, **381**, 404–7.

Calkins, J. & Thordardøttir, T. R. (1980). The ecological significance of solar UV radiation on aquatic organisms. *Nature*, **283**, 563–6.

Carder, K. L., Steward, R. G., Harvey, G. R. & Ortner, P. B. (1989). Marine humic and fulvic acids: their effects on remote sensing of ocean chlorophyll. *Limnology and Oceanography*, **34**, 68–81.

Carlson, C. A., Ducklow, H. W. & Sleeter, T. D. (1996). Stocks and dynamics of bacterioplankton in the northwestern Sargasso Sea. *Deep Sea Research II*, **43**, 491–516.

Carlucci, A. F. & Williams, P. M. (1978). Simulated *in situ* growth rates of pelagic marine bacteria. *Naturwissenschaften*, **65**, 541–2.

Chen, R. F. & Bada, J. L. (1992). The fluorescence of dissolved organic matter in seawater. *Marine Chemistry*, **37**, 191–221.

Chen, S., Inskeep, W. P., Williams, S. A. & Callis, P. R. (1994). Fluorescence lifetime measurements of fluoranthene, 1-naphthol, and napropamide in the presence of dissolved humic acid. *Environmental Science and Technology*, **28**, 1582–8.

Chen, Y., Khan, S. U. & Schnitzer, M. (1978). Ultraviolet irradiation of dilute fulvic acid solutions. *Soil Science Society of America Journal*, **42**, 292–6.

Cherrier, J. (1997). Carbon flow through bacterioplankton in eastern North Pacific surface waters. PhD thesis, Florida State University.

Cherrier, J., Bauer, J. E., Druffel, E. R. M., Coffin, R. B. & Chanton, J. P. (1999). Radiocarbon in oceanic bacteria: evidence for the age of assimilated organic matter. *Limnology and Oceanography*, **44**, 730–6.

Conrad, R., Seiler, W., Bunse, G. & Giehl, H. (1982). Carbon monoxide in seawater (Atlantic Ocean). *Journal of Geophysical Research*, **87**, 8839–52.

Cooper, W. J., Zika, R. G., Petasne, R. G. & Fischer, A. M. (1989). Sunlight-induced photochemistry of humic substances in natural waters: major reactive species. In *Aquatic Humic Substances: Influence on Fate and Treatment of Pollutants*, ed. I. H. Suffet & P. MacCarthy, pp. 333–62. American Chemical Society, Washington, DC.

Creac'h, M. P. (1955). Some components of organic material in littoral seawater. Solar oxidation in the marine environment. *Comptes Rendus de l'Académie des Sciences, Paris*, **241**, 437–9.

Cullen, J. J. & Lesser, M. P. (1991). Inhibition of photosynthesis by ultraviolet radiation as a function of dose and dosage rate: results for a marine diatom. *Marine Biology*, **111**, 183–90.

De Haan, H. (1993). Solar UV-light penetration and photodegradation of humic substances in peaty lake water. *Limnology and Oceanography*, **38**, 1072–6.

del Giorgio, P. A., Cole, J. J. & Cimbleris, A. (1997). Respiration rates in bacteria exceed phytoplankton production in unproductive aquatic systems. *Nature*, **385**, 148–51.

Deuser, W. G. (1988). Whither organic carbon? *Nature*, **332**, 396–7.

Döhler, G. (1990). Impact of UV-B (290–320 nm) radiation on metabolic processes of marine phytoplankton. In *Effects of Solar Ultraviolet Radiation on Biogeochemical Dynamics in Aquatic Environments*, Woods Hole Oceanographic Institute Technical Report, WHOI 90-09, pp. 133–4.

Druffel, E. R. M., Williams, P. M. & Suzuki, Y. (1989). Concentrations and radiocarbon signatures of dissolved organic matter in the Pacific Ocean. *Geophysical Research Letters*, **16**, 991–4.

Ehrhardt, M. (1984). Marine Gelbstoff. In *The Handbook of Environmental Chemistry*, ed. O. Hutzinger, vol. 2, Part A, pp. 63–77. Springer-Verlag, Berlin.

Faust, B. C. (1994). A review of the photochemical redox reactions of iron(III) species in atmospheric, oceanic, and surface waters: influences on geochemical cycles and oxidant formation. In *Aquatic and Surface Photochemistry*, ACS Symposium Series, ed. G. R. Helz, R. G. Zepp & D. G. Crosby, pp. 3–37. Lewis Publishers, Boca Raton, FL.

Faust, B. C. & Hoigné, J. (1987). Sensitized photooxidation of phenols by fulvic acid and in natural waters. *Environmental Science and Technology*, **21**, 957–64.

Ferek, R. J. & Andreae, M. O. (1984). Photochemical production of carbonyl sulphide in marine surface waters. *Nature*, **307**, 148–50.

Fischer, A. M., Kliger, D. S. Winterle, J. S. & Mill, T. (1985). Direct observation of phototransients in natural waters. *Chemosphere*, **14**, 1299–306.

Fitzgerald, W. F., Gill, G. A. & Kim J. P. (1984). An equatorial Pacific Ocean source of atmospheric mercury. *Science*, **224**, 597–9.

Flöck, O. R. & Andreae, M. O. (1996). Photochemical and non-photochemical formation and destruction of carbonyl sulfide and methyl mercaptan in ocean waters. *Marine Chemistry*, **54**, 11–26.

Franko, D.A. and Heath, R.T. (1982). UV-sensitive complex phosphorus: Association with dissolved humic material and iron in a bog lake. *Limnology and Oceanography*, **27**, 564–9.

Geller, A. (1985). Degradation and formation of refractory DOM by bacteria during simultaneous growth on labile substrates and persistent lake water constituents. *Schweizerische Zeitschrift für Hydrologie*, **47**, 27–44.

Geller, A. (1986). Comparison of mechanisms enhancing biodegradability of refractory lake-water constituents. *Limnology and Oceanography*, **31**, 755–64.

Gieskes, W. W. C. and Kraay, G. W. (1982). Effect of enclosure in large plastic bags on diurnal change in oxygen concentration in tropical ocean water. *Marine Chemistry*, **70**, 99–104.

Granéli, W., Lindell, M. J. & Tranvik, L. J. (1996). Photo-oxidative production of dissolved inorganic carbon in lakes of different humic content. *Limnology and Oceanography*, **41**, 698–706.

Guiliano, M., El Anba-Lurot, F., Doumenq, P., Mille, G. and Rontani, J. F. (1997). Photo-oxidation of *n*-alkanes in simulated marine environmental conditions. *Journal of Photochemistry and Photobiology A: Chemistry*, **102**, 127–32.

Haag, W. R. & Hoigné, J. (1985). Photo-sensitized oxidation in natural water via OH radicals. *Chemosphere*, **14**, 1659–71.

Haag, W. R. & Hoigné, J. (1986). Singlet oxygen in surface water. 3. Photochemical formation and steady-state concentrations in various types of waters. *Environmental Science and Technology*, **20**, 341–8.

Häder, D.-P. & Worrest, R. C. (1991). Effects of enhanced solar ultraviolet radiation on aquatic ecosystems. *Photochemistry and Photobiology*, **53**, 717–25.

Hansell, D. A., Bates, N. R. & Gundersen, K. (1995). Mineralization of dissolved organic carbon in the Sargasso Sea. *Marine Chemistry*, **51**, 201–12.

Hansell, D. A. & Waterhouse, T. Y. (1997). Controls on the distribution of organic carbon and nitrogen in the eastern Pacific Ocean. *Deep Sea Research I*, **44**, 843–57.

Hansen, H. P. (1975). Photochemical degradation of petroleum hydrocarbon surface films on seawater. *Marine Chemistry*, **3**, 183–95.

Harvey, G. R. & Boran, D. A. (1985). Geochemistry of humic substances in seawater. In *Humic Substances in Soil, Sediment and Water: Geochemistry, Isolation and Characterization*, ed. G. R. Aiken, D. M. McKnight, R. L. Wershaw & P. McCarthy, pp. 233–47. John Wiley & Sons, New York.

Harvey, G. R., Boran, D. A., Chesal, L. A. & Tokar, J. M. (1983). The structure of marine fulvic and humic acids. *Marine Chemistry*, **12**, 119–32.

Harm, W. (1980). *Biological Effects of Ultraviolet Radiation*, IUPAB Biophysics Series I. Cambridge University Press, Cambridge.

Hedges, J. I. (1992). Global biogeochemical cycles: progress and problems. *Marine Chemistry*, **39**, 67–93.

Hedges, J. I., Hatcher, P. G., Ertel, J. R. & Meyers-Schulte, K. J. (1992). A comparison of dissolved humic substances from seawater with Amazon River counterparts by 13C-NMR spectrometry. *Geochimica et Cosmochimica Acta*, **56**, 1753–7.

Heissenberger, A. & Herndl, G. J. (1994). Formation of high molecular weight material by free-living marine bacteria. *Marine Ecology Progress Series*, **111**, 129–35.

Hellebust, J. A. (1965). Excretion of some organic compounds by marine phytoplankton. *Limnology and Oceanography*, **10**, 192–206.

Herndl, G. J., Muller-Niklas, G. & Frick, J. (1993). Major role of ultraviolet-B in controlling bacterioplankton growth in the surface layer of the ocean. *Nature*, **361**, 717–19.

Hoge, F. E., Vodacek, A. & Blough, N. V. (1993). Inherent optical properties of the ocean: retrieval of the absorption coefficient of chromophoric dissolved organic matter from fluorescence measurements. *Limnology and Oceanography*, **38**, 1394–402.

Hoigné, J., Faust, B. C., Haag, W. R., Scully, Jr, F. E. & Zepp, R. G. (1989). Aquatic humic substances as sources and sinks of photochemically produced transient reactants. In *Aquatic Humic Substances: Influence on Fate and Treatment of Pollutants*, ed. I. H. Suffet & P. MacCarthy, pp. 363–81. American Chemical Society, Washington, DC.

Højerslev, N. K. (1982). Yellow substance in the sea. In *The Role of Solar Ultraviolet Radiation in Marine Ecosystems*, ed. J. Calkins, pp. 263–81. Plenum Press, New York.

Ishiwatari, R. (1992). Macromolecular material (humic substance) in the water column and sediments. *Marine Chemistry*, **39**, 151–66.

Jagger, J. (1985). *Solar-UV Actions on Living Cells*, Praeger Publishing, New York.

Jones, R. D. (1991). Carbon monoxide and methane distribution and consumption in the photic zone of the Sargasso Sea. *Deep Sea Research*, **38**, 625–35.

Jones, R. D. & Amador, J. A. (1993). Methane and carbon monoxide production, oxidation and turnover times in the Caribbean Sea as influenced by the Orinoco River. *Journal of Geophysical Research*, **98**, 2353–9.

Karabashev, G. S. (1992). On the influence of dissolved organic matter on remote-sensing of chlorophyll in the straits of Skagerrak and Kattegat. *Oceanologica Acta*, **15**, 255–9.

Karentz, D. (1991). Ecological considerations of Antarctic ozone depletion. *Antarctic Science*, **3**, 3–11.

Karentz, D. (1994). Ultraviolet tolerance mechanisms in Antarctic marine organisms. In *Ultraviolet Radiation in Antarctica: Measurements and Biological Effects*, Antartic Research Series, **62**, ed., C. S. Weiler & P. A. Penhale, pp. 93–110. American Geophysical Union, Washington, DC.

Kieber, D. J. & Blough, N. V. (1990). Determination of carbon-centered radicals in aqueous solution by high performance liquid chromatography with fluorescence detection. *Analytical Chemistry*, **62**, 2275–83.

Kieber, D. J., Jiao, J., Kiene, R. P. & Bates, T. S. (1996). Impact of dimethyl sulfide photochemistry on methyl sulfur cycling in the equatorial Pacific Ocean. *Journal of Geophysical Research*, **101(C)**, 3715–22.

Kieber, D. J., McDaniel, J. A. & Mopper, K. (1989). Photochemical source of biological substrates in seawater: implications for carbon cycling. *Nature*, **341**, 637–9.

Kieber, D. J. & Mopper, K. (1987). Photochemical formation of glyoxylic and pyruvic acids in seawater. *Marine Chemistry*, **21**, 135–49.

Kieber, R. J., Zhou, X. & Mopper, K. (1990). Formation of carbonyl compounds from UV-induced photodegradation of humic substances in natural waters: fate of riverine carbon in the sea. *Limnology and Oceanography*, **35**, 1503–15.

Kirchman, D. L., Rich, J. H. & Barber, R. T. (1995). Biomass and biomass production of heterotrophic bacteria along 140°W in the equatorial Pacific: effect of temperature on the microbial loop. *Deep Sea Research II*, **42**, 603–19.

Kirchman, D. L., Suzuki, Y., Garside, C. & Ducklow, H. W. (1991). High turnover rates of dissolved organic carbon during a spring phytoplankton bloom. *Nature*, **352**, 612–14.

Kotzias, D., Herrmann, M., Zsolnay, A., Beyerle-Pfnür, R., Parlar, H. & Korte, F. (1987). Photochemical aging of humic substances. *Chemosphere*, **16**, 1463–8.

Kouassi, A. M. & Zika, R. G. (1990). Light-induced alteration of the photophysical properties of dissolved organic matter in seawater. Part I. Photoreversible properties of natural water fluorescence. *Netherlands Journal of Sea Research*, **27**, 25–32.

Kouassi, A. M. & Zika, R. G. (1992). Light-induced destruction of the absorbance property of dissolved organic matter in seawater. *Toxicology and Environmental Chemistry*, **35**, 195–211.

Kouassi, A. M., Zika, R. G. & Plane, J. M. C. (1990). Light-induced alteration of the photophysical properties of dissolved organic matter in seawater. Part II. Estimates of the environmental rates of the natural water fluorescence. *Netherlands Journal of Sea Research*, **27**, 33–41.

Kramer, C. J. M. (1979). Degradation by sunlight of dissolved fluorescing substances in the upper layers of the eastern Atlantic Ocean. *Netherlands Journal of Sea Research*, **13**, 325–9.

Laane, R. W. P. M., Gieskes, W. W. C., Kraay, G. W. & Eversdijk, A. (1985). Oxygen consumption from natural waters by photo-oxidizing processes. *Netherlands Journal of Sea Research*, **19**, 125–8.

Larson, R. A. (1995). Singlet oxygen in the environmental sciences. Preprint of paper presented at the 210th ACS National Meeting, Chicago, IL, 20–24 August, *American Chemical Society Division of Environmental Chemistry*, vol 35(2), pp. 398–9. American Chemical Society, Washington, DC.

Lee, C. & Wakeham, S. G. (1988). Organic matter in seawater: biogeochemical processes. In *Chemical Oceanography*, vol. 9, ed. J. P. Riley, pp. 1–49. Academic Press, San Diego, CA.

Lee, C. & Wakeham, S. G. (1992). Organic matter in the water column: future research challenges. *Marine Chemistry*, **39**, 95–118.

Lee, R. F. & Baker, J. (1992). Ethylene and ethane production in an estuarine river: formation from the decomposition of polyunsaturated fatty acids. *Marine Chemistry*, **38**, 25–36.

Leifer, A. (1988). *The Kinetics of Environmental Aquatic Photochemistry*. American Chemical Society, Washington, DC.

Lichtenthaler, R. G., Haag, W. R. & Mill, T. (1989). Photooxidation of probe compounds sensitized by crude oils in toluene and as an oil film on water. *Environmental Science and Technology*, **23**, 39–45.

Lindell, M. J., Granéli, W. & Tranvik, L. J. (1995). Enhanced bacterial growth in response to photochemical transformation of dissolved organic matter. *Limnology and Oceanography*, **40**, 195–9.

Lindell, M. J., Granéli, W., & Tranvik, L. J. (1996). Effects of sunlight on bacterial growth in lakes of different humic content. *Aquatic Microbial Ecology*, **11**, 135–41.

Lindell, M. J. & Rai, H. (1994). Photochemical oxygen consumption in humic waters. *Ergebnisse der Limnologie (Archiv für Hydrobiologie. Beiheft)*, **43**, 145–55.

Mabury, S. A. & Crosby, D. G. (1995). The carbonate radical in natural water. Preprint from a paper presented before the Division of Environmental Chemistry, American Chemical Society, Chicago, IL, 21–16 August, symposium volume, pp. 499–502.

MacKinnon, M. D. (1981). The measurement of organic carbon in sea water. In *Marine Organic Chemistry*, Elsevier Oceanography Series, 31, ed. E. K. Duursma & R. Dawson, pp. 415–43. Elsevier Scientific Publishing, Amsterdam.

Mague, T. H., Friberg, E., Hughes, D. J. & Morris, I. (1980). Extracellular release of carbon by marine phytoplankton: a physiological approach. *Limnology and Oceanography*, **25**, 262–79.

Malloy, K. D., Holman, M. A., Mitchell, D. & Detrich, III, H. W. (1997). Solar UVB-induced DNA damage and photoenzymatic DNA repair in antarctic zooplankton. *Proceedings of the National Academy of Sciences, USA*, **94**, 1258–63.

Mantoura, R. F. C. & Woodward, E. M. S. (1983). Conservative behavior of riverine dissolved organic carbon in the Severn Estuary: chemical and geochemical implications. *Geochimica et Cosmochimica Acta*, **47**, 1293–309.

Martin, J. H. & Fitzwater, S. E. (1988). Iron deficiency limits phytoplankton growth in the north-east Pacific Subarctic. *Nature*, **331**, 341–3.

Martin, J. H., Knauer, G. A., Karl, D. M. & Broenkow, W. W. (1987). VERTEX: carbon cycling in the northeast Pacific. *Deep Sea Research*, **34**, 267–85.

Mee, L. K. (1987). Radiation chemistry of biopolymers. In *Radiation Chemistry: Principles and Applications*, ed. Farhataziz & M. A. J. Rodgers, pp. 477–99. VCH Publishers, New York.

Menzel, D. W. (1974). Primary productivity, dissolved and particulate organic matter, and the sites of oxidation of organic matter. In *The Sea*, vol. 5, ed. E. D. Goldberg, pp. 659–78. John Wiley & Sons, New York, NY.

Meyers-Schulte, K. J. & Hedges, J. I. (1986). Molecular evidence for a terrestrial component of organic matter dissolved in ocean water. *Nature*, **321**, 61–3.

Miles, C. J. & Brezonik, P. L. (1981). Oxygen consumption in humic-colored waters by a photochemical ferrous–ferric catalytic cycle. *Environmental Science and Technology*, **15**, 1089–95.

Mill, T. (1980). Chemical and photo oxidation. In *The Handbook of Environmental Chemistry*, ed. O. Hutzinger, vol. 2(A), pp. 77–105. Springer-Verlag, Berlin.

Mill, T., Haag, W. & Karentz, D. (1990). Estimated effects of indirect photolysis on marine organisms. In *Effects of Solar Ultraviolet Radiation on Biogeochemical Dynamics in Aquatic Environments.* Woods Hole Oceanographic Institute Technical Report, WHOI 90-09, pp. 89–93.

Mill, T., Hendry, D. G. & Richardson, H. (1980). Free-radical oxidants in natural waters. *Science,* **207**, 886–7.

Miller, W. L. (1994). Recent advances in the photochemistry of natural dissolved organic matter. In *Aquatic and Surface Photochemistry*, ed. G. R. Helz, R. G. Zepp & D. G. Crosby, pp. 111–27. Lewis Publishers, Boca Raton, FL.

Miller, W. L., King, D. W., Lin, J. & Kester, D. R. (1995). Photochemical redox cycling of iron in coastal seawater. *Marine Chemistry,* **50**, 63–77.

Miller, W. L. & Moran, M. A. (1997). Interaction of photochemical and microbial processes in the degradation of refractory dissolved organic matter from a coastal marine environment. *Limnology and Oceanography,* **42**, 1317–24.

Miller, W. L. & Zepp, R. G. (1995). Photochemical production of dissolved inorganic carbon from terrestrial organic matter: significance to the oceanic organic carbon cycle. *Geophysical Research Letters,* **22**, 417–20.

Miller, W. L., Zepp, R. G., Moran, M. A., Sheppard, E. S. & Hodson, R. E. (1994). Organic carbon cycling in water from a Georgia salt marsh: photochemical and biological transformations. *Transactions of the American Geophysical Union,* **75**, 327.

Moffett, J. W. & Zika, R. G. (1983). Oxidation kinetics of Cu (I) in seawater: implications for its existence in the marine environment. *Marine Chemistry,* **13**, 239–51.

Moffett, J. W. & Zika, R. G. (1987). Photochemistry of copper complexes in seawater. In *Photochemistry of Environmental Aquatic Systems,* ACS Symposium Series, vol. 327, ed. R. G. Zika & W. J. Cooper, pp. 116–30. American Chemical Society, Washington, DC.

Moffett, J. W. & Zika R. G. (1988). Measurement of copper(I) in surface waters of the subtropical Atlantic and Gulf of Mexico. *Geochimica et Cosmochimica Acta,* **52**, 1849–57.

Momzikoff, A., Santus, R. & Giraud, M (1983). A study of the photosensitizing properties of seawater. *Marine Chemistry,* **12**, 1–14.

Moore, C. A., Farmer, C. T. & Zika, R. G. (1993). Influence of the Orinoco River on hydrogen peroxide distribution and production in the eastern Caribbean. *Journal of Geophysical Research,* **98(C2)**, 2289–98.

Moore, R. M. & Zafiriou, O. C. (1994). Photochemical production of methyl iodide in seawater. *Journal of Geophysical Research,* **99**, 16415–20.

Mopper, K. & Degens, E. T. (1979). Organic carbon in the ocean: Nature and cycling. In *The Global Carbon Cycle,* ed. B. Bolin, E. T. Degens, S. Kempe & P. Ketner, pp. 293–316. John Wiley & Sons, New York.

Mopper, K., Sarpal, R. S. & Kieber, D. J. (1995). Protein and humic fluorescence of dissolved organic matter in antarctic seawater. *Antarctic Journal of the USA,* **30(5)**, 137–9.

Mopper, K. & Schultz, C. A. (1993). Fluorescence as a possible tool for studying the nature and water column distribution of DOC components. *Marine Chemistry,* **41**, 229–38.

Mopper, K. & Stahovec, W. L. (1986). Sources and sinks of low-molecular-weight organic carbonyl compounds in seawater. *Marine Chemistry,* **19**, 305–21.

Mopper, K. & Zhou, X. (1990). Hydroxyl radical photoproduction in the sea and its potential impact on marine processes. *Science,* **250**, 661–4.

Mopper, K., Zhou, X,. Kieber, R. J., Kieber, D. J., Sikorski, R. J. & Jones, R. D. (1991). Photochemical degradation of dissolved organic carbon and its impact on the oceanic carbon cycle. *Nature*, **353**, 60–2.

Moran, M. A. & Zepp, R. G. (1997). Role of photoreactions in the formation of biologically labile compounds from dissolved organic matter. *Limnology and Oceanography*, **42**, 1307–16.

Moran, S. B. & Buessler, K. O. (1992). Short residence time of colloids in the upper ocean estimated from $^{238}U-^{214}Th$ disequilibria. *Nature*, **359**, 221–3.

Morris, D. P. & Hargreaves, B. R. (1997). The role of photochemical degradation of dissolved organic carbon in regulating the UV transparency of three lakes on the Pocono Plateau. *Limnology and Oceanography*, **42**, 239–49.

Najjar, R. G., Erickson, III, D. J. & Madronich, S. (1995). Modeling the air–sea fluxes of gases formed from the decomposition of dissolved organic matter: carbonyl sulfide and carbon monoxide. In *Role of Nonliving Organic Matter in the Earth's Carbon Cycle*, ed. R. G. Zepp & Ch. Sonntag, pp. 107–32. John Wiley & Sons, New York.

Neale, P. J., Davis, R. F. & Cullen, J. J. (1998). Interactive effects of ozone depletion and vertical mixing on photosynthesis of Antarctic phytoplankton. *Nature*, **392**, 585–9.

Nelson, N. B., Siegel, D. A. & Michaels, A. F. (1998). Seasonal dynamics of colored dissolved material in the Sargasso Sea. *Deep Sea Research I*, **45**, 931–57.

Opsahl, S. & Benner, R. (1997). Distribution and cycling of terrigenous dissolved organic matter in the ocean. *Nature*, **386**, 480–2.

Paerl, H. W. (1991). Ecophysiological and trophic implications of light-stimulated amino acid utilization in marine picoplankton. *Applied Environmental Microbiology*, **57**, 473–9.

Palenik, B., Price, N. M. & Morel, F. M. M. (1991). Potential effects of UV-B on the chemical environmental of marine organisms: a review. *Environmental Pollution*, **70**, 117–30.

Petasne, R. G. & Zika, R. G. (1987). Fate of superoxide in coastal sea water. *Nature*, **325**, 516–18.

Plane, J. M. C., Zika, R. G., Zepp, R. G. & Burns, L. A. (1987). Photochemical modeling applied to natural waters. In *Photochemistry of Environmental Aquatic Systems*, ACS Symposium Series, vol. 327, ed. R. G. Zika and W. J. Cooper, pp. 250–67. American Chemical Society, Washington, DC.

Pos, W. H., Milne, P. J., Riemer, D. & Zika, R. G. (1997). Photoinduced oxidation of H_2S species: a sink for sulfide in seawater. *Journal of Geophysical Research*, **102**, 12831–7.

Porter, N. A. & Wujek, D. G.(1988). The autooxidation of polyunsaturated lipids. In *Reactive Oxygen Species in Chemistry, Biology, and Medicine*, ed. A. Quintanilha, pp. 55–79. Plenum Press, New York.

Riemer, D. D., Pos, W. H. & Zika, R. G. (1997). Nonmethane hydrocarbons (NMHCs) in seawater: photochemical production from dissolved organic matter (DOM). In *Program and Abstracts from the 1997 ASLO Ocean Sciences Meeting*, 10–14 February. Santa Fe, NM, p. 282.

Salonen, K. & Vähätalo (1994). Photochemical mineralisation of dissolved organic matter in Lake Skjervatjern. *Environment International*, **20**, 307–12.

Santschi, P. H., Guo, L., Baskaran, M., Trumbore, S., Southon, J., Bianchi, T. S., Honeyman, B. & Cifuentes, L. (1995). Isotopic evidence for the contemporary origin of high-molecular weight organic matter in oceanic environments. *Geochimica et Cosmochimica Acta*, **59**, 625–31.

Sarpal, R. S., Mopper, K. & Kieber, D. J. (1995). Absorbance properties of dissolved organic matter in antarctic seawater. *Antarctic Journal of the USA*, **30**, 139–40.

Scully, Jr, F. E. & Hoigné, J. (1987). Rate constants for reactions of singlet oxygen with phenols and other compounds in water. *Chemosphere*, **16**, 681–94.

Sellers, P., Kelly, C. A., Rudd, J. W. M. & MacHutchon, A. R. (1996). Photodegradation of methylmercury in lakes. *Nature*, **380**, 694–7.

Sholkovitz, E. R. (1976). Flocculation of dissolved organic and inorganic matter during the mixing of river water and seawater. *Geochimica et Cosmochimica Acta*, **40**, 831–45.

Smith, R. C. & Baker, K. S. (1989). Stratospheric ozone, middle ultraviolet radiation and phytoplankton productivity. *Oceanography*, **2**, 4–10.

Smith, R. C., Prézelin, B. B., Baker, K. S., Bidigare, R. R., Boucher, N. P., Coley, T. *et al.* (1992). Ozone depletion: ultraviolet radiation and phytoplankton biology in Antarctic waters. *Science*, **255**, 952–9.

Spokes, L. J. & Liss, P. S. (1995). Photochemically induced redox reactions in seawater I. Cations. *Marine Chemistry*, **49**, 201–13.

Stewart, A. J. & Wetzel, R. G. (1981). Dissolved humic materials: photodegradation, sediment effects, and reactivity with phosphate and calcium carbonate precipitation. *Archiv für Hydrobiologie*, **92**, 265–86.

Strome, D. J. & Miller, M. C. (1978). Photolytic changes in dissolved humic substances. *Verhandlungen-Internationale Vereinigung für Theoretische und Angewandte Limnologie*, **20**, 1248–54.

Stuiver, M., Quay, P. D. & Ostlund, H. G. (1983). Abyssal water carbon-14 distribution and the age of the world oceans. *Science*, **219**, 849–51.

Sulzberger, B. (1990). Photoredox reactions at hydrous metal oxide surfaces: a surface coordination chemistry approach. In *Aquatic Chemical Kinetics – Reaction Rates of Processes in Natural Waters*, ed. W. Stumm, pp. 401–29. John Wiley & Sons, New York.

Sunda, W. G. & Gillespie, P. A. (1979). The response of a marine bacterium to cupric ion and its use to estimate cupric ion activity in seawater. *Journal of Marine Research*, **37**, 761–77.

Sunda, W. G., Huntsman, S. A. & Harvey, G. R. (1983). Photoreduction of manganese oxides in seawater and its geochemical and biological implications. *Nature*, **301**, 234–6.

Thominette, F. & Verdu, J. (1984). Photo-oxidative behavior of crude oils relative to sea pollution. Part I. Comparative study of various crude oils and model systems. *Marine Chemistry*, **15**, 91–104.

Thompson, A. M. & Zafiriou, O. C. (1983). Air–sea fluxes of transient atmospheric species. *Journal of Geophysical Research*, **88(C11)**, 6696–708.

Tranvik, L. J. (1993). Microbial transformation of labile dissolved organic matter into humic-like matter in seawater. *FEMS Microbiological Ecology*, **12**, 177–83.

Truesdale, V. W. & Moore, R. M. (1992). Further studies on the chemical reduction of molecular iodine added to seawater. *Marine Chemistry*, **40**, 199–213.

Uher, G. & Andreae, M. O. (1997). Photochemical production of carbonyl sulfide in North Sea water: a process study. *Limnology and Oceanography*, **42**, 432–42.

Valentine, R. L. & Zepp, R. G. (1993). Formation of carbon monoxide from the photodegradation of terrestrial dissolved organic carbon in natural waters. *Environmental Science and Technology*, **27**, 409–12.

Vaughan, G. M. (1989). Determination of nanomolar levels of formate in natural waters based on a luminescence enzymatic assay. M.S. thesis, University of Miami.

Vodacek, A., Blough, N. V., DeGrandpre, M. D., Peltzer, E. T. & Nelson, R. K. (1997). Seasonal variation of CDOM and DOC in the middle Atlantic bight: terrestrial inputs and photooxidation. *Limnology and Oceanography*, **42**, 674–86.

Vodacek, A., Hoge, F. E., Swift, R. N., Yungel, J. K., Peltzer, E. T. & Blough, N. V. (1995). The use of in situ and airborne fluorescence measurements to determine UV absorption coefficients and DOC concentrations in surface waters. *Limnology and Oceanography*, **40**, 411–15.

Voelker, B. M., Morel, F. M. M. & Sulzberger, B. (1997). Iron redox cycling in surface waters: effects of humic substances and light. *Environmental Science and Technology*, **31**, 1004–11.

Waite, T. D. & Morel, F. M. M. (1984). Photoreductive dissolution of colloidal iron oxides in natural waters. *Environmental Science and Technology*, **18**, 860–8.

Weiss, P. S., Andrews, S. S., Johnson, J. E. & Zarfiriou, O. C. (1995). Photoproduction of carbonyl sulfide in south Pacific Ocean as a function of irradiation wavelength. *Geophysical Research Letters*, **22**, 215–18.

Wetzel, R. G., Hatcher, P. G. & Bianchi, T. S. (1995). Natural photolysis by ultraviolet irradiance of recalcitrant dissolved organic matter to simple substrates for rapid bacterial metabolism. *Limnology and Oceanography* **40**, 1369–80.

Williams, P. M. & Carlucci, A. F. (1976). Bacterial utilization of organic matter in the deep sea. *Nature*, **262**, 810–11.

Williams, P. M. & Druffel, E. R. M. (1987). Radiocarbon in dissolved organic matter in the Central North Pacific Ocean. *Nature*, **330**, 246–8.

Williamson, C. E. (1995). What role does UV-B radiation play in freshwater ecosystems? *Limnology and Oceanography*, **40**, 386–92.

Wilson, D. F., Swinnerton, J. W. & Lamontagne, R. A. (1970). Production of carbon monoxide and gaseous hydrocarbons in seawater: relation to dissolved organic carbon. *Science*, **168**, 1577–9.

Zafiriou, O. C. (1974). Sources and reactions of hydroxyl and daughter radicals in seawater. *Journal of Geophysical Research*, **79**, 4491–7.

Zafiriou, O. C. (1977). Marine organic chemistry previewed. *Marine Chemistry*, **5**, 497–522.

Zafiriou, O. C. (1983). Natural water photochemistry. In *Chemical Oceanography*, 2nd edn, vol. 8, ed. J. P. Riley & R. Chester, pp. 339–79. Academic Press, London.

Zafiriou, O. C., Blough, N. V., Micinski, E., Dister, B., Kieber, D. & Moffett, J. (1990). Molecular probe systems for reactive transients in natural waters. *Marine Chemistry*, **30**, 45–70.

Zafiriou, O. C. & Dister, B. (1991). Photochemical free radical production rates: Gulf of Maine and Woods Hole-Miami transect. *Journal of Geophysical Research*, **96**, 4939–45.

Zafiriou, O. C., Joussot-Dubien, J., Zepp, R. G. & Zika, R. G. (1984). Photochemistry of natural waters. *Environmental Science and Technology*, **18**, 358A–371A.

Zafirou, O. C & McFarland, M. (1981). Nitric oxide from nitrite photolysis in the central equatorial Pacific. *Journal of Geophysical Research*, **86(C4)**, 3173–82.

Zafiriou, O. C. & True, M. B. (1979). Nitrate photolysis in seawater by sunlight. *Marine Chemistry*, **8**, 33–42.

Zepp, R. G. (1988). Environmental photoprocesses involving natural organic matter. In *Humic Substances and Their Role in the Environment*, ed. F. H. Frimmel & R. F. Christman, pp. 193–214. John Wiley & Sons, New York.

Zepp, R. G. (1991). Photochemical conversion of solar energy in the environment. In *Photochemical Conversion of Solar Energy*, ed. E. Pelizzetti & M. Schiavello, pp. 497–515. Kluwer Academic, Dordrecht.

Zepp, R. G. & Andreae, M. O. (1994). Factors affecting the photochemical production of carbonyl sulfide in seawater. *Geophysical Research Letters*, **21**, 2813–16.

Zepp, R. G., Braun, A. M., Hoigné, J. & Leenheer, J. A. (1987a). Photoproduction of hydrated electrons from natural organic solutes in aquatic environments. *Environmental Science and Technology*, **21**, 485–90.

Zepp, R. G., Callaghan, T. V. & Erickson, D. J. (1995). Effects of increased solar ultraviolet radiation on biogeochemical cycles. *Ambio*, **24**, 181–7.

Zepp, R. G., Hoigné, J. & Bader, H. (1987). Nitrate-induced photooxidation of trace organic chemicals in water. *Environmental Science and Technology*, **21**, 443–50.

Zepp, R. G., Schlotzhauer, P. F. & Sink, R. M. (1985). Photosensitized transformations involving electronic energy transfer in natural waters: role of humic substances. *Environmental Science and Technology*, **19**, 74–81.

Zepp, R. G., Wolfe, N. L., Baughman, G. L. & Hollis, R. C. (1977). Singlet oxygen in natural waters. *Nature*, **267**, 421–3.

Zhou, X. & Mopper, K. (1990). Determination of photochemically produced hydroxyl radicals in seawater and freshwater. *Marine Chemistry*, **30**, 71–88.

Zhou, X. & Mopper, K. (1997). Photochemical production of low molecular weight carbonyl compounds in seawater and surface microlayer and their air-sea exchange. *Marine Chemistry*, **56**, 201–14.

Zika, R. G. (1981). Marine organic photochemistry. In *Marine Organic Chemistry*, ed. E. K. Duursma & R. Dawson, pp. 299–325. Elsevier, Amsterdam.

Zika, R. G. (1987). Advances in marine photochemistry 1983–1987. *Reviews of Geophysics*, **25**, 1390–4.

Zlotnik, I. & Dubinsky, Z. (1989). The effect of light and temperature on DOC excretion by phytoplankton. *Limnology and Oceanography*, **34**, 831–9.

Zuo, Y. & Jones, R. D. (1995). Formation of carbon monoxide by photolysis of dissolved organic material and its significance in the carbon cycling of the oceans. *Naturwissenschaften*, **82**, 472–4.

5

○ ○ ○ ○ ○ ○ ○ ○ ○ ○ ○ ○ ○ ○ ○ ○ ○ ○ ○ ○

Photochemical production of biological substrates

David J. Kieber

5.1 Introduction

Solar radiation initiates a wide variety of photochemical transformations in natural waters, ranging from primary photolyses involving well-defined chromophores to complicated photoreactions involving ill-defined chromophores, such as humic substances. The broad spectrum of photochemical transformations that have been documented attest to the importance of photochemistry in aquatic biogeochemical cycles (Zafiriou, 1977; Zika, 1981; Larson & Berenbaum, 1988; Cooper et al., 1989; Hoigné et al., 1989; Blough & Zepp, 1990; Helz, Zepp & Crosby, 1994; Zepp, Callaghan & Erickson, 1995; Blough, 1997; Moran & Zepp, 1997). These transformations are driven largely by the absorption of light by dissolved organic matter (DOM), principally in the ultraviolet (UV) region of the solar spectrum ($\lambda < 400$ nm).

Currently, it is not possible to assess quantitatively the effects of various photochemical processes on plankton in marine and freshwater environments. A comprehensive understanding of the nature of photochemical–biological interactions in aquatic systems is lacking, at both the cellular and the ecosystems levels. There is evidence for short term deleterious effects of UV radiation including DNA damage (Karentz, Cleaver & Mitchell, 1991), inhibition of photosynthesis (Cullen, Neale & Lesser, 1992; Smith et al., 1992) and decreased nutrient uptake (Döhler et al., 1991). It is also clear that organisms have mechanisms to repair damage and screen incident UV radiation (Vincent & Roy, 1993; Karentz, 1994; and see Chapter 8). As may be expected, there is considerable interspecies sensitivity to short and long term exposure to UV radiation that may lead to changes in the community structure favouring UV-resilient species (Calkins & Thordardóttir, 1980; Worrest, Thomson & Van Dyke, 1981; Worrest, 1983). However, some of the changes that are observed are counter-intuitive and make sense only through examination of trophic level interactions. For example, Boothwell, Sherbot & Pollock (1994) observed that the benthic algal population increased upon exposure to UV-B due to a differential effect of UV-B on herbivores and the algae. Presumably, all of these observed 'UV effects' are the result of photochemical transformations that occur intracellularly or in the seawater surrounding the plankton.

The driving force for UV-related environmental research is the recognition that UV-B fluxes are increasing at the earth's surface due to the depletion of stratospheric ozone (Crutzen, 1992; Harris *et al.*, 1995). UV-B radiation may also penetrate deeper in the water column due to increased water clarity stemming from lake water acidification and global warming (Schindler *et al.*, 1996). Potential effects of photochemistry on aquatic ecosystems in response to increasing UV-B fluxes have been highlighted in several reviews. Larson & Berenbaum (1988) discussed the role of aquatic photochemistry in altering the toxicity of xenobiotics, while Palenik, Price & Morel (1991) focused their review on potential impacts of increased UV-B levels on free radical damage, and trace metal toxicity and bioavailability. More recently, Madronich *et al.* (1995) discussed trends in UV radiation reaching the earth's surface and calculated radiation amplification factors for a variety of biological and photochemical processes. In the same issue of *Ambio*, Zepp *et al.* (1995) reviewed the effects of changes in UV fluxes on the biogeochemical cycles of carbon, nitrogen, oxygen and sulfur both in terrestrial and aquatic ecosystems; and Häder *et al.* (1995) examined the effects of increased UV radiation on aquatic ecosystems. This chapter complements these and other earlier reviews (e.g. Williamson, 1995) by examining a burgeoning area of research on the effects of extracellular photochemical transformations of naturally occurring organic matter on the growth of aquatic plankton. A related review that focuses on the photochemical aspects of this process has recently been reported (Moran & Zepp, 1997). Results from both freshwater and marine studies are discussed, since much of the research on the photochemical production of biological substrates has been conducted in freshwater systems.

5.2   Photochemical sources of biological substrates

The photochemical breakdown of dissolved organic matter (DOM) in natural waters is a potentially important source of substrates for plankton, either by releasing nutrients and producing specific low molecular weight

(LMW) species, or by rendering the high molecular weight fraction of DOM more labile to microbial attack (Figure 5.1). Strome & Miller (1978) made some of the earliest observations that the photolysis of humic compounds, isolated by base extraction from soil or Arctic tundra ponds, greatly increased the ability of the humic compounds to support bacterial growth. Geller (1986) had a similar finding using DOM that was isolated from a mesoeutrophic lake by gel filtration through a G-15 Sephadex column. This DOM was added to an inorganic media in a Pyrex flask ($<1\%$ transmission ≤ 320 nm) and exposed to sunlight for over six weeks through a north-facing window. Bacterial isolates added to the photolysed solutions readily utilised the DOM, whereas little growth was observed in the dark controls. Loss of DOM was observed only in the first week,

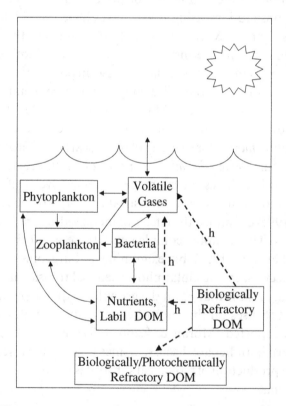

Figure 5.1. Impact of photochemistry on the cycling of organic matter in the photic zone. The solid arrows denote known inputs and removal pathways for organic matter in the photic zone (excluding mixing), while dashed arrows depict pathways involving photochemistry. Photochemical modification of dissolved organic matter (DOM) may also result in the formation of a pool of DOM that is photochemically stable and biologically recalcitrant.

however, with no change in DOM detected thereafter. These initial qualitative observations were largely unrecognised in the literature until the early 1990s, when numerous, more detailed studies were conducted that showed a correlation between the exposure of natural water to sunlight and its increased ability to support bacterial growth.

Lindell, Granéli & Tranvik (1995) conducted an experiment using sterilised water collected from a mesohumic lake (Sjättesjön), with 12 mg DOC (dissolved organic carbon) l^{-1} and amended with millimolar levels of nutrients. The water was exposed to simulated sunlight for 0–100 h, and then inoculated with a natural bacterial assemblage and incubated in the dark. Both bacterial numbers and cell volumes increased substantially with increasing exposure of the lake water to UV radiation for up to 8–10 h of light exposure. However, bacterial numbers and volumes were not affected by longer periods of irradiation, implying that the microbes could no longer utilise the remaining DOM (or that nutrients were limiting). Although this study showed a correlation between photolysis of DOM and bacterial growth, the quantitative importance of this process was not determined nor were photochemical changes in DOM examined (e.g. photoproducts, optical properties). Assuming that the increase in bacterial biomass (*circa* 2.5 mg C l^{-1}) is due to the utilisation of DOM photoproducts, then the bacteria were able to use approximately 20% more of the carbon in the light-exposed lake water relative to the dark controls. This is a conservative estimate because the contribution of bacterial respiration was not included, as it was not measured by Lindell *et al.* (1995). Even with this uncertainty, this estimate suggests that the photochemical production of biological substrates may have a significant, but not exhaustive, contribution to bacterial growth.

Repeated cycles of sunlight exposure and microbial attack should degrade the DOM even further than was observed by these investigators, especially if higher energy UV-B radiation is not partially or totally excluded (Figure 5.2). Evidence for enhanced degradation of DOM through alternating cycles of light exposure and dark microbial degradation was demonstrated for seawater from a salt marsh (Sapelo Island, Georgia, USA), with and without added humic acids (Miller & Moran, 1997). These investigators found that the fraction of DOM degraded through multiple photochemical/microbial cycles was significantly higher than the fractional loss that was achieved through a single cycle (Figure 5.2). This idea of alternating cycles fits well with the notion that DOM (and plankton) is continually mixed in and out of the UV-active photic zone in the oceans due to wind-driven mixing. Thus, even if there is photoinhibition at or

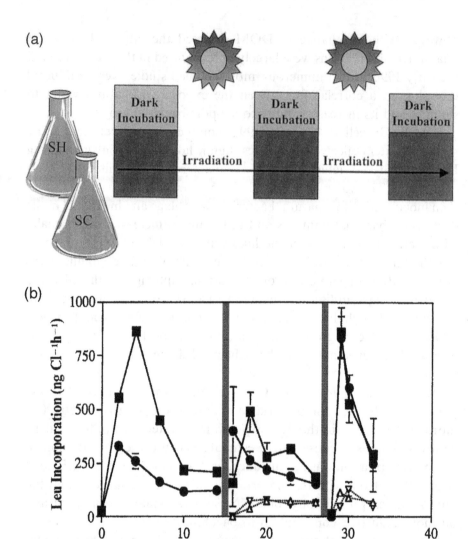

Figure 5.2.   (a) Experimental design to study alternating biological and photochemical degradation of coastal dissolved organic matter (DOM). A 20 ml sample of a bacterioplankton inoculum was added to 1.6 l of 0.2 µm filtered Sapelo Island salt marsh water containing either added humic substances (SH) or added control water (SC). Samples were incubated in the dark at 24 °C for 14 days, with ³H-leucine added to estimate bacterial protein production. The inoculum that was used for the biological incubations was obtained by 0.6 µm filtration of the marsh water to remove the bacteriovores. After the 14 day incubation in the dark, samples were filtered through a 0.2 µm Nuclepore filter and exposed to sunlight for the equivalent of 8 h of midday sunlight at 34° N latitude. The water was then amended with the bacterial inoculum and incubated in the dark for 10 days. After this second biological incubation, samples were refiltered through a 0.2 µm filter and again irradiated for approximately 8 h of midday sunlight exposure. This was followed by a third addition of bacterial inoculum and

near the sea surface, photochemically derived substrates will be utilised deeper in the water column.

Photolysis of DOM does not necessarily lead to higher bacterial productivity in marine waters. Benner & Biddanda (1998) observed that bacterial production rates decreased by 75% when 0.2 µm-filtered surface seawater from the Gulf of Mexico was exposed to sunlight and subsequently amended with unfiltered seawater and incubated in the dark. In contrast to the surface seawater results, exposure of filtered, deep seawater increased bacterial productivity by 40% relative to the controls. This result is not surprising, given that deep seawater contains a large fraction of biologically refractory DOM. Sunlight exposure of this water probably resulted in a net increase in biological substrates. Surface seawater, by contrast, has lower absorbances, lower DOM concentrations and a much higher proportion of biomolecules in the DOM. Therefore, although it is possible that the decreased productivity observed in surface seawater resulted from the photochemical formation of bacterial inhibitors, a decrease in limiting nutrient concentrations, or the photochemical alteration of DOM to form biologically refractory compounds, it is much more likely that decreased bacterial productivity resulted from a net photochemical loss of substrates. This interpretation is consistent with the finding that the solar photolysis of virally released biomolecular organic matter from a culture of *Aureococcus anophagefferens* resulted in a decrease in the amount of carbon available to bacteria (Gobler *et al.*, 1997). It can be concluded from these studies that photolysis of DOM will increase bacterial productivity in relatively high DOM, humic-rich environments (e.g. coastal waters and many freshwater lakes), and decrease productivity in oligotrophic, photobleached surface waters of the open ocean. Further studies are needed to test this hypothesis.

Studies by Strome & Miller (1978), Geller (1986), Lindell *et al.* (1995), Miller & Moran (1997), and Jørgensen *et al.* (1998) show that there is an increase in the microbial use of DOM that has been exposed to sunlight.

Caption for fig. 5.2 (*cont.*)

dark incubation for 7 days at 24 °C. Controls were not exposed to sunlight. (b) Biological degradation of DOM as measured by rates of bacterial protein production. Vertical grey bars indicate when the bacteria were removed by filtration for the photochemical degradation phase of the study, and then reinoculated. Error bars denote the standard deviation of triplicate samples. Symbol notation: (■) SH light treatment, (●) SC light treatment, (▽) SH dark control and (△) SC dark control. (Reprinted with permission from Miller & Moran, 1997. Copyright 1997. American Society of Limnology and Oceanography, Inc.)

One consequence of the photolysis of the DOM that is consistent with this hypothesis is the release of substrates that were previously bound (e.g. as a complex or through hydrophobic interactions) to DOM in a manner that rendered them resistant to microbial attack. For example, Francko & Heath (1982) observed that exposure of DOM to sunlight in an acid bog lake (pH 5.2–5.8) caused the release of phosphate complexed to high molecular weight DOM. The rate of release of the bound phosphate was further shown to be highly correlated to the reduction of Fe(III) to Fe(II). The mechanism underlying this process is not well understood (Francko, 1990), and there appears to be a seasonality in the extent of phosphate release (Cotner & Heath, 1990). Photochemically induced increases in dissolved phosphorus levels have not been observed in other systems, including a culture of the marine diatom *A. anophagefferens* (Gobler *et al.*, 1997) and a freshwater lake (Jørgensen *et al.*, 1998). Given these differences, it is not evident that the photochemical release of phosphate will be important in marine systems. Nonetheless, the work of Francko & Heath (1982) demonstrates the potential for DOM photolysis to release bound species that are substrates for marine plankton (*vide infra*).

The photolysis of DOM may also be important for nitrogen speciation and availability. Indirect evidence suggests that DOM, and humic compounds in particular, can be an important source of inorganic nitrogen to phytoplankton. Carlsson & Granéli (1993) conducted a microcosm study in the Gullmar Fjord by suspending 150 litre polyethylene enclosures from the sea surface and adding humic acids as the only nitrogen source. They found that, with daily additions of humic acids, the concentrations of chlorophyll *a*, nitrate and ammonia increased, especially when the enclosures were supplemented with phosphate. Concomitantly, they observed an increase in phytoplankton growth rates, as well as bacterial numbers. On the basis of these results, Carlsson & Granéli (1993) concluded that the humic acid additions stimulated this growth, and that the humic acids were the source of nitrogen for the plankton. Presumably the humic-bound nitrogen was the source of the inorganic nitrogen species that were detected, possibly through the photolysis of the humic acids, although this hypothesis was not explicitly tested.

Bushaw *et al.* (1996) isolated fulvic acids from a boreal pond and added them to laboratory water. The resultant solution was irradiated with sunlight (or artificial light). They observed that irradiation of the fulvic acids resulted in an appreciable increase in the concentration of ammonia in solution. Ammonia production was also observed in several other natural water samples, including high DOM coastal waters from the Satilla River Estuary (Georgia, USA) and humic isolates. The mechanism for ammonia photoproduction was not investigated, but it may involve either the release of bound ammonia or the photochemical breakdown of the DOM to form ammonia (e.g. deammification of peptides or amino acids). Kinetic results support the second hypothesis (R. G. Zepp *et al.*, unpublished results). Ammonia photoproduction was confirmed for the Satilla River sample that was diluted 1:15 in Gulf of Mexico seawater (Figure 5.3). There was no

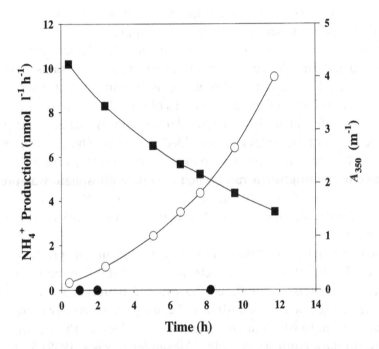

Figure 5.3. Production of ammonia ($\bigcirc$) in 0.2 μm-filtered Satilla River water and decrease in the absorbance coefficient (*A*) at 350 nm ($\blacksquare$) plotted as a function of irradiation time. The dark control for ammonia production is denoted by $\bullet$. All Satilla River water samples were diluted 1:15 in Gulf of Mexico seawater. The pH of the diluted samples was 7.9. No ammonia production was detected when the Gulf of Mexico seawater was irradiated alone (data not shown). Ammonia production rates were normalised to the absorption coefficient at 350 nm. Samples were not corrected for screening, and therefore are an underestimation of the true production rate by approximately 30%. Samples were irradiated at 25 °C in a 5 cm Teflon-screw capped quartz cell with constant stirring. Sample irradiations were with a 300 W solar simulator, with an intensity of 130 mW cm^{-2} at the front surface of the quartz cell, as determined with an IL 1700 Research Radiometer fitted with a calibrated broadband silicon detector (SUD 0380), quartz diffuser and neutral density filter (International Light Incorporated). Absorbance measurements were made in a 5 cm quartz cell with a Hewlett Packard model 8453 diode array spectrophotometer. Spectra were referenced against high purity laboratory water.

evidence for amino acid production (at the nanomolar level) in this sample, as determined by high performance liquid chromatography (Mopper & Furton, 1991), even after 11 h of irradiation (D. J. Kieber & W. L. Miller, unpublished results). In contrast, Jørgensen et al. (1998) observed no change in ammonium concentrations and an increase in amino acids (and carbohydrates) when water from Lake Skärshult was exposed to sunlight. The increase in amino acid and carbohydrate concentrations was presumably due to their photochemical release from organically bound DOM (vide infra). The reasons for these contrasting results are not known, but are suggestive of inherent differences in the DOM composition between these water samples.

Bushaw et al. (1996) also examined the growth of a natural bacterial inoculum in the irradiated fulvic acid water in dark incubations, both with and without added nutrients (phosphate, nitrate and glucose). A substantial increase in the bacterial growth rate was observed in the irradiated water with phosphate added when compared to the non-irradiated controls and all other irradiated samples without added phosphate. Thus, photochemical sources of carbon and nitrogen, originating from DOM, were sufficient to increase bacterial growth rates when adequate phosphate was present. The importance of this process in seawater at low DOM levels has not been demonstrated, but it probably involves limiting nutrients other than phosphate.

Solar irradiation of DOM not only results in the production (or release) of nutrients but also affects the bioavailability of bound organic matter. Photolysis of humic-bound organic material, such as amino acids and aromatic compounds have been shown to greatly increase the rate and extent to which micro-organisms can degrade these compounds relative to dark controls. Amador, Alexander & Zika (1989) found that in the dark only 20% of ^{14}C-labelled glycine in a humic complex could be utilised over a 60 day period. Most of the glycine (circa 80%) was bound to intermediate and high molecular weight humic substances (>5000 relative molecular mass) that could not be utilised by the microbes. When humic acid solutions were exposed to sunlight, the percentage of glycine that was remineralised by the heterotrophic bacteria in the dark increased significantly to 40–60% of the total (Amador et al., 1989). This increase parallels the increase in the percentage of glycine present in the LMW humic acid fraction and is consistent with the finding that only the photochemically formed LMW fraction is remineralised by the bacteria. However, not all organic compounds that are bound to humic acids show increased biological lability upon exposure to sunlight. For example, sunlight had no affect on the microbial remineralisation of humic-bound aniline. This was attributed to the formation of photochemically and biologically resistant N-heterocycles in the humic acids (Amador *et al.*, 1991).

In Amador *et al.*'s (1989) study, there remained a fraction of the bound glycine (*circa* 40%) that could not be utilised by the microbes (despite extended exposure to sunlight), and which was photochemically stable in the high molecular weight humic acid fraction. A similar finding was observed for humic-bound phenol (Amador *et al.*, 1991). Likewise, Lindell *et al.* (1995) observed that a natural assemblage of bacteria in lake water could utilise only a fraction of the DOM when exposed to simulated sunlight. These results suggest that the photolysis of DOM not only enhances the breakdown of the DOM and its biological lability but also results in the production of DOM that is photochemically stable and recalcitrant to microbial attack (Figure 5.1). This photochemically stable DOM is likely to be transparent to solar radiation, and it may well form the basis for the isotopically 'old' carbon ($>4000\,yr$) that has been observed in the oceans (Williams & Druffel, 1987). Further studies are warranted to examine this hypothesis more closely. However, careful controls will be needed to account for possible nutrient limitation or the build-up of inhibitory substances.

One of the prominent features in the photolysis of DOM is its decrease in absorbance (Blough, 1997). This 'bleaching' is a direct consequence of the loss of extended resonance and aromaticity in the chromophores and the oxidative cleavage of bonds within DOM. As a result, the photo-oxidised DOM shows a general increase in its aliphatic character and a decrease in its average molecular weight (e.g. Kulovaara & Backlund, 1993). A number of LMW products is formed in the process, including carbon monoxide, carbon dioxide (for a review, see Chapter 4), hydrocarbons (Wilson, Swinnerton & Lamontagne, 1970), aldehydes, ketones (Mopper *et al.*, 1991) and organic acids (Vaughan, 1989; Kieber, McDaniel & Mopper, 1989; Wetzel, Hatcher & Bianchi, 1995). Micro-organisms readily utilise some of these products. It is, therefore, not surprising that there is a direct correlation between the rate of photoproduction of LMW products and their rate of uptake by bacteria (Kieber *et al.*, 1989) (Figure 5.4). Clearly, these substrates alone will not have an impact on microbial food web dynamics, since their fluxes are too small to affect secondary production (Mopper *et al.*, 1991). It is reasonable to expect, however, on the basis of the work of Strome & Miller (1978), Geller (1986), Miller & Moran (1997) and others, that a variety of chemical species will be formed, and that together they may represent an important carbon (and possibly nitrogen/phosphorus) input into the food web. At present, though, this research has been limited to only a few substrates because of analytical constraints. We simply do not know, nor have the methods to detect, many of the LMW species that may form in seawater. For example, it is currently not possible to detect LMW alcohols (e.g. methanol, ethanol) and some LMW organic acids (e.g. acetate) at nanomolar levels that are expected in oligotrophic, low DOM oceanic waters. Despite these limitations, the photochemical formation of substrates has the potential to be quite important in food web dynamics, as shown by Carlsson & Granéli (1993), Lindell *et al.* (1995), and Bushaw *et al.* (1996). It is also reasonable to expect, from the knowledge that biological niches do not go unexploited, that there are plankton that have adapted to live at the surface of the photic zone (i.e. in the sea surface microlayer) to exploit this photochemical food source (as well as atmospheric nutrient inputs, e.g. iron).

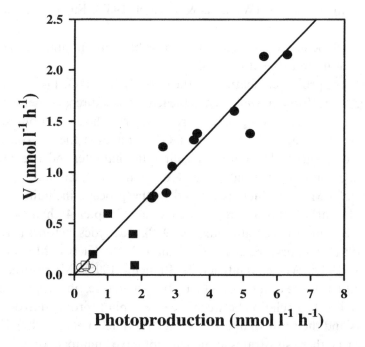

Figure 5.4. Biological uptake rate (*V*) of pyruvate plotted against its midday rate of photochemical production. The slope of the best fit line (±SE) is 0.35±0.03, as determined from linear correlation analysis. ○ Sargasso Sea, □ Gulf Stream, ■ Gulf of Maine and ● Biscayne Bay, FL. (Reprinted with permission from Kieber *et al.*, 1989. Copyright 1989. *Nature*, Macmillan Publishers, Ltd.)

It is also possible that in some instances carbon or other nutrients do not limit the growth of plankton. Under these conditions, increased photochemical inputs of substrates are not expected to increase productivity. This was the observation in a study conducted in the Amazon River system (Amon & Benner, 1996). In this study, bacterial growth was not stimulated by the exposure of Pyrex-enclosed Rio Negro water to sunlight. This finding is consistent with results from nutrient addition experiments where it was shown that the bacteria in this water were not carbon limited. Interestingly, the same result was obtained in sunlight-exposed water from the Rio Solimes, where bacteria were carbon limited (Amon & Benner, 1996). No effect may have been observed due to co-limitation from another nutrient (e.g. phosphate), offsetting stimulatory and inhibitory effects of UV to the bacteria, or the attenuation of more than 50% of the UV-B and 20% of the UV-A by the Pyrex bottles. Attenuation of this high energy, short wavelength radiation will substantially decrease the amount of biological substrates that are produced, since most of the photoproduction that has been observed in seawater occurs within this spectral region (Blough, 1997; Moran & Zepp, 1997). Photoinhibition is also quite possible in these bottle experiments, and has been observed previously for both bacteria (Herndl, Müller-Niklas & Frick, 1993) and viruses (Suttle & Chen, 1992). Results that are obtained in these simple bottle experiments may not be relevant to natural waters. These experiments do not address the critical questions: (a) 'Does photoinhibition occur to an appreciable extent when vertical mixing is taken into account?' and (b) 'Does photochemical oxidation of DOM increase the amount of substrates available to plankton?' It may be that substrates are photochemically formed in the upper few metres, whereas the utilisation of these substrates occurs primarily deeper in the water column.

Implicit assumptions of most photolysis/uptake studies are that humic substances are the major component of DOM, and that humic substances are biologically recalcitrant or are only very slowly degraded by heterotrophic bacteria. Photolysis of the humic substances may increase their lability with respect to microbial attack, thereby representing a source of substrates that are otherwise not available to the bacteria. The basic premise that humic substances are biologically refractory is borne out in several studies in both seawater (for a review, see Mopper & Degens, 1979) and freshwater (for a review, see Jones, 1992). However, there is evidence that this is not completely accurate, at least for some natural waters (Jones, 1992; Moran & Hodson, 1994).

The discrepancy in these results surely reflects the chemical heterogeneity of humic substances studied from a wide diversity of aquatic environments, as well as inherent uncertainties in the methods of collection and concentration of humic substances. It is possible, for example, that the microbial utilisation of a small percentage of humic substances (*circa* 10%) that has been observed by some investigators is solely the result of the uptake of trace quantities of biomolecules that are isolated along with the humic substances, possibly due to: (a) adsorption of the biomolecules onto the humic substances during the isolation procedure, (b) chemical reactions that occur during the isolation of humic substances (e.g. aldol condensation), (c) direct adsorption of the biomolecules onto the resin (XAD or C_{18}), or (d) co-precipitation of a small fraction of the biomolecules with the humic substances at pH 2. Amador and co-workers' results with glycine (Amador *et al.*, 1989) suggest that binding of biomolecules to humic substances is not trivial.

Irrespective of this controversy, the pertinent issue from the perspective of food web dynamics is not one of biological inertness of the humic substances under dark conditions; it is rather a question of how fast (and to what extent) do the humic substances breakdown in the presence and absence of sunlight. In this respect, there is little disagreement that for any of these studies, the rate of microbial utilisation of the humic substances is accelerated when they are exposed to sunlight.

In most natural waters, excluding humic-rich waters, the photochemical production of substrates from DOM is expected to occur primarily at wavelengths less than 400 nm, corresponding to the region of maximum absorption and photobleaching of DOM (Blough, 1997; Moran & Zepp, 1997; and see Chapter 4). The UV-B radiation (280–320 nm) may be particularly important. Action spectra for the production of carbonyl compounds, determined in the Everglades (Florida) and Biscayne Bay seawater (Florida) were confined to wavelengths of < 320 nm (Kieber, Zhou & Mopper, 1990). Similarly, the production of glyoxylate and pyruvate in Antarctic waters was confined entirely to the UV-B (Wu, 1996). The release of humic-bound glycine only occurred at wavelengths < 380 nm (Amador *et al.*, 1989). If substrate production is limited to UV radiation, then the photochemical production of substrates will occur in the upper 20–30 m of optically clear marine waters to as little as a few centimetres in humic-rich waters (Smith & Baker, 1979; Williamson, 1995). This depth range reflects the trend of decreasing penetration of UV radiation with increasing DOM. Offsetting this is the trend of increasing substrate production rates with increasing DOM (Kieber *et al.*, 1990). For example, when organic-rich Everglades (Florida) water was exposed to midday sunlight, net photochemical production rates of formaldehyde, acetaldehyde and glyoxylate were 77, 62 and 127 $nmol\,l^{-1}\,h^{-1}$, respectively. By comparison, production rates in oligotrophic waters in the Sargasso Sea were less than 1 $nmol\,l^{-1}\,h^{-1}$ at the sea surface. One might expect, *a priori*, that the photochemical production of substrates would be quite important in shallow littoral environments, such as streams or some lakes and rivers. However, the impact of this process on food web dynamics is not as evident in pelagic environments where the depth of the photic zone is much deeper than the penetration of UV-B. In the latter case, the only way to determine the importance of photochemically produced substrates in microbial food web dynamics is to conduct a comprehensive field study, and combine this effort with the development of an integrated mixing model that incorporates photochemistry, optics and biological activity. This is a multidisciplinary problem that will require a large collaborative effort to evaluate.

On the basis of published action spectra, it is not unreasonable to expect that increased UV-B fluxes, resulting from depletion of stratospheric ozone, will increase the rate of input of substrates that are available to algae or bacteria. This may have important consequences in environments where carbon (or other nutrients) may be limiting growth such as in the Sargasso Sea (Fuhrman *et al.*, 1989; Hansell, Bates & Gundersen, 1995), the North Atlantic (Kirchman *et al.*, 1994), the equatorial Pacific (Kirchman & Rich, 1997) and the subarctic Pacific (Kirchman, 1990). A change in the trophic structure may also be expected, favouring organisms that can utilise photochemically produced substrates while coping with increasing doses of UV-B radiation.

5.3 Summary

Photolysis of DOM in natural waters results in the formation of LMW products and the release of adsorbed species, such as phosphate, that can be utilised by plankton. At the same time, photolysis of DOM results in the remineralisation of DOM forming carbon monoxide and carbon dioxide, thereby circumventing the microbial loop (see Chapter 4). Thus, the photolysis of DOM is expected to play an important role in its remineralisation. One end result of photolysis may be the formation of DOM that is photochemically stable and biologically recalcitrant. This process may represent an important mechanism for the formation of isotopically old carbon in natural waters. The optimal wavelengths for the photochemical decomposition of DOM in natural waters are in the ultraviolet region of the solar spectrum, as determined by action spectra. Therefore, an increase in the photo-oxidation rate of DOM in surface waters is expected with an increase in UV-B fluxes due to stratospheric ozone depletion. The net effect of this increase on the depth integrated algal and bacterial productivity is not known owing to offsetting stimulatory and inhibitory processes. The unknown effect of changes in UV-B on productivity addresses three central questions that still have not been answered. Specifically, what is the total flux of photochemically generated biological substrates to plankton in natural waters under ambient conditions? Is this total flux significant with respect to oceanic food web dynamics? And under what conditions is it important?

Given the current state of knowledge and uncertainties regarding photochemical remineralisation and substrate/nutrient production in aquatic systems, the best approach to determine the importance of photochemically produced substrates in microbial food web dynamics is to conduct a large, multidisciplinary field study, and combine this effort with the development of a mixing model that includes photochemistry, optics and biological activity.

Acknowledgements

The author gratefully acknowledges the National Science Foundation Office of Polar Programs (OPP-96-10173, D.J.K.) for their support of this work. The author also thanks Drs R. G. Zepp and K. Mopper for discussions pertaining to this manuscript and G. W. Miller for the ammonia results.

References

Amador, J. A., Alexander, M. & Zika, R. G. (1989). Sequential photochemical and microbial degradation of organic matter bound to humic acid. *Applied and Environmental Microbiology*, **55**, 2843–9.

Amador, J. A., Alexander, M. & Zika, R. G. (1991). Degradation of aromatic compounds bound to humic acid by the combined action of sunlight and microorganisms. *Environmental Toxicology and Chemistry*, **10**, 475–82.

Amon, R. M. W. & Benner, R. (1996). Photochemical and microbial consumption of dissolved organic carbon and dissolved oxygen in the Amazon River system. *Geochimica et Cosmochimica Acta*, **60**, 1783–92.

Benner, R, & Biddanda, B. (1998). Photochemical transformations of surface and deep marine dissolved organic matter: effects on bacterial growth. *Limnology and Oceanography*, **43**, 1373–8.

Blough, N. V. (1997). Photochemistry in the sea-surface microlayer. In *The Sea Surface and Global Change*, ed. P. S. Liss & R. A. Duce, pp. 383–424. Cambridge University Press, Cambridge.

Blough, N. V. & Zepp, R. G. (eds.) (1990). *Effects of Solar Ultraviolet Radiation on Biogeochemical Dynamics in Aquatic Environments*. Woods Hole Oceanographic Institution Technical Report, WHOI 90-09.

Boothwell, M. L., Sherbot, D. M. J. & Pollock, C. M. (1994). Ecosystem response to solar ultraviolet-B radiation: influence of trophic level interactions. *Science*, **265**, 97–100.

Bushaw, K. L., Zepp, R. G., Tarr, M. A., Schulz-Lander, D., Bourbonnière, R. A., Hodson, R. E., Miller, W. L., Bronk, D. A. & Moran, M. A. (1996). Photochemical release of biologically available nitrogen from dissolved organic matter. *Nature*, **381**, 404–7.

Calkins, J. & Thordardøttir, T. (1980). The ecological significance of solar UV radiation on aquatic organisms. *Nature*, **283**, 563–6.

Carlsson, P. & Granéli, E. (1993). Availability of humic bound nitrogen for coastal phytoplankton. *Estuarine, Coastal and Shelf Science*, **36**, 433–47.

Cooper, W. J., Zika, R. G., Petasne, R. G. & Fischer, A. M. (1989). Sunlight-induced photochemistry of humic substances in natural waters: major reactive species. In *Aquatic Humic Substances: Influence on Fate and Treatment of Pollutants*, ed. I. H. Suffet & P. MacCarthy, pp. 333–62. American Chemical Society, Washington, DC.

Cotner, Jr, J. B. & Heath, R. T. (1990). Iron redox effects on photosensitive phosphorus release from dissolved humic materials. *Limnology and Oceanography*, **35**, 1175–81.

Crutzen, P. J. (1992). Ultraviolet on the increase. *Nature*, **356**, 104–5.

Cullen, J. C., Neale, P. J. & Lesser, M. P. (1992). Biological weighting function for the inhibition of phytoplankton photosynthesis by ultraviolet radiation. *Science*, **258**, 646–50.

Döhler, G., Hagmeier, E., Grigoleit, E. & Krause, K. D. (1991). Impact of solar UV radiation on uptake of ^{15}N-ammonia and ^{15}N-nitrate by marine diatoms and natural phytoplankton. *Biochemie und Physiologie der Pflanzen*, **187**, 293–303.

Francko, D. A. (1990). Alteration of bioavailability and toxicity by phototransformation of organic acids. In *Organic Acids in Aquatic Ecosystems: Report of the Dahlem Workshop on Organic Acids in Aquatic Ecosystems*, ed. E. M. Perdue & E. T. Gjessing, pp. 167–77. John Wiley & Sons, New York.

Francko, D. A. & Heath, R. T. (1982). UV-sensitive complex phosphorus: association with dissolved humic material and iron in a bog lake. *Limnology and Oceanography*, **27**, 564–9.

Fuhrman, J. A., Sleeter, T. D., Carlson, C. A. & Proctor, L. M. (1989). Dominance of bacterial biomass in the Sargasso Sea and its ecological implications. *Marine Ecology Progress Series*, **57**, 207–17.

Geller, A. (1986). Comparison of mechanisms enhancing biodegradability of refractory lake water constituents. *Limnology and Oceanography*, **31**, 755–64.

Gobler, C. J., Hutchins, D. A., Fisher, N. S., Cosper, E. M. & Sañudo-Wilhelmy, S. A. (1997). Release and bioavailability of C, N, P, Se and Fe following viral release of a marine chrysophyte. *Limnology and Oceanography*, **42**, 1492–504.

Häder, D.-P., Worrest, R. C., Kumar, H. D. & Smith, R. C. (1995). Effects of increased solar ultraviolet radiation on aquatic ecosystems. *Ambio*, **24**, 174–80.

Hansell, D. A., Bates, N. R. & Gundersen, K. (1995). Mineralization of dissolved organic carbon in the Sargasso Sea. *Marine Chemistry*, **51**, 201–12.

Harris, R. P., Albritton, D. L., Allaart, M., Alyea, F. N., Ancellet, G., Andreae, M. O. *et al.*

(1995). Ozone measurements. In *Scientific Assessment of Ozone Depletion*, ed. C. A. Ennis. Global Ozone Research Monitoring Project-Report No. 37, World Meteorological Organization, 1994, Chapter 1.

Helz, G. R., Zepp, R. G. & Crosby, D. G. (eds.) (1994). *Aquatic and Surface Photochemistry*. Lewis Publishers, Ann Arbor, MI.

Herndl, G. J., Müller-Niklas, G. & Frick, J. (1993). Major role of ultraviolet-B in controlling bacterioplankton growth in the surface layer of the ocean. *Nature*, **361**, 717–19.

Hoigné, J., Faust, B. C., Haag, W. R., Scully, Jr F. E. & Zepp, R.G. (1989). Aquatic humic substances as sources and sinks of photochemically produced transient reactants. In *Aquatic Humic Substances: Influence on Fate and Treatment of Pollutants*, ed. I. H. Suffet & P. MacCarthy, pp. 363–81. American Chemical Society, Washington, DC.

Jones, R. I. (1992). The influence of humic substances on lacustrine planktonic food chains. *Hydrobiologia*, **229**, 73–91.

Jørgensen, O. G. N., Tranvik, L., Edling, H., Granéli, W. & Lindel, M. (1998). Effects of sunlight on occurrence and bacterial turnover of specific carbon and nitrogen compounds in lake water. *FEMS Microbiology Ecology*, **25**, 217–27.

Karentz, D. (1994). Ultraviolet tolerance mechanisms in antarctic marine organisms. In *Ultraviolet Radiation in Antarctica: Measurements and Biological Effects*, Antartic Research Series, ed. C. S. Weiler & P. A. Penhale, pp. 93–110. American Geophysical Union, Washington, DC.

Karentz, D., Cleaver, J. E. & Mitchell, D. L. (1991). Cell survival characteristics and molecular responses of antarctic phytoplankton to ultraviolet-B radiation. *Journal of Phycology*, **27**, 326–41.

Kieber, D. J., McDaniel, J. & Mopper, K. (1989). Photochemical source of biological substrates in seawater. Implications for carbon cycling. *Nature*, **341**, 637–9.

Kieber, R. J., Zhou, X. & Mopper, K. (1990). Formation of carbonyl compounds from UV-induced photodegradation of humic substances in natural waters: fate of riverine carbon in the sea. *Limnology and Oceanography*, **35**, 1503–15.

Kirchman, D. L. (1990). Limitation of bacterial growth by dissolved organic matter in the subarctic Pacific. *Marine Ecology Progress Series*, **62**, 47–54.

Kirchman, D. L., Ducklow, H. W., McCarthy, J. J. & Garside, C. (1994). Biomass and nitrogen uptake by heterotrophic bacteria during the spring phytoplankton bloom in the North Atlantic Ocean. *Deep Sea Research*, **41**, 879–95.

Kirchman, D. L. & Rich, J. H. (1997). Regulation of bacterial growth rates by dissolved organic carbon and temperature in the equatorial Pacific Ocean. *Microbial Ecology*, **33**, 11–20.

Kulovaara, M. & Backlund, P. (1993). Effects of simulated sunlight on aquatic humic matter. *Vatten*, **49**, 100–3.

Larson, R. A. & Berenbaum, M. R. (1988). Environmental phototoxicity. *Environmental Science and Technology*, **22**, 354–60.

Lindell, M. J., Granéli, W. & Tranvik, L. J. (1995). Enhanced bacterial growth in response to photochemical transformation of dissolved organic matter. *Limnology and Oceanography*, **40**, 195–9.

Madronich, S., McKenzie, R. L., Caldwell, M. M. & Björn, L. O. (1995). Changes in ultraviolet radiation reaching the earth's surface. *Ambio*, **24**, 143–52.

Miller, W. L. & Moran, M. A. (1997). Interaction of photochemical and microbial processes in the degradation of refractory dissolved organic matter from a coastal marine environment. *Limnology and Oceanography*, **42**, 1317–24.

Mopper, K. & Degens, E. T. (1979). Organic carbon in the ocean: Nature and cycling. In *The Global Carbon Cycle*, SCOPE 13, ed. B. Bolin, E. T. Degens, S. Kempe & P. Ketner, pp. 293–316. John Wiley & Sons, New York.

Mopper, K. & Furton, K. G. (1991). Extraction and analysis of polysaccharides, chiral amino acids, and SFE-extractable lipids from marine POM. In *Marine Particles: Analysis and Characterization*, Geophysical Monograph, **63**, pp. 151–61. American Geophysical Union, Washington, DC.

Mopper, K., Zhou, X., Kieber, R. J., Kieber, D. J., Sikorski, R. J. & Jones, R. D. (1991). Photochemical degradation of dissolved organic carbon and its impact on the oceanic carbon cycle. *Nature*, **353**, 60–2.

Moran, M. A. & Hodson, R. E. (1994). Support of bacterioplankton production by dissolved humic substances from three marine environments. *Marine Ecology Progress Series*, **110**, 241–7.

Moran, M. A. & Zepp, R. G. (1997). Role of photoreactions in the formation of biologically labile compounds from dissolved organic matter. *Limnology and Oceanography*, **42**, 1307–16.

Palenik, B., Price, N. M. & Morel, F. M. M. (1991). Potential effects of UV-B on the chemical environment of marine organisms: a review. *Environmental Pollution*, **70**, 117–30.

Schindler, D. W., Curtis, P. J., Parker, B. R. & Stainton, M. P. (1996). Consequences of climate warming and lake acidification for UV-B penetration in North American boreal lakes. *Nature*, **379**, 705–8.

Smith, R. C. & Baker, K. S. (1979). Penetration of UV-B and biologically effective dose rates in natural waters. *Photochemistry and Photobiology*, **29**, 311–23.

Smith, R. C., Prézelin, B. B., Baker, K. S., Bidigare, R. R., Boucher, N. P., Coley, T., Karentz, D., MacIntyre, S., Matlick, H. A., Menzies, D., Ondrusek, M., Wan, Z. & Waters, K. J. (1992). Ozone depletion: ultraviolet radiation and phytoplankton biology in Antarctic waters. *Science*, **255**, 952–9.

Strome, D. J. & Miller, M. C. (1978). Photolytic changes in dissolved humic substances. *Verhandlungen-Internationale Vereinigung für Theoretische und Angewandte Limnologie*, **20**, 1248–54.

Suttle, C. A. & Chen, F. (1992). Mechanisms and rates of decay of marine viruses in seawater. *Applied and Environmental Microbiology*, **58**, 3721–9.

Vaughan, G.M. (1989). Determination of nanomolar levels of formate in natural waters based on a luminescence enzymatic assay. MS thesis, University of Miami, Rosenstiel School of Marine and Atmospheric Science, Miami, FL.

Vincent, W. F. & Roy, S. (1993). Solar ultraviolet-B radiation and aquatic primary production: damage, protection and recovery. *Environmental Reviews*, **1**, 1–12.

Wetzel, R. G., Hatcher, P. G. & Bianchi, T. S. (1995). Natural photolysis by ultraviolet irradiance of recalcitrant dissolved organic matter to simple substrates for rapid bacterial metabolism. *Limnology and Oceanography*, **40**, 1369–80.

Williams, P. M. & Druffel, E. R. M. (1987). Radiocarbon in dissolved organic matter in the central North Pacific Ocean. *Nature*, **330**, 246–8.

Williamson, C. E. (1995). What role does UV-B radiation play in freshwater ecosystems? *Limnology and Oceanography*, **40**, 386–92.

Wilson, D. F., Swinnerton, J. W. & Lamontagne, R. A. (1970). Production of carbon monoxide and gaseous hydrocarbons in seawater: relation to dissolved organic carbon. *Science*, **168**, 1577–9.

Worrest, R. C. (1983). Impact of solar ultraviolet-B radiation (290–320 nm) upon marine microalgae. *Physiologia Plantarum*, **58**, 428–34.

Worrest, R. C., Thomson, B. E. & Van Dyke, H. (1981). Impact of UV-B radiation upon estuarine microcosms. *Photochemistry and Photobiology*, **33**, 861–7.

Wu, J. (1996). Photochemical production of glyoxylate and pyruvate in antarctic seawater. MS thesis, State University of New York, College of Environmental Science and Forestry, Syracuse, NY.

Zafiriou, O. C. (1977). Marine organic photochemistry previewed. *Marine Chemistry*, **5**, 497–522.

Zepp, R. G., Callaghan, T. V. & Erickson, D. J. (1995). Effects of increased solar ultraviolet radiation on biogeochemical cycles. *Ambio*, **24**, 181–7.

Zika, R. G. (1981). Marine organic photochemistry. In *Marine Organic Chemistry*, ed. E. K. Duursma & R. Dawson, pp. 299–325. Elsevier, Amsterdam.

6

○ ○ ○ ○ ○ ○ ○ ○ ○ ○ ○ ○ ○ ○ ○ ○ ○ ○ ○

Mechanisms of UV damage to aquatic organisms

Warwick F. Vincent* and Patrick J. Neale

6.1 Introduction

UV radiation is the photochemically most reactive waveband of the incident solar radiation field and causes a broad spectrum of genetic and cytotoxic effects in aquatic organisms. In the natural environment, these responses are offset by various protection strategies such as avoidance, screening, photochemical quenching and repair. The net stress imposed by UV exposure thus reflects a balance between damage, repair and the energetic costs of protection, and may be manifested in terms of increased energy demand, changes in cell composition, and decreased growth and survival rates (Figure 6.1). In this chapter, we focus on the damaging effects of UV exposure, while Chapter 7 examines the protection mechanisms that allow organisms to avoid, reduce or recover from such effects.

At the cellular level the toxic effects of UV radiation are initiated by one of two photochemical pathways (Figure 6.1). Firstly, certain biomolecules such as proteins and nucleic acids have chromophores that absorb in the UV region of the spectrum. Under high UV fluxes these molecules are photochemically degraded or transformed, resulting in impairment or even complete loss of biological function. The magnitude of damage caused by these so-called direct or primary mechanisms is determined by the amount of radiation absorbed (optical cross-section in the UV range) and the quantum yield of photodestruction (molecules damaged per photon absorbed). A second class of UV toxicity effects is caused by a series of indirect mechanisms. UV is absorbed by some intermediate compound (photosensitising agent) either inside or outside the cell to produce reactive oxygen species (ROS). The resulting high energy oxidants such as hydrogen peroxide, superoxide or hydroxyl radicals can then diffuse and react with other cellular components with sites of damage that can be well away from the site of photoproduction. The quantum yields of damage by these secondary effects are more variable and difficult to quantify because of the multiple pathways of ROS generation, quenching and reaction.

This chapter first examines the molecular and cellular sites of direct UV damage and the indirect mechanisms operating via ROS production. Then some of the general issues and unknowns in the site-specific effects of UV radiation, with emphasis on the causes of variability in response, are considered. Much of the data concerning UV toxicity effects have been derived from laboratory studies under artificial irradiance regimes that bear little resemblance to the spectral composition and intensity of incident solar radiation. This review concludes by examining the relevance of such studies towards understanding UV effects in the natural environment, and proposes a set of criteria to aid and evaluate the transfer of information from laboratory experiments to the field.

6.2   Direct effects

UV radiation has many potential targets in the cellular environment but there is still considerable debate as to the primary reactions leading to impairment of specific biological activities. DNA and RNA have the

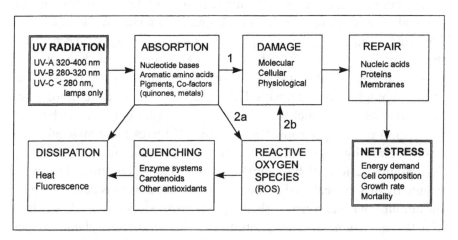

Figure 6.1.   The pathways from UV radiation exposure to cellular stress. Damage can occur directly by photochemical degradation of biomolecules (pathway 1) or indirectly via the production of reactive oxygen species such as hydrogen peroxide and superoxide radicals (pathway 2a), which then cause more widespread oxidative damage within the cell (2b). The net stress is manifested in terms of: the increased energy demands of protection and repair; compositional changes (e.g. lipid content), which may affect the nutritional quality of the cells for higher trophic levels; an impairment of growth rate resulting from the photochemical damage and from the increased energy requirements; and, under severe exposures, an increased rate of mortality.

highest absorbance coefficients for short wavelength UV among all cellular components, and these molecules are therefore especially sensitive to UV exposure (see Figure 3.1, p. 76). Aromatic amino acids such as tryptophan, tyrosine, phenylalanine and histidine also absorb UV, although much less so than do nucleic acids, and the resultant excess excitation energy can result in protein breakdown and the loss of specific enzymatic or other biological functions. For example, protein-based pigments, in particular phycobiliproteins, absorb and are readily damaged by UV exposure (Lao & Glazer, 1996; Rajagopal *et al.*, 1998). Other pigments are also bleached by UV, although such effects may be more the result of indirect rather than of direct mechanisms.

6.2.1 Genetic damage

Nucleic acid bases absorb maximally in the UV-C range, with peak absorbance around 260 nm, and exhibit a tail that extends well into the UV-B (Figure 6.2). This absorbed energy results in the first excited singlet state, with a lifetime of only a few picoseconds. Most of this energy is dissipated by radiationless processes inside the molecule, but a small fraction is available for a variety of chemical reactions (Table 6.1). This can result in the photodestruction of nucleotides (Görner, 1994), with a two- to four-fold greater effect on pyrimidines (thymine and cytosine) relative to purines (adenine and guanine). In addition, three principal photoproducts are formed by the UV-induced reactions (Figure 6.3): (a) 5,6-dipyrimidines, which are cyclobutane-type dimers, generally referred to simply as pyrimidine dimers; (b) photohydrates; and (c) pyrimidine (6–4) pyrimidones, often referred to as (6–4) photoproduct.

Pyrimidine dimers are the dominant products and the action spectrum for their photoproduction from UV irradiation of purified DNA follows that of absorbance (Figure 6.2). Their production has two effects. Firstly, a small fraction of the dimers results in mutation. Secondly, and more importantly in the short term, pyrimidine dimers cause the RNA polymerase to stall during transcription; even a single dimer is sufficient to eliminate completely the expression of a transcriptional unit. There is also evidence that the polymerase remains bound at the dimer; UV exposure thereby has the additional effect of reducing the overall concentration of free RNA polymerase (Britt, 1996, and references therein).

Photohydrates are formed by the photoaddition of H_2O across the 5,6-ethylenic bond of a pyrimidine base. These photoproducts are produced with a quantum yield approaching that of pyrimidine dimers, but they gradually dehydrate at a time scale of hours and appear to have little toxicological effect (Görner, 1994).

The (6–4) photoproduct is produced at less than one-tenth of the rate of pyrimidine dimers; however, this lesion has more damaging effects on the genome and cellular function. The (6–4) dimer is 300 times more effective at blocking transcription by DNA polymerase than are pyrimidine dimers (Mitchell & Nairn, 1989). Furthermore, unlike the pyrimidine dimers, the (6–4) photoproduct cannot be excised and repaired by enzymatic photoreactivation (Brash *et al.*, 1985), causing long term effects on transcription and replication, and ultimately mutation or even death. This photoproduct absorbs maximally in the range 310–320 nm and is converted to its photoisomer, the Dewar pyrimidone. From studies on phage replicated in bacteria, this isomeric form appears to be more lethal but less mutagenic than the (6–4) product from which it is derived (Sage, 1993).

DNA damage has been observed in a wide range of aquatic species exposed to UV-B radiation, including microalgae (Buma *et al.*, 1996a),

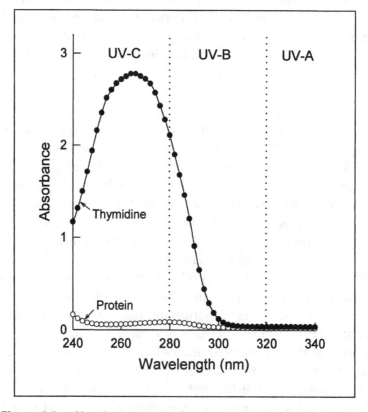

Figure 6.2. Absorbance spectra for a protein standard (serum albumin) and a constituent of DNA (thymidine), both at a concentration of 100 μg ml⁻¹. Although both spectra peak in the UV-C, their absorbance curves extend into the solar UV waveband (280–400 nm).

zooplankton (Malloy *et al.*, 1997) and bacterioplankton (Jeffrey *et al.*, 1996). One of the observed consequences of such effects in diatoms is an increase in cell size, and a concomitant increase in the amount of pigment, protein and carbohydrate per cell (Karentz *et al.*, 1991; Behrenfeld *et al.*, 1992; A. Veen cited in Buma *et al.*, 1995; Buma *et al.*, 1996b). This appears to be associated with an arrested or prolonged cell cycle in which the S (DNA synthesis) phase is extended while the UV-induced dimers are removed by DNA repair mechanisms (Buma *et al.*, 1996a).

The longer-term effects of UV-induced DNA damage on mutation and carcinogenesis has been studied extensively in human skin cells, but are less well understood in aquatic organisms. Studies on hybrid fish of the genus *Xiphophus* have shown that melanomas can be induced by single exposures of fry to UV radiation (Nairn *et al.*, 1996), lending support to the 'one bad day' hypothesis (Vetter, 1997) that even brief exposure to high levels of UV-B at a critical stage in the life cycle of marine organisms could have devastating effects on survival and recruitment. The genetic analysis of melanoma formation in *Xiphophus* indicates that both UV and blue light have carcinogenic effects, but via independent mechanisms (Nairn *et al.*, 1996).

6.2.2 Protein damage

UV radiation at high irradiance can have a destructive effect on total cellular protein in aquatic organisms. For example, UV-induced changes

Table 6.1. The quantum yields for photochemical end-products derived from UV-irradiation of purified double-stranded DNA

Photoproduct or process	Quantum yield at 254 nm (10^{-4} products per photon absorbed)
Chromophore loss	31
Pyrimidine dimers	24
Destruction of nucleotides	21
Photohydrates	10
(6–4) products	1.4
Single-strand breaks	0.4
Base release	0.23
Double-strand breaks	0.014
Interstrand cross-links	0.01
Locally denatured sites	0.01

Modified from Görner, 1994.

in the total protein profile of the cyanobacterium *Anabaena* were examined by polyacrylamide gel electrophoresis. Almost every band showed a gradual decrease in protein content with increasing time of exposure to a high flux of UV-B (500 μW cm^{-2} unweighted). Within 2 h there was nearly complete elimination of proteins in the relative molecular mass 12 000–45 000, including phycocyanin monomers (Sinha *et al.*, 1996). UV-B irradiation of two cultivars of rice (*Oryza sativa*) caused a reduction in total leaf nitrogen, soluble protein and the concentration of ribulose 1,5-bisphosphate carboxylase (Rubisco) (Hidema, Kang & Kumagai, 1996). These results imply that wetland species, including emergent plants, may also be prone to UV-B effects. Such effects may also operate on sensitive animal tissues; for example, several of the proteins within the lens of the eye break down in response to UV irradiation, and this process may contribute to cataract formation (Weinreb & Dovrat, 1996).

The mechanism of protein damage involves the absorption of UV radiation by a chromophore within the protein, the production of a

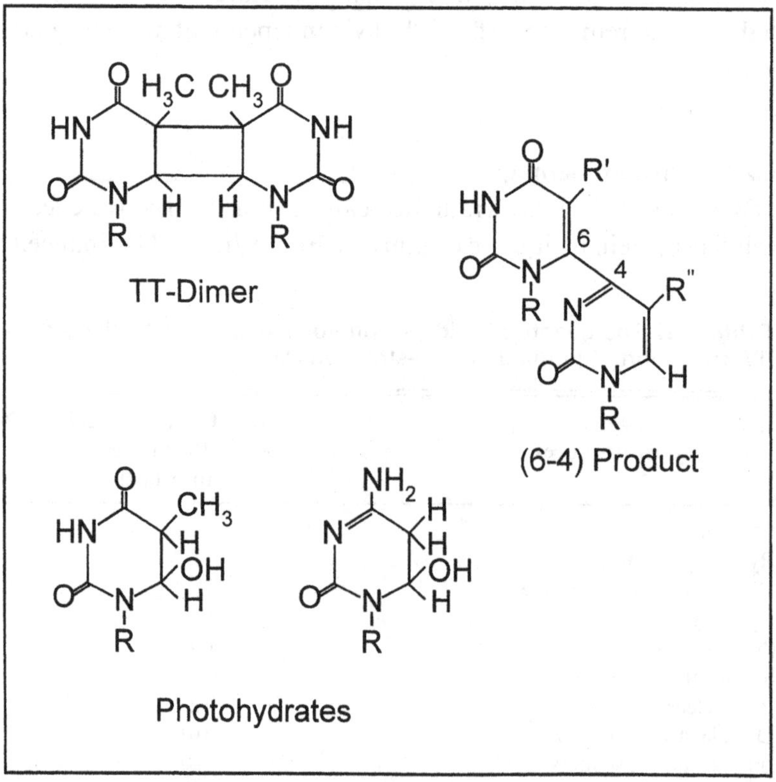

Figure 6.3. Photoproducts from the interaction between UV radiation and nucleic acids (redrawn from Görner, 1994).

photosensitiser, and subsequent reactions that cause photo-oxidative breakdown or cross-linking of amino acids. For example, the toxic effects of UV-B radiation on the photosynthetic enzyme Rubisco is hypothesised to involve the following sequence (Wilson *et al.*, 1995, and references therein):

1. A UV-B photon is absorbed by the aromatic amino acid tryptophan (Trp), which is then transformed to the excited state Trp*.
2. Trp* undergoes intersystem crossing to form a triplet state, and the absorbed energy is passed to oxygen to form $^1O_2$ (singlet oxygen).
3. $^1O_2$ then oxidises Trp, yielding the photoproduct N-formyl-kynurenine (N-fku).
4. N-fku acts as a photosensitiser and in the presence of UV-A radiation reacts with a nearby amino acid residue to form a covalent bond.

Trp occurs at the active site of Rubisco and at interfaces between the subunits of this enzyme, and photo-oxidative damage to these residues is therefore likely to have strong inhibitory effects on enzyme function. Additionally, Trp can be photo-oxidised to the radical cation $Trp^{+\cdot}$. This is associated with intramolecular transfer of electrons from Trp to a nearby amino acid residue resulting in direct damage or cross-linking of residues.

6.2.3 Photosynthesis

Direct inhibition of photosynthesis is one of the most sensitive responses to UV radiation, but the initial target of such effects is a subject of ongoing experiments and debate. The evidence to date indicates a primary response at two levels, both involving damage to a protein complex: (a) on the light reactions of photosynthesis at photosystem 2 (PS II); and (b) on the dark reactions of photosynthesis, specifically on Rubisco, the first enzyme of the reductive pentose phosphate pathway (Calvin cycle).

Several studies on UV-damage of the photosynthetic apparatus have analysed effects occurring in isolated systems. Jones & Kok (1966) measured electron transport by isolated spinach chloroplasts and lyophilised cells of *Synechococcus* 6301 (formerly *Anacystis nidulans*) after exposure to monochromatic UV. This exposure affected electron transport primarily through PS II, leading to an early suggestion that PS II could be an important target for UV effects. Subsequent studies using biophysical assays of primary photochemistry have indicated a variety of possible sites of damage in the PS II reaction centre complex. These include the primary electron acceptor, Q_A (Melis, Nemson & Harrison, 1992), the primary charge separation at P680 (Iwanzik et al., 1983), the primary donor to PS II, Z, which is a tyrosine residue (Renger et al., 1989), and the water-oxidising complex (Vass, 1997). Other studies have been consistent with multiple effects (Bornmann, Björn & Åkerlund, 1984). More recently, Post et al. (1996) examined the effects of UV-A plus UV-B on thylakoid membrane fragments by flash-induced absorbance measurements and concluded that the primary effect was on the donor side of PS II. A recent review of this literature is provided by Vass (1997).

These sophisticated biophysical studies provide an interesting picture of how UV can affect a large molecular complex such as PS II, but the environmental significance of these studies is unknown. The studies have used a variety of broadband UV sources, e.g. mercury vapour lamps (Melis et al., 1992) and xenon arc lamps (Renger et al., 1989), with spectral output more or less different from solar UV-B. The localisation of UV effects at a specific site may depend on the match between the UV treatment spectrum and the action spectrum for effects at that site. Vass's (1997) review of the absorption spectra of PS II functional components showed several that align with the action spectrum for inactivation of PS II electron transport (see Chapter 3), but only over limited spectral ranges. Thus, damage induced under some types of lamp may not occur to any significant degree under solar exposure, and different lamps may favour different damage mechanisms. The latter may help to explain the lack of consensus on a primary site of UV damage in PS II.

Another way that in vivo conditions contrast with the isolated photosynthetic apparatus is that PS II activity in plants is maintained through rapid turnover of certain components of the reaction centre complex. A single protein (D1) within PS II plays a pivotal role in binding all primary donors and acceptors active in electron transport. Indeed, all the previously mentioned sites of UV damage are connected with the D1 protein; in addition, the electron donor Z is a tyrosine residue of D1. Damage to the D1 protein seems to occur in all irradiance regimes and to be a 'cost-of-business' for photosynthesis (for a review, see Long, Humphries & Falkowski, 1994). This damage is efficiently repaired by proteolytic breakdown of damaged D1 and replacement with a newly synthesised protein (Matoo & Edelman, 1987) resulting in turnover of active PS II. This process, referred to as the PS II repair cycle (Guenther & Melis, 1990), ensures that there is little net loss of PS II activity under a wide range of irradiance conditions. Greenberg et al. (1989) reported that D1 turnover can occur under UV-B irradiation alone and that illumination with both UV-B and phosynthetically available radiation (PAR) enhanced turnover rates beyond what was measured under PAR alone. This indicated the possibility of a separate sensitiser for UV-induced PS II repair; the difference spectrum for D1 degradation suggested that the plastosemiquinone anion radical may mediate this response (see Figure 3.2, p. 77). In cyanobacteria, enhanced turnover of D1 is coupled with UV-B induced expression of D1 (*psbA*) genes (Campbell *et al.*, 1998; Máté *et al.*, 1998). It is important to note that enhanced PS II turnover was evident during UV exposure of intact plants, even when there was no net loss of photosynthetic activity (Wilson & Greenberg, 1993), implying an energetic cost to the plants that would not be apparent from measurements of photosynthesis.

Consistent with an active role for the PS II repair cycle, studies of UV effects on the photophysiology of intact plants have found minor or negligible damage to PS II activity compared with effects on other components. Strid, Chow & Anderson (1990) reported a decrease in functional PS II reaction centers during exposure of an UV-B sensitive variety of pea (*Pisum sativum* cv. Greenfeast) to UV-B lamps, but a greater decrease was displayed in the primary carbon fixation enzyme Rubisco. Solar UV exposure of Antarctic diatoms in outside enclosures resulted in no significant decrease in the maximum quantum yield of PS II as measured by fluorescence, but did result in detectable decreases in Rubisco. In these experiments, cultures were gradually adapted to 17% incident light (without UV) and then transferred to tanks with and without natural UV. Over a two week period, overall irradiance was further increased from 17% to 41% of incident. The solar UV exposure resulted in a 22% decrease in Rubisco polypeptides (normalised to chlorophyll *a*) which was similar to the magnitude of decrease in light-saturated rates of photosynthesis. Similarly, in another pea study, exposure to UV-B lamps led to decreases in light-saturated rates of carbon assimilation without any significant decrease in the quantum yield of PS II electron transport (Nogués & Baker, 1995). This latter study also suggested that Rubisco could have a key role in the mechanism(s) of UV radiation inhibition of plant photosynthesis.

Quantifying Rubisco may be a useful approach to distinguish between UV inhibition and PAR-induced photoinhibition because the latter phenomenon involves damage to PS II, with no apparent effects on Rubisco (Neale, 1987; Neale *et al.*, 1993). The enzyme consists of four catalytic dimers of large subunits (LSUs) and eight small subunits (SSUs). Studies on the angiosperms *Brassica*, *Pisum*, *Lycopersicon* and *Nicotiana* have shown that UV-B exposure causes the conversion of the 54 kDa LSU to a novel 64 kDa particle, probably from LSU–SSU cross-linking via the molecular mechanisms described above in Section 6.2.2 (Wilson *et al.*, 1995). However, the cellular content of Rubisco is affected by a number of environmental factors (see e.g. Pichard, Frischer & Paul, 1993), so another possibility is that Rubisco is downregulated as a result of chronic damage to other photosynthetic components. Finally, even though net damage to PS II may not be evident during UV exposure, UV damage to PS II could take its toll through the additional metabolic overhead incurred due to faster turnover of D1 (cf. Raven & Samuelsson, 1986).

6.2.4   Carbon allocation

There is mounting evidence that UV exposure can have a substantial downstream influence on the relative allocation of fixed carbon to specific macromolecular pools in addition to its direct inhibitory effect on photosynthetic rates. For example, in a laboratory assay with a phytoflagellate, a cyanobacterium and a diatom, Arts & Rai (1997) found that UV-B affected the relative allocation of photosynthate to lipid, protein, polysaccharide and low molecular weight compounds. The extent of this effect varied greatly between species. The most sensitive fraction appeared to be protein, suggesting that UV exposure could lead to unbalanced growth and a change in the nutritional value of microalgae to grazers. Changes in the lipid and fatty acid composition of phytoplankton have also been observed in cultures exposed to UV radiation (Döhler & Bierman, 1994; Goes *et al.*, 1994; Wang & Chai, 1994), again with implications for the quality of food available to higher trophic levels (Hessen, de Lange & Van Donk, 1997).

6.2.5   Pigment damage

UV radiation has a strong bleaching effect on pigmented cells, although most experiments reporting this effect have been conducted at high UV fluxes. The evidence of UV-induced pigment damage comes from a great variety of biological systems, including the light-harvesting apparatus of diatoms (Buma *et al.*, 1996b; Lohmann *et al.*, 1998), cyanobacteria (Quesada, Mouget & Vincent, 1995; Quesada & Vincent, 1997; Araoz & Häder, 1997), flagellates (Häder & Häder, 1990; Döhler & Buchmann, 1995), macroalgae (Döhler, Hagmeier & David, 1995) and higher plants, and also the light-sensing organs or organelles of many organisms, including the eyespot of euglenoids (Häder & Worrest, 1991). These effects, however, may be secondary to some of the other mechanisms previously described. For example, Jones & Kok (1966) noted that pigment bleaching in chloroplasts occurred at a much slower rate than did loss of photosynthetic capacity.

There are large differences between species in the qualitative pattern of UV response by specific pigments, and a strong spectral dependence. For example, UV-B irradiation (305 nm radiation and above) of the brown macroalga *Fucus vesiculosus* for 5 h caused a 20% to 40% decrease in almost all of its pigments. The same treatment of the green macroalga *Ulva lactuca* caused a more severe pigment loss, with up to 100% decline in some carotenoids, but there was also an apparent increase in some pigments, notably chlorophyll *a*. Under shorter wavelength UV-B (down to 295 nm) both chlorophylls *a* and *b* declined, but certain other pigments such as violaxanthin showed a marked increase. Alternatively, under UV-A chlorophyll *a* and most of the carotenoids declined and chlorophyll *b* synthesis was apparently stimulated (Döhler *et al.*, 1995). A similar variability in pigment response has been observed in the marine flagellate *Pavlova* (Döhler & Buchmann, 1995).

Pigment bleaching can result from one or more of three mechanisms: (i) *Direct absorbance and damage.* Certain protein-based pigments absorb UV energy directly and undergo photochemical degradation via the mechanisms described in Section 6.2.2. Cyanobacteria are ubiquitous components of aquatic ecosystems and the light-harvesting phycobili-proteins of these organisms appear to be especially sensitive to UV-B (Quesada *et al.*, 1995). As with other pigments, there are large variations in the extent of damage even between closely related species. For example, two mat-forming species of the family Oscillatoriaceae isolated from an ice shelf habitat in Antarctica differed greatly in their dose response to UV-B radiation, although both showed the same qualitative effect of a marked UV-induced decline in C-phycocyanin relative to chlorophyll *a* (Figure 6.4).

Photochemical studies on the effects of UV-B (295 nm) on the filamentous cyanobacterium *Anabaena* have shown that damage occurs by destruction of individual phycobiliproteins, and is compounded *in vivo* by dissociation of the phycobilisome complex (Lao & Glazer, 1996). The quantum yield for the photodestruction of C-phycocyanin was 1.9×10^{-3} molecules damaged per photon of 295 nm radiation absorbed; this quantum yield dropped by two orders of magnitude at 380 nm, and by another two orders of magnitude in the visible. Similar photodestruction yields were obtained for other proteins such as Rubisco and other aromatic substances such as 2,4-dinitrotoluene.

Phycobiliprotein damage is caused by photon absorbance and reactions by aromatic amino acids in the protein moiety, as well as by the associated bilins. Phycobiliproteins account for a large percentage of the total protein content in cyanobacteria and the UV absorbance of individual phycobilins is relatively high; for example, the absorbance by C-phycocyanin at 320 nm is about 10% of its absorbance maximum at 614 nm. This, in combination with the high photodestruction quantum yield, suggests that the phycobilisomes are early sites of damage in cyanobacteria exposed to UV radiation, and that these UV effects perhaps play a greater role than DNA damage in these organisms (Lao & Glazer, 1996).

A more recent study on the cyanobacterium *Spirulina platensis* confirmed that phycobilisomes are a primary target for UV-B-induced damage of the photosynthetic apparatus (Rajagopal *et al.*, 1998). The UV-B irradiation caused photobleaching of the phycobiliproteins, specific loss of the 85.5 kDa anchor protein of phycobilisomes and an alteration of energy transfer processes within these light-harvesting complexes.

(ii) *Photosensitiser action.* Some pigments (P) absorb radiation, are excited to a higher energy level and then transfer this energy to oxygen, which in turn causes pigment and other damage to the cell. The tryptophan mechanism described above in Section 6.2.2 provided an

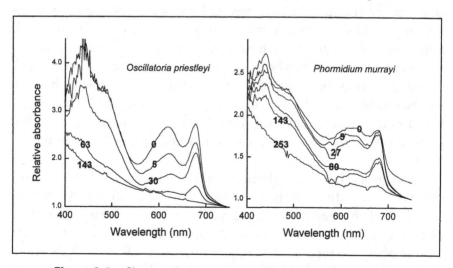

Figure 6.4. Changes in the *in vivo* absorption spectra of two filamentous species of cyanobacteria grown for 5 days under PAR with UV-B radiation (values in $\mu W\,cm^{-2}$ are given on the curves). The phycocyanin-C peak at 615 nm is gradually reduced relative to chlorophyll *a* with increasing UV radiation. (Redrawn from Quesada & Vincent, 1997.)

introduction to this process. The general mechanism is (Larson, 1978):

$$^1P + h\nu \rightarrow *^1P$$
$$*^1P \rightarrow ISC \rightarrow *^3P$$

where ^1P, $*^1P$ and $*^3P$ are the pigment ground state, excited singlet state and excited triplet state, respectively, and ISC is intersystem crossing, a process of radiationless conversion of one photochemically excited spin state to another. In the presence of ground state oxygen (the triplet state $^3O_2$) the energy of the excited pigment triplet can be transferred by physical quenching to singlet oxygen ($^1O_2$), which in turn can have a wide range of destructive effects:

$$*^3P + {}^3O_2 \rightarrow {}^1P + {}^1O_2 + \text{heat}$$

Many natural pigments can act as photosensitisers, including tetrapyrroles such as chlorophylls, phaeophytins, protoporphyrin, zinc porphyrins and some haems. Other cellular compounds that can play this role include flavins, retinal and the intermediate in tryptophan metabolism, N-formylkynurenine. A variety of pollutants can also act as triplet sensitisers for the production of singlet oxygen, including aromatic hydrocarbons, ketones and naphthols.

(iii) *ROS production in addition to singlet oxygen.* These are generated at various sites within the cell during exposure to UV and cause oxidative degradation of pigments, as well as other biomolecules (see Section 6.3 below). Many pigment complexes are within or near lipid membranes, and the production of lipid peroxides may be especially damaging. One class of pigments, the carotenoids, reacts with ROS and can then be regenerated by natural reduction processes. Carotenoids thus appear to play an important role as natural quenching agents of ROS and are an important cellular defense against UV damage.

In addition to the three short term effects described above, UV can have a controlling influence on pigment accumulation during cellular development. In part, this may reflect photoprotective and acclimation responses to lessen and mitigate the effects of continuing UV exposure. For example, carotenoids are effective ROS scavengers, and concentrations of specific carotenoid pigments in algal and cyanobacterial cells commonly increase during UV exposure (Quesada & Vincent, 1997, and references therein). However, there can be a variety of negative responses to UV during development including photo-oxidation of newly synthesised pigments, UV-inhibition of gene expression and pigment biosynthesis, and damage to photosystem proteins disrupting the subsequent assembly of light-harvesting pigment complexes (Jordan et al., 1991; Strid, Chow & Anderson, 1994). Experiments with *Spirodela oligorrhiza* have shown that UV-B exposure can result in fewer and/or smaller photosystems during chloroplast development (Marwood & Greenberg, 1996).

6.2.6   Respiration

Little attention has been given to the physiological effects of UV damage on respiration, but the results so far available indicate considerable variation between species, even for the same UV exposure. For example, Beardall et al. (1997) found that UV-A + UV-B inhibited respiration in *Aphanizomenon* but stimulated oxygen uptake in *Selenastrum*. The latter effect could be the result of increased biosynthesis associated with the repair of UV damage, while UV-inhibition of respiration could result from a great variety of molecular mechanisms, including lipid membrane damage or degradation of key proteins such as the cytochrome oxidases. UV-induced impairment of respiratory activity has also been observed in experiments with animals, including fish (Winckler & Fidhiany, 1996).

6.2.7   Nutrient uptake

UV exposure has a wide ranging impact on the nutrient physiology of cells via transport (e.g. caused by damage to membrane lipids or ATPases) and assimilation (e.g. through degradation of key enzymes). Experiments on a unicellular marine prasinophyte have shown that such effects on nitrogen transport and assimilation can result in symptoms that mimic nitrogen deficiency (Goes et al., 1995). For these experiments, nitrogen-replete and nitrogen-deficient cultures were grown in the presence and absence of UV-B radiation. The UV treatment resulted in: (a) reduced rates of amino acid production; (b) a much slower rate of peptide biosynthesis; (c) a larger pool size of total amino acids, the net effect of (a) plus (b); (d) decreased transamination after the initial nitrogen-assimilation step resulting in increased cellular concentrations of glutamate; and (e) less production of downstream amino acids such as alanine. These five effects were also characteristic of nitrogen deficiency and were most strongly manifested in cultures exposed to the combined extreme of nitrogen limitation plus UV-B.

[15]N-ammonium uptake experiments with the marine diatom *Lithodesmium variabile* have similarly shown that UV-B affects the pool size of amino acids (Döhler & Kugel-Anders, 1994). UV exposure caused an increased pool size of glutamine associated with enhanced activity of glutamine synthetase and inhibition of glutamate synthetase, and a

decrease in the alanine pool size associated with UV-B inhibition of alanine aminotransferase. The inhibitory effects of UV on nitrogen assimilation have also been demonstrated in the marine macroalgae *Fucus*, *Laminaria*, *Ulva* and *Rhodomela* (Döhler *et al.*, 1995). All species showed a decrease in ammonium uptake and changes in the intracellular pool size of amino acids. There were large differences between species in the magnitude of effect and the qualitative pattern of change in amino acids.

Phosphorus uptake by phytoplankton (Hessen, Van Donk & Andersen, 1995) and aquatic microbial growth responses to phosphorus enrichment (Bergeron & Vincent, 1997) can also be impaired by UV exposure. These effects on phosphorus uptake in combination with those on nitrogen uptake (see above) and carbon allocation (Section 6.2.4) have the potential to alter substantially the C:N:P stoichiometry of cells, and thereby to change the nutritional quality of food available for higher trophic levels (Hessen *et al.*, 1997).

6.2.8 Nitrogen fixation

The enzyme responsible for nitrogen fixation, nitrogenase, is also susceptible to UV damage. When cultures of the cyanobacterium *Anabaena* were exposed to strong UV-B fluxes there was a major decrease in nitrogenase activity over a period of 2 h (Sinha *et al.*, 1996). The analysis of isolated heterocysts (the cells responsible for nitrogen fixation) by sodium dodecyl sulfate–polyacrylamide gel electrophoresis (SDS–PAGE) showed that there were three prominent peptides in the range 26–55 kDa. Two of these decreased during the 2 h UV-B exposure while the third was completely eliminated after 1 h. These authors suggest that, apart from the direct photodegradation of peptides, nitrogenase inactivation could operate via UV-disruption of the thick protein-rich cell wall of heterocysts. This would then allow oxygen to diffuse in and cause inactivation of this O_2-sensitive enzyme.

6.2.9 Cell motility

UV exposure is known to inhibit cellular motility in a variety of organisms and such effects are likely to impair the growth and survival of planktonic, as well as benthic, micro-organisms in the aquatic environment (Häder, 1997, and references therein). For example, the gliding motility of cyanobacterial trichomes completely ceased when the cultures were exposed to ambient UV radiation in the tropics (Donkor, Damian & Häder, 1993). Similarly, the motility in four species of flagellate in culture was inhibited by UV-B, but to differing extents between species (Ekelund,

1994). The exact molecular mechanisms of these effects are unknown, and could operate through photochemical damage of specific proteins (e.g. microtubulin, dynein) or via downstream physiological effects on the energy status of the cells. For some species, the effects operate in part through a disruption of the photoreceptors. For example, experiments with *Euglena gracilis* have shown that UV-B radiation selectively destroys proteins within the paraflagellar body, the photoreceptor organelle (Häder & Brodhun, 1991). In chlamydomonads, complete loss of flagella has been observed after exposure to UV radiation in laboratory, as well as field, experiments (Hessen *et al.*, 1995; Van Donk & Hessen, 1996).

6.2.10 Other effects

Prolonged exposure to UV radiation is known to result in an impaired immune response in mammalian systems, including humans (Garssen *et al.*, 1998). This effect appears to be caused by a variety of biochemical and histological mechanisms operating in parallel including induction of interleukins by keratinocytes, reduced numbers of Langerhans cells in the skin and a change in their response to antigens, and possibly DNA damage. Urocanic acid, a substance found in the epidermal layer of skin, has also been implicated in immunosuppression. The *trans* form of this molecule can be converted by UV radiation into the *cis*-isomer, which suppresses hypersensitivity reactions that play a role in the immune response (van der Leun, 1996). It is not known, however, to what extent these responses might operate in the aquatic environment, for example, in the skin of marine mammals.

There is a large literature on the role of UV in carcinogenesis (e.g. Gilchrest, 1995), but little attention has been directed to marine and freshwater organisms in this regard. Tumours can be induced in certain fish species, for example by single exposures of fry to UV radiation (Nairn *et al.*, 1996). The species *Xiphophorus* has been of particular interest as a model system for the study of sunlight carcinogenesis because the resultant melanomas closely resemble human melanomas (Friend, 1993).

The erythemal ('sunburn') effect of UV radiation has also been the subject of considerable study in mammals and particularly in human skin (see Gilchrest, 1995), but little is known about similar effects on aquatic animals that live in the near-surface waters of lakes or the sea. At least some fish may have natural sunscreens in their epidermal layer (Fabacher & Little, 1995), but skin lesions characteristic of sunburn have been observed in fish exposed to high solar irradiance (e.g. Ramos *et al.*, 1994).

6.3 Indirect effects

The production and subsequent activity of ROS is a central element of UV toxicology. Although these highly reactive oxidants are produced by cellular processes under PAR and even in the dark by a variety of physiological mechanisms, their rate of production both inside the cell and in the surrounding aquatic medium is greatly accelerated in the presence of UV radiation. Once an ROS is formed, various fast reactions can then follow, and the resultant photoproducts may be more reactive and cytotoxic than either their precursors or the direct effects of UV radiation. The nature of these fast reactions is determined by chemical properties of the cell in the vicinity of ROS production, in particular pH, metal concentrations and the presence of lipids.

ROS generation within the cell may be more important than external production, and given the short half-life of these products, concentrations in the surrounding medium may be a misleading guide to cellular levels. Here we consider some of the production mechanisms and cellular effects of four types of ROS. More detailed information about ROS photochemistry is given in Chapter 4.

6.3.1 Hydrogen peroxide and hydroxyl radicals

Hydrogen peroxide (H_2O_2) is produced photochemically by a variety of mechanisms both internally within the cells of aquatic organisms and also externally in their surrounding environment. The main photochemical pathway is by dismutation from superoxide:

$$O_2 + hv \rightarrow O_2^{-\cdot} \rightarrow H_2O_2$$

H_2O_2 has long been used as an algicide and bactericide, although usually at concentrations in the millimolar range (e.g. 0.25 to 1.5 mmol l^{-1} against filamentous cyanobacteria; Barroin & Feuillade, 1986), substantially above the maximum concentrations that are produced in marine and fresh waters (e.g. up to 900 nmol l^{-1} in lakewater, Scully & Vincent, 1997). However, much higher concentrations can occur in rainfall (up to 10 mmol l^{-1}; Shtamm, Purmal' & Skurlatov, 1991), and stratospheric ozone depletion could have biologically significant effects on the input of H_2O_2 to shallow or stratified surface waters (Scully et al., 1997). High concentrations may also be generated locally by photochemical reactions within the cell. H_2O_2 is less reactive than many other ROS, and therefore has a relatively long half-life that allows it to diffuse well away from the site of its photoproduction.

H_2O_2 is a powerful oxidant capable of damaging nucleic acids, proteins and pigments. Although this damage may be incurred by direct reaction between biomolecules and the H_2O_2, many of these effects are mediated by the Fenton reaction, whereby more reactive oxidants are generated, in particular hydroxyl radicals:

$$H_2O_2 + Fe^{2+} \rightarrow OH^{\cdot} + OH^{-} + Fe^{3+}$$

The extreme reactivity of the hydroxyl radical is indicated by its short half-life, typically nanoseconds by comparison with minutes to hours for H_2O_2. Hydroxyl radicals attack nucleic acids, proteins and membranes and consequently few parts of the cell are immune from their effects; even catalases and peroxidases are attacked and degraded by this highly reactive ROS.

Certain types of genetic damage have now been attributed to the Fenton reaction, which proceeds at faster rates in the presence of UV (photoFenton reaction). Most importantly, the photoFenton reaction causes a breakdown in DNA supercoiling or tertiary structure via the production of 'nicks' or breaks in the sugar polyphosphate backbone of the double helix. This tertiary structure influences the affinity constant, as well as the number of ligands that can bind to DNA, thereby affecting gene expression (Giacomoni, 1995). Such UV effects were initially puzzling because their action spectrum extends well into the UV-A region, where direct photon absorption by DNA as well as the quantum yield of dimer formation are minimal (Figure 6.2). More recent experiments (Shih & Hu, 1996) have shown that moderate exposure to UV-A (30 min of 365 nm radiation at $200\,\mu W\,cm^{-2}$) in the presence or absence of H_2O_2 has no damaging effect on DNA, and the combination of Fe^{2+} and H_2O_2 (in the absence of UV) has only a small effect. However the frequency of strand breaks and unwinding increases four- to five-fold when all components of the photoFenton reaction are present, i.e. $UV\text{-}A + H_2O_2 + Fe^{2+}$.

6.3.2   Superoxide radicals

Superoxide radicals ($O_2^{-\cdot}$) are generated by the interaction between UV and oxygen and can subsequently react with and damage biomolecules and cells via at least four mechanisms (Fridovich, 1986):

1. *Direct effects.* $O_2^{-\cdot}$ reacts directly with many small molecules including bile pigments, ascorbate, diols, α-tocopherol and sugars such as triose.
2. *Protonation.* In intracellular microenvironments that are rich in

protons, such as at the surface of biological membranes, O_2^- can undergo the following reaction to form a hydroperoxyl radical:

$$O_2^- + H^+ \rightleftharpoons HO_2^-; \; pK_a = 4.8$$

HO_2 is a much stronger oxidant than O_2^- and will directly attack polyunsaturated fatty acids in membranes.

3. *Cation association.* Superoxide radicals can associate with other cations, such as vanadate (VO_4^{3-}) and cause, for example, the oxidation of NADPH, resulting in loss of reducing power in biological membranes.
4. *Generation of hydroxyl radical.* The reaction between superoxide, H_2O_2 plus Fe(III) or Cu(II) can produce OH, which as noted in the previous section can have devastating effects on the cell.

6.3.3 Membrane lipid peroxidation

Lipid peroxides are a general indicator of oxidative tissue damage and can be produced by UV exposure via the production of radicals (R) such as hydroxyl ions. The general initiation reaction is the removal of a hydrogen atom from a methylene carbon of an unsaturated fatty acid (LH):

$$LH + R \rightarrow L + RH$$

The resultant carbon-centred radical then reacts rapidly with oxygen to produce a peroxy radical, which in turn attacks an unsaturated fatty acid, in the process regenerating a lipid radical and a lipid hydroperoxide (LOOH). This set of steps is referred to as chain branching or propagation:

$$L + O_2 \rightarrow LO_2$$
$$LH + LO_2 \rightarrow LOOH + L$$

These reactions continue until two free radicals react with each other and terminate the chain:

$$L + L \rightarrow LL$$
$$L + LO_2 \rightarrow LOOL$$
$$LO_2 + LO_2 \rightarrow LOOL + O_2$$

These photochemical effects on lipids may be particularly damaging for membrane structure and function (Murphy, 1983).

6.3.4 Singlet oxygen and other photochemical reactions

A variety of other toxic photoproducts can also be generated by the interaction between UV radiation and cellular components. These include singlet oxygen (see above) and hydrated electrons (for details, see Chapter 2). Photochemical reactions beyond those discussed above and which might also be important in the aqueous cellular environment include the production of hydroxyl radical from nitrate (a form of nitrogen stored in high concentration in the vacuole of some phytoplankton such as diatoms), hydroxyl and hydroperoxyl radicals from transition metals, and hydrogen peroxide from flavins (Shtamm *et al.*, 1991).

6.4   General issues

6.4.1   Variability in response

The high UV reactivity of several major classes of biomolecules – nucleic acids, proteins and lipids – means that all cell types and organisms are susceptible to the toxic effects of UV radiation. The magnitude of damage, however, is highly variable between life stages, tissues and species. This variability in response is the result of differences to be expected at many levels, particularly the following:

(i) *Efficiency of protection and repair strategies.* An example noted above is the increased rate of repair of D1 protein in *Spirodela*, resulting in no apparent inhibitory effect of UV radiation on photosynthesis (Wilson & Greenberg, 1993). The increased carotenoid concentration of cells may increase their ability to remove ROS as they form, while UV-absorbing pigments such as mycosporine-like amino acids may screen the cells from damaging UV irradiance (see Chapter 7).

(ii) *Pre-acclimation state of the cell.* Many of the protection-repair mechanisms operating in cells are inducible, and cells gradually exposed to bright light and/or UV are much less prone or even tolerant of UV effects (e.g. Roos & Vincent, 1998).

(iii) *Intensity of exposure.* Some target-specific effects obey reciprocity, meaning that the amount of damage increases as a linear function of UV flux. However, changes in the UV dose rate can also cause qualitative shifts in the nature of UV effects, which result in deviations from this linear pattern, or the sudden onset of damage once a threshold tolerance is exceeded. For example, moderate UV-B dosages cause an increase in cell size and cellular concentration of proteins in marine diatoms, whereas higher dosages may cause a general breakdown of cells. High UV fluxes have been found to cause PS II damage in *Pisum sativum*, but these effects are absent from short term or moderate UV exposures (Nogués & Baker, 1995). Increasing UV dose rates will eventually swamp the ROS-quenching and other protection–repair responses of even the most resistant cells and cause general photo-oxidative damage and death.

(iv) *Spectral irradiance.* The severity of UV effects depends on the spectral composition of the incident radiation relative to the action spectrum for damage. This spectral dependence can be expressed as a biological weighting function (BWF), essentially a series of multiplier factors for the dosage rate at each wavelength (for details, see Chapter 3). BWFs vary greatly between cell types depending on the primary target; for example the UV-inhibition of bacterial production seems to follow the BWF for DNA (Jeffrey *et al.*, 1996), whereas UV inhibition of phytoplankton production includes a strong UV-A component consistent with PS II (and/or Rubisco) protein damage (Cullen & Neale, 1994). The incident spectrum also influences the efficiency of protection and repair mechanisms that are activated by long wavelength UV and/or short wavelength PAR (Quesada *et al.*, 1995, and references therein).

(v) *Duration of exposure.* At long time scales the effects of UV may deviate completely from any recognisable dose–response function. Organisms may acclimate to prolonged exposure by improving their cellular protection against UV effects (Lesser, Neale & Cullen, 1996). In cyanobacteria, the cellular concentration of UV-screening pigments (Garcia-Pichel, Wingard & Castenholz, 1993) and ROS-quenching carotenoids (Quesada & Vincent, 1997) markedly increases with increasing UV exposure, and UV-acclimated cells can have growth rates approaching those achieved in the absence of UV (Roos & Vincent, 1998).

(vi) *Delayed effects and recovery.* At time intervals of several generations, UV exposure may have cumulative genetic effects that ultimately impair biological performance. There has been little attention given to such effects in the aquatic environment, but delayed and/or long term cumulative responses to UV exposure are well known from the skin cancer literature, and increasingly from studies with higher plants. For example, in a multiyear study of loblolly pine (*Pinus taeda*), UV-B exposures simulating 25% ozone depletion caused twice the decrease in biomass after three years than after one year (Sullivan & Teramura, 1992). The germination success of seeds from the annual plant *Dimorphotheca sinuata* was markedly reduced for cultivars grown under UV-B, and the second generation of UV-exposed plants showed a broader range and increased magnitude of toxicity responses; these effects indicated a cumulative UV-B effect on DNA integrity (Musil, 1996).

(vii) *Interactions with other variables.* The physiological stress imposed by other aspects of the environment, for example nutrient limitation, inadequate energy supply, scarcity of refuges or the presence of toxic materials, may lessen the ability of cells to escape the damaging effects of UV radiation via screening, quenching, avoidance or repair (Vincent & Roy, 1993). Low temperatures could reduce the effectiveness of UV defences that depend on biosynthetic activity; for example, the UV-inhibition of growth by Antarctic cyanobacteria has been found to increase with decreasing temperature, although these effects may be offset by cellular acclimation to the UV field (Roos & Vincent, 1998).

6.4.2   Evaluating the literature

Concern about the effects of stratospheric ozone depletion has stimulated a rapid increase in research effort towards understanding the impact of solar UV radiation on natural systems. Much of the literature produced to date, however, is derived from experimental manipulations that do not accurately reproduce the UV exposures and other conditions found in the environment. Before any direct extrapolation is made from laboratory and field experiments to aquatic ecosystems the appropriateness of the assay must be assessed with respect to UV dosage rates (fluxes), doses (flux × duration of exposure), spectral regime, pre-acclimation state of the organisms and the opportunities for recovery or protection that may exist in the natural environment.

By these assessment criteria, most reports in the literature of UV damage appear exaggerated relative to natural systems. Most of the UV studies reviewed in this chapter have been conducted in the laboratory under artificial lamps with spectral outputs and biologically effective dosages that differ markedly from those of natural sunlight. Many of the field studies have been performed under experimental regimes where the organisms are artificially held under bright solar irradiance (for example, on the deck of ships or at the surface of the sea or lakes) for much longer periods than they would normally experience in Nature. Some UV assay designs allow little opportunity for the natural avoidance, protection and recovery mechanisms to operate; for example by not providing the UV-A or blue wavelengths required to activate repair mechanisms.

On the more positive side, the studies to date have provided invaluable mechanistic information about how UV radiation can interact with and impair biological systems, and the potential effects to look for in the natural environment. Of special interest to ecologists, this work has revealed UV damage signatures that can now be used to identify and assess UV effects in aquatic ecosystems. The prime example of the latter is the use of DNA dimers as indices of UV-B damage, for example in phytoplankton populations using flow cytometry (Buma *et al.*, 1995), in Antarctic zooplankton communities during periods of increased UV-B (Malloy *et al.*, 1997) and in marine bacterial communities to determine how UV effects are modified by the hydrodynamic regime (Jeffrey *et al.*, 1996). Other UV-B effects, such as the photodestruction of Rubisco (as seen in Antarctic phytoplankton communities exposed to natural UV; Lesser *et al.*, 1996) and the differential bleaching of phycobilin pigments relative to chlorophyll *a* in cyanobacteria (Lao & Glazer, 1995; Quesada & Vincent, 1997) have the potential to be applied in similar ways. An important goal for future studies is to understand better how such indices can be translated into ecologically meaningful effects, such as in terms of productivity and growth.

Acknowledgements

Our research on UV effects on aquatic biota is funded by the Natural Sciences and Engineering Research Council (W.F.V.), les Fonds pour la formation de chercheurs et l'aide à la recherche (W.F.V.) and the National Science Foundation (P.J.N.). We thank Drs Michael T. Arts, Hakumat Rai, Hendika de Lange and Fumie Kasai for their helpful review comments.

References

Araoz, R. & Häder, D.-P. (1997). Ultraviolet radiation induces both degradation and synthesis of phycobilisomes in *Nostoc* sp.: a spectroscopic and biochemical approach. *FEMS Microbiological Ecology*, **23**, 301–13.

Arts, M. T. & Rai, H. (1997). Effects of ultraviolet-B radiation on the production of lipid, polysaccharide and protein in three freshwater algal species. *Freshwater Biology*, **38**, 597–610.

Barroin, G. & Feuillade, M. (1986). Hydrogen peroxide as a potential algicide for *Oscillatoria rubescens* D.C. *Water Research*, **20**, 619–23.

Beardall, J., Berman, T., Markager, S., Martinez, R. & Montecino, V. (1997). The effects of ultraviolet radiation on respiration and photosynthesis in two species of microalgae. *Canadian Journal of Fisheries and Aquatic Science*, **54**, 687–96.

Behrenfeld, M. J., Hardy, J. T. & Lee II., H. (1992). Chronic effects of ultraviolet-B radiation on growth and cell volume of *Phaeodactylum tricornutum* (Bacillariophyceae). *Journal of Phycology*, **28**, 757–60.

Bergeron, M. & Vincent, W. F. (1997). Microbial food web responses to phosphorus and solar UV radiation in a subarctic lake. *Aquatic Microbial Ecology*, **12**, 239–49.

Bornman, J. F., Björn, L. O. & Åkerlund, H.-E. (1984). Action spectrum for inhibition by ultraviolet radiation of photosystem II activity in spinach thylakoids. *Photobiochemistry and Photobiophyics*, **8**, 305–13.

Brash, D. E., Franklin, W. A., Sancar, G. B., Sancar, A. & Haseltine, W. A. (1985). *Escherichia coli* DNA photolyase reverses cyclobutane dimers but not pyrimidine–pyrimidone (6–4) photoproducts. *Journal of Biological Chemistry*, **260**, 11438–41.

Britt, A. B. (1996). DNA-damage and repair in plants. *Annual Reviews of Plant Physiology and Plant Molecular Biology*, **47**, 75–100.

Buma, A. G. J., Van Hannen, E. J., Roza, L., Veldhuis, M. J. W. & Gieskes, W. W. C. (1995). Monitoring ultraviolet-B induced DNA damage in individual diatom cells by immunofluorescent thymine dimer detection. *Journal of Phycology*, **31**, 314–21.

Buma, A. G. J., Van Hannen, E. J., Veldhuis, M. J. W. & Gieskes, W. W. C. (1996a). UV-B induces DNA-damage and DNA-synthesis delay in the marine diatom *Cyclotella* sp. *Scientia Marina*, **60**, 101–6.

Buma, A. G. J., Zemmelink, H. J., Sjollema, K. & Gieskes, W. W. C. (1996b). UVB radiation modifies protein, photosynthetic pigment content, volume and ultrastructure of marine diatoms. *Marine Ecological Progress Series*, **142**, 47–54.

Campbell, D., Erikson, M.-J., Oquist, G., Gustafsson, P. & Clarke, A. K. (1998). The cyanobacterium *Synechococcus* resists UV-B by exchanging photosystem II reaction-center D1 proteins. *Proceedings of the National Academy of Science, USA*, **95**, 364–9.

Cullen, J. J. & Neale, P. J. (1994). Ultraviolet radiation, ozone depletion, and marine photosynthesis. *Photosynthesis Research*, **39**, 303–20.

Döhler, G. & Bierman, T. (1994). Impact of UV-B radiation on the lipid and fatty acid composition of synchronised *Ditylum brightwellii* (West) Grunow. *Naturforschung*, **49**, 607–14.

Döhler, G. & Buchmann, T. (1995). Effects of UV-A and UV-B irradiance on pigments and ^{15}N-assimilation of the haptophycean *Pavlova*. *Journal of Plant Physiology*, **146**, 29–34.

Döhler, G., Hagmeier, E. & David, C. (1995). Effects of solar and artificial UV radiation on pigments and assimilation of ^{15}N ammonium and ^{15}N nitrate by macroalgae. *Journal of Photochemistry and Photobiology B. Biology*, **30**, 179–87.

Döhler, G. & Kugel-Anders, A. (1994). Assimilation of ^{15}N-ammonium by *Lithodesmium variabile* Takano during irradiation with UV-B (300–320 nm) and complemental monochromatic light. *Photochemistry and Photobiology B. Biology*, **24**, 55–60.

Donkor, V. A., Damian, H. A. K. & Häder, D. P. (1993). Effects of tropical solar radiation on the motility of filamentous cyanobacteria. *FEMS Microbiological Ecology*, **12**, 143–8.

Ekelund, N. G. A. (1994). Influence of UV-B radiation on photosynthetic light response curves, absorption spectra and motility of four phytoplankton species. *Physiologia Plantarum*, **91**, 696–702.

Fabacher, D. L. & Little, E. E. (1995). Skin component may protect fish from ultraviolet-B radiation. *Environmental Science and Pollution Research*, **2**, 30–2.

Fridovich, I. (1986). Biological effects of the superoxide radical. *Archives of Biochemistry and Biophysics*, **247**, 1–11.

Friend, S. H. (1993). Genetic models for studying cancer susceptibility. *Science*, **259**, 774–5.

Garcia-Pichel, F., Wingard, C. E. & Castenholz, R. W. (1993). Evidence regarding the UV sunscreen role of a mycosporine-like compound in the cyanobacterium *Gloeocapsa* sp. *Applied Environmental Microbiology*, **59**, 170–6.

Garssen, J., Norval, M., El-Ghorr, A., Gibbs, N. K., Jones, C. D., Cerimele, D., De Simone, C., Caffieri, S., Dall'Acqua, F., De Gruijl, F. R., Sontag, Y. & Van Loveren, H. (1998). Estimation of the effect of increasing UVB exposure on the human immune system and related resistance to infectious diseases and tumours. *Journal of Photochemistry and Photobiology B. Biology*, **42**, 167–79.

Giacomoni, P. U. (1995). Open questions in photobiology. II. Induction of nicks by UV-A. *Journal of Photochemistry and Photobiology*, **29**, 83–5.

Gilchrest, B. A. (ed.) (1995). *Photodamage*. Blackwell Science, Oxford.

Goes, J. I., Handa, N., Taguchi, S. & Hama, T. (1995). Changes in the patterns of biosynthesis and composition of amino acids in a marine phytoplankter exposed to ultraviolet-B radiation: nitrogen limitation implicated. *Photochemistry and Photobiology*, **62**, 703–10.

Goes, J. I., Nobuhiko, H., Taguchi, S. & Hama, T. (1994). Effect of UV-B radiation on the fatty acid composition of the marine phytoplankter *Tetraselmis* sp: relation to cellular pigments. *Marine Ecological Progress Series*, **114**, 259–74.

Görner, H. (1994). Photochemistry of DNA and related biomolecules: quantum yields and consequences of photoionization. *Journal of Photochemistry and Photobiology*, **26**, 117–39.

Greenberg, B. M., Gaba, V., Canaani, O., Malkin, S., Matoo, A. K. & Edelman, M. (1989). Separate photosensitizers mediate degradation of the 32-kDa photosystem II reaction center protein in the visible and UV spectral regions. *Proceedings of the National Academy of Sciences, USA*, **86**, 6617–20.

Guenther, J. E. & Melis, A. (1990). The physiological significance of photosystem II heterogeneity in chloroplasts. *Photosynthesis Research*, **23**, 105–9.

Häder, D.-P. (1997). Effects of UV radiation on phytoplankton. *Advances in Microbial Ecology*, **15**, 1–26.

Häder, D.-P. & Brodhun, B. (1991). Effects of ultraviolet radiation on the photoreceptor proteins and pigments in the paraflagellar body of the flagellate *Euglena gracilis*. *Journal of Plant Physiology*, **137**, 641–6.

Häder, D.-P. & Häder, M. (1990). Effects of UV radiation on motility, photo-orientation and pigmentation in a freshwater *Cryptomonas*. *Journal of Photochemistry and Photobiology B. Biology*, **5**, 105–14.

Häder, D.-P. & Worrest, R. C. (1991). Effects of enhanced solar ultraviolet radiation on aquatic ecosystems. *Photochemistry and Photobiology*, **53**, 717–25.

Hessen, D. O., de Lange, H. J. & Van Donk, E. (1997). UV-induced changes in phytoplankton cells and its effects on grazers. *Freshwater Biology*, **38**, 513–24.

Hessen, D. O., Van Donk, E. & Andersen, T. (1995). Growth response, P-uptake and loss of flagella in *Chlamydomonas reinhardtii* exposed to UV-B. *Journal of Plankton Research*, **17**, 17–27.

Hidema, J., Kang, H.-S. & Kumagai, T. (1996). Differences in the sensitivity to UVB radiation of two cultivars of rice (*Oryza sativa*). *Plant Cell Physiology*, **37**, 742–7.

Iwanzik, W., Tevini, M., Dohnt, G., Voss, M., Weiss, W., Gräber, P. & Renger, G. (1983). Action of UV-B radiation on photosynthetic primary reactions in spinach chloroplasts. *Physiologia Plantarum*, **58**, 401–7.

Jeffrey, W. H., Aas, P., Lyons, M. M., Coffin, R. B., Pledger, R. J. & Mitchell, D. L. (1996). Ambient solar radiation-induced photodamage in marine bacterioplankton. *Photochemistry and Photobiology*, **64**, 419–27.

Jones, L. W. & Kok, B. (1966). Photoinhibition of chloroplast reactions. I. Kinetics and action spectra. *Plant Physiology*, **41**, 1037–43.

Jordan, B. R., Chow, W. S., Strid, A. & Anderson, J. M. (1991). Reduction in cab and psb A RNA transcripts in response to supplementary ultraviolet-B radiation. *FEBS Letters*, **284**, 5–8.

Karentz, D., Cleaver, J. E. & Mitchell, D. L. (1991). Cell survival characteristics and molecular responses of Antarctic phytoplankton to ultraviolet-B radiation. *Journal of Phycology*, **27**, 326–41.

Lao, K. & Glazer, A. N. (1996). Ultraviolet-B photodestruction of a light-harvesting complex. *Proceedings of the National Academy of Sciences, USA*, **93**, 5258–63.

Larson, R. A. (1978). Environmental chemistry of reactive oxygen species. *CRC Critical Reviews in Environmental Control*, pp. 197–246.

Lesser, M. P., Neale, P. J. & Cullen, J. J. (1996). Acclimation of Antarctic phytoplankton to ultraviolet radiation: ultraviolet absorbing compounds and carbon fixation. *Molecular Marine Biology and Biotechnology*, **5**, 314–25.

Lohmann, M., Döhler, G., Huckenbeck N. & Verdini S. (1998). Effects of UV radiation of different wavebands on pigmentation, ^{15}N-ammonium uptake, amino acid pools and adenylate contents of marine diatoms. *Marine Biology*, **130**, 501–7.

Long, S. P., Humphries, S. & Falkowski, P. (1994). Photoinhibition of photosynthesis in nature. *Annual Reviews of Plant Physiology and Molecular Biology*, **45**, 633–62.

Malloy, K. D., Holman, M. A., Mitchell, D. & Detrich, H. W. (1997). Solar UVB-induced DNA damage and photoenzymatic DNA repair in Antarctic zooplankton. *Proceedings of the National Academy of Sciences, USA*, **94**, 1258–63.

Marwood, C. A. & Greenberg, B. M. (1996). Effect of supplementary UVB radiation on chlorophyll synthesis and accumulation of photosystems during chloroplast development in *Spirodela oligorrhiza*. *Photochemistry and Photobiology*, **64**, 664–70.

Máté, Z., Sass, L., Szekeres, M., Vass, I. & Nagy, F. (1998). UV-B-induced differential transcription of psbA genes encoding the D1 protein of photosystem II in the cyanobacterium *Synechocystis* 6803. *Journal of Biological Chemistry*, **273**, 17439–44.

Matoo, A. K. & Edelman, M. (1987). Intramembrane translocation and post-translational palmitylation of the chloroplast 32-kDa herbicide-binding protein. *Proceedings of the National Academy of Sciences, USA*, **84**, 1497–501.

Melis, A., Nemson, J. A. & Harrison, M. A. (1992). Damage to functional components and partial degradation of photosystem II reaction center proteins upon chloroplast exposure to ultraviolet-B radiation. *Biochimica et Biophysica Acta*, **1100**, 312–20.

Mitchell, D. L. & Nairn, R. S. (1989). The biology of the (6–4) photoproduct. *Annual Reviews of Photochemistry and Photobiology*, **49**, 805–19.

Murphy, T. M. (1983). Membranes as targets of ultraviolet radiation. *Physiologia Plantarum*, **58**, 381–8.

Musil, C. F. (1996). Accumulated effect of elevated ultraviolet-B radiation over multiple generations of the arid-environment annual *Dimorphotheca sinuata* DC (Asteraceae). *Plant Cell Environment*, **19**, 1017–27.

Nairn, R. S., Morizot, D. C., Kazianis, S., Woodhead, A. D. & Setlow, R. B. (1996). Nonmammalian models for sunlight carcinogenesis: genetic analysis of melonoma formation in *Xiphophorus* hybrid fish. *Photochemistry and Photobiology*, **64**, 440–8.

Neale, P. J. (1987). Algal photoinhibition and photosynthesis in the aquatic environment. In *Photoinhibition*, ed. D. J. Kyle, C. B. Osmond & C. J. Arntzen, pp. 35–65. Elsevier, Amsterdam.

Neale, P. J., Cullen, J. J., Lesser, M. P. & Melis, A. (1993). Physiological bases for detecting and predicting photoinhibition of aquatic photosynthesis by PAR and UV radiation. In *Photosynthetic Responses to the Environment*, ed. H. Yamamoto & C. M. Smith, pp. 61–77. American Society of Plant Physiology, Washington, DC.

Nogués, S. & Baker, N. K. (1995). Evaluation of the role of damage to photosystem II in the inhibition of CO_2 assimilation in pea leaves on exposure to UV-B radiation. *Plant Cell Environment*, **18**, 781–7.

Pichard, S. L., Frischer, M. E. & Paul, J. H. (1993). Ribulose bisphosphate carboxylase gene expression in subtropical marine phytoplankton populations. *Marine Ecological Progress Series*, **101**, 55–65.

Post, A., Lukins, P. B., Walker, P. J. & Larkum, A. W. D. (1996). The effects of ultraviolet irradiation on P680 + reduction in PS II core complexes measured for individual S-states and during repetitive cycling of the oxygen-evolving complex. *Photosynthesis Research*, **49**, 21–7.

Quesada, A., Mouget, J.-L. & Vincent, W. F. (1995). Growth of Antarctic cyanobacteria under ultraviolet radiation: UVA counteracts UVB radiation. *Journal of Phycology*, **31**, 242–8.

Quesada, A. & Vincent, W. F. (1997). Strategies of adaptation by Antarctic cyanobacteria to ultraviolet radiation. *European Journal of Phycology*, **32**, 335–42.

Rajagopal, S., Jha, I. B., Murthy, S. D. S. & Mohanty, P. (1998). Ultraviolet-B effects on *Spirulina platensis* cells: modification of chromophore–protein interaction and energy transfer characteristics of phycobilisomes. *Biochemical and Biophysical Research Communications*, **249**, 172–7.

Ramos, K. T., Fries, L. T., Berkhouse, C. S. & Fries, J. N. (1994). Apparent sunburn of juvenile paddlefish. *Progressive Fish Culturist*, **56**, 214–16.

Raven, J. A. & Samuelsson, G. (1986). Repair of photoinhibitory damage in *Anacystis nidulans* 625 (*Synechococcus* 5301): relating catalytic capacity for, and energy supply to, protein synthesis, and implications for P_{max} and the efficiency of light-limited growth. *New Phytologist*, **103**, 625–43.

Renger, G., Volker, M., Eckert, H. J., Fromme, R., Hohm-Veit, S. & Graber, P. (1989). On the mechanism of photosystem II deterioration by UV-B irradiation. *Photochemistry and Photobiology*, **49**, 97–105.

Roos, J. C. & Vincent, W. F. (1998). Temperature dependence of UV radiation effects on Antarctic cyanobacteria. *Journal of Phycology*, **34**, 188–25.

Sage, E. (1993). Distribution and repair of photolesions in DNA: genetic consequences and the role of sequence context. *Photochemistry and Photobiology*, **57**, 163–74.

Scully, N. & Vincent, W. F. (1996). Hydrogen peroxide: a natural tracer of stratification and mixing processes in subarctic lakes. *Ergebnisse de Limnologie (Archiv für Hydrobiologie. Beihefte)*, **139**, 1–15.

Scully, N., Vincent, W. F., Lean, D. R. S. & Cooper, W. J. (1997). Implications of ozone depletion for surface water photochemistry: sensitivity of clear lakes. *Aquatic Sciences*, **59**, 260–74.

Shih, M. K. & Hu, M. L. (1996). UVA-potentiated damage to calf thymus DNA by Fenton reaction system and protection by para-aminobenzoic acid. *Photochemistry and Photobiology*, **3**, 286–91.

Shtamm, E. V., Purmal', A. P. & Skurlatov, Yu I. (1991). The role of hydrogen peroxide in natural aquatic media. *Russian Chemical Reviews*, **60**, 1228–48.

Sinha, R. P., Singh, N., Kumar, H. D., Häder, H. D., Häder, M. & Häder, D.-P. (1996). Effects of UV radiation on certain physiological and biochemical processes in cyanobacteria. *Journal of Photochemistry and Photobiology*, **32**, 107–13.

Strid, A., Chow, W. S. & Anderson, J. M. (1990). Effects of supplementary ultraviolet-B radiation on photosynthesis in *Pisum sativum*. *Biochimica et Biophysica Acta*, **1020**, 260–8.

Strid, A., Chow, W. S. & Anderson, J. M. (1994). UV-B damage and protection at the molecular level in plants. *Photosynthesis Research*, **39**, 475–89.

Sullivan, J. H. & Teramura, A. H. (1992). The effects of ultraviolet-B radiation on loblolly pines. 2. Growth of field-grown seedlings. *Trees*, **6**, 115–20.

van der Leun, J. C. (1996). UV radiation from sunlight: summary, conclusions and recommendations. *Journal of Photochemistry and Photobiology B. Biology*, **35**, 237–44.

Van Donk, E. & Hessen, D. O. (1996). Loss of flagella in the green alga *Chlamydomonas reinhardtii* due to *in situ* UV-exposure. *Scientia Marina*, **60**, 107–12.

Vass, I. (1997). Adverse effects of UV-B light on the structure and function of the photosynthetic apparatus, In *Handbook of Photosynthesis* ed. M. Pessarakli, pp. 931–49. Marcel Dekker, New York.

Vetter, R. D. (1997). UV effects on pelagic fish eggs and larvae: the 'one bad day' hypothesis. *Limnology and Oceanography*, **42**. 432 (abstract).

Vincent, W. F. & Roy, S. (1993). Solar UV-B effects on aquatic primary production: damage, repair and recovery. *Environmental Reviews*, **1**, 1–12.

Wang, K. S. & Chai, T.-J. (1994). Reduction in omega-3 fatty acids by UV-B radiation in microalgae. *Journal of Applied Phycology*, **6**, 415–21.

Weinreb, O. & Dovrat, A. (1996). Transglutaminase involvement in UV-A damage to the eye lens. *Experimental Eye Research*, **63**, 591–7.

Wilson, M. I., Ghosh, S., Gerhardt, K. E., Holland, N., Sudhakar Babu T., Edelman, M., Dumbroff, E. B. & Greenberg, B. M. (1995). *In vivo* photomodification of ribulose-1,5-bisphosphate carboxylase/oxygenase holoenzyme by ultraviolet-B radiation. *Plant Physiology*, **109**, 221–9.

Wilson, M. I. & Greenberg, B. M. (1993). Protection of the D1 photosystem II reaction center protein from degradation in ultraviolet radiation following adaptation of *Brassica napus* L. to growth in ultraviolet-B. *Photochemistry and Photobiology*, **57**, 556–63.

Winckler K. & Fidhiany L. (1996). Significant influence of UVA on the general metabolism in the growing cichlid fish, *Cichlasoma nigrofasciatum. Journal of Photochemistry and Photobiology B – Biology*, **33**, 131–5.

7

○ ○ ○ ○ ○ ○ ○ ○ ○ ○ ○ ○ ○ ○ ○ ○ ○ ○ ○

Strategies for the minimisation of UV-induced damage

Suzanne Roy

7.1 Introduction

When faced with a stressful situation, an organism can show four basic responses (Table 7.1):

avoid the stress (e.g. by moving away from it),
reduce the stress by some physiological or behavioural mechanism,
repair damages produced by the stress, and
acclimate to the stress, either physiologically or genetically, over a
longer period of time.

These choices are not mutually exclusive. UV radiation, and particularly the high energy, low wavelength radiation (UV-B: 280–320 nm), represents a stress for most living organisms because of its highly absorptive character. A number of recent reviews (as well as Chapter 6) give additional information to interested readers about UV effects (e.g. Karentz *et al.*, 1994; Häder *et al.*, 1995; Booth *et al.*, 1997). This chapter considers the various strategies for minimising UV-induced damage used by marine organisms, particularly the autotrophs at the base of marine food webs.

7.2 UV avoidance mechanisms

Autotrophic organisms need light for photosynthesis. Hence they can generally control to some extent their movement towards light or the water surface. The control mechanisms can involve the presence of flagella or cilia, or buoyancy regulation. When light becomes excessive and potentially dangerous for the photosynthetic system, other mechanisms control movement away from the light. Light is generally perceived by

Table 7.1. The four major types of strategy used by marine organisms to minimise UV-induced damage

Avoidance	Screening	Repair		Acclimation
		of direct UV damage	of indirect UV damage	
Move away from the light	Extracellular (in skin, cell walls, cuticles, sheaths)	DNA repair: photoreactivation	Antioxidant enzymes	Short-term physiological changes (e.g. fluorescence)
Circadian rhythm (avoid surface at mid-day)	Intracellular: MAAs and others	DNA repair: excision repair	Lipid-soluble antioxidants (e.g. carotenoids)	Long-term physiological modifications (e.g. biosynthesis of screening compounds)
Habitat selection	Inorganic coverings	Protein repair (stress proteins)	Water-soluble antioxidants (e.g. ascorbic acid)	UV sensitivity: community changes
Life in a vertically mixed environment	Organic coverings (e.g. epiphytes)			

MAA, mycosporine-like amino acids.

photoreceptors. Green light and UV-A/blue light photoreceptors are prevalent in aquatic organisms, but their exact nature is not well known (Rüdiger & López-Figueroa, 1992). The carotenoid zeaxanthin has been suggested as a blue light photoreceptor in higher plants (Quiñones & Zeiger, 1994). Although not yet found in marine organisms (Häder & Worrest, 1991) some freshwater invertebrates have been shown to possess photoreceptors specific to UV-B (Pennington & Emlet, 1986; Smith & Macagno, 1990). Consequently, cells can sense an increase in light intensity in the UV-A (320–400 nm) and photosynthetically available radiation (PAR) (400–700 nm) ranges and respond to it, but they cannot sense an increase in the UV-B range unaccompanied by concurrent increases in other wavelengths, as is the case with stratospheric ozone depletion.

UV-B light can reduce motility in many micro-organisms, including cyanobacteria, chlorophytes, cryptophytes, dinoflagellates and ciliates (Häder *et al.*, 1995). Moreover, loss of flagella has been reported for some chlorophytes (Hessen, Van Donk & Andersen, 1995). UV-B light can also affect photo- and graviorientation (Häder *et al.*, 1995). Thus changes in the spectral balance (= ratio of UV-B over PAR + UV-A + UV-B) caused by ozone depletion can potentially interfere with the mechanisms controlling motility and avoidance of excessive light.

Non-planktonic pelagic organisms can generally control their position in the water. However, without a UV-B specific sensor, they are likely to be unable to move away when UV-B increases due to ozone depletion. Benthic organisms that live intertidally or in shallow, illuminated waters can generally move in and out of the sediment layer for photoprotection. Sessile organisms are more at risk. However, they can generally develop protective mechanisms, such as the synthesis of UV screening compounds (see below).

Avoidance of excessive light can be a targeted direct movement in response to high light (cf. Vincent & Quesada, 1997). It can also be built into an endogenous, circadian response to daily light patterns: e.g. many phytoplankton species move away from the surface at mid-day, when light intensities are strongest (e.g. dinoflagellates, Tilzer, 1973; cyanobacteria, Reynolds, Oliver & Walsby, 1987). Alternatively, avoidance can be more passive, taking the form of habitat selection. For example, many tropical corals harbour symbiotic dinoflagellates (zooxanthellae) that benefit from the light coverage afforded by the host (e.g. Banaszak & Trench, 1995a).

Cells that live in highly mixed surface waters are generally thought to benefit passively from this mixing, because it reduces the average light intensity impinging upon the cells. Thus, UV-induced DNA damage in bacterioplankton (Jeffrey *et al.*, 1996) and photosynthetic activity in Antarctic phytoplankton (Ferreyra, 1995) show reduced UV effects under a mixed regime. However, mixing can also cause an increase in UV-induced damage, depending on repair mechanisms and the kinetics of damage and repair. If the rate of repair of UV-induced damage is slow or inefficient, then the amount of damage increases as a function of UV flux (the 'reciprocity' concept, cf. Chapter 3) and mixing provides a way of bringing undamaged cells to the surface increasing the total number of cells damaged by UV. Alternatively, if repair is efficient, removing cells from the UV-damaging surface layer allows them time to recover, such that mixing becomes beneficial in this case (cf. Chapter 3). Understanding the effects of mixing thus requires knowledge of the vertical circulation and of the kinetics of both repair and UV-induced damage (Neale, Davis & Cullen, 1998).

7.3   Reduction of effective UV radiation

Another strategy for minimisation of UV-induced impairment is to reduce the effective UV radiation that penetrates the cell. This can be achieved in a number of ways: by synthesising intra- or extracellular compounds that can absorb the damaging wavelengths ('screening agents'), or by adopting a specific behaviour that helps to reduce the amount of UV radiation reaching UV-sensitive targets such as DNA or photosynthetic reaction centres. A variety of compounds with UV-screening abilities have been described. These are examined below.

7.3.1   Extracellular screening agents

Extracellular screening compounds can be found in cell walls where they block incoming UV radiation. Sporopollenin is a cell wall biopolymer with proposed UV-B screening abilitites (Xiong *et al.*, 1996, 1997). Melanin is another UV-screening compound found in the skin of aquatic vertebrates (Lowe & Goodman-Lowe, 1996) or in the cuticle of zooplankton (Hessen & Soerensen, 1990). Cyanobacteria that form colonies or mats often possess a different kind of extracellular screening compound. A yellow-brown, lipid-soluble pigment found in extracellular sheaths from more than 30 species of cyanobacteria presents an absorption maximum around 370 nm (with substantial absorption also in the UV-B range). This pigment, called scytonemin by Nägeli & Schwenderer (1877) plays a photoprotective role against excessive UV-A radiation and its synthesis is promoted by increases in UV-A (Garcia-Pichel & Castenholz, 1991). This compound can absorb > 85% of incoming UV-A radiation. Removal results in a loss of resistance to photobleaching by UV-A. Its chemical structure, determined in 1993 (Proteau *et al.*, 1993), suggests that scytonemin is formed from a condensation of tryptophan- and phenyl-propanoid-derived subunits (Figure 7.1). Another water-soluble pigment has been found in the extracellular polysaccharide matrix of cells from colonies of *Nostoc commune*, a terrestrial cyanobacterium. It absorbs maximally at 312 nm and is induced by increases in UV-B radiation but has not been described chemically (Scherer, Chen & Böger, 1988). Its role is thought to be photoprotective. Gloeocapsin is another proposed UV-screening compound found in extracellular sheaths of colonies of some marine cyanobacteria (Lukas & Hoffman, 1984).

7.3.2 Intracellular screening agents
Various intracellular substances absorb in the UV range and are thought to play a role in photoprotection. Among these, the mycosporine-like amino acids (MAAs), have elicited much interest in recent UV research.

7.3.2.1 Mycosporine-like amino acids
MAAs (Figure 7.2) are found in most marine organisms, from cyanobacteria to fish (Karentz *et al.*, 1991b; Dunlap & Shick, 1998). These compounds were originally discovered in fungi and were associated with increased sporulation, hence their name (Leach, 1965; Favre-Bonvin, Arpin & Brevard, 1976). They are biosynthesised via the shikimate pathway (Favre-Bonvin *et al.*, 1987), in a manner similar to the biosynthesis of UV-screening flavonoids in terrestrial plants (Figure 7.3), which is controlled by UV-B-inducible enzymes (e.g. phenylalanine ammonia lyase, Tevini, Braun & Fieser, 1991). Only bacteria, fungi and algae can synthesise MAAs; other marine organisms acquire MAAs by diet transfer,

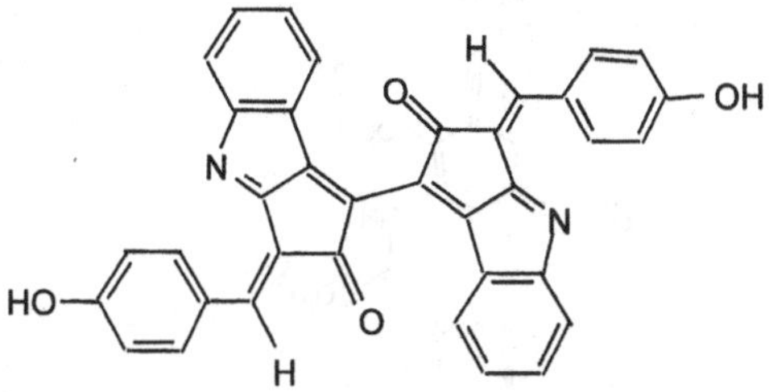

Scytonemin

Figure 7.1. The chemical structure of scytonemin, an extracellular screening agent common in cyanobacteria.

symbiotic or bacterial associations (Shick *et al.*, 1992; Stochaj, Dunlap & Shick, 1994; Carroll & Shick, 1996). The evolution of UV-B-absorbing phenolic compounds seems to have played an important role in the evolution of plants. An increasing degree of polymerisation and complexity of polyphenolics is observed from algae (MAAs) to non-vascular and vascular plants (flavonoids, Rozema, van de Staaij & Björn, 1997).

The sunscreen role of MAAs has been inferred from:

> their high absorptivity in the range 309–360 nm,
> their increase in cellular concentration with increased UV exposure (Dunlap, Chalker & Oliver, 1986; Wood, 1989),
> correlation with reduced photoinhibition (Neale, Lesser & Cullen, 1994; Vernet *et al.*, 1994; Lesser, 1996a) and decreased photodamage (Wood, 1987),

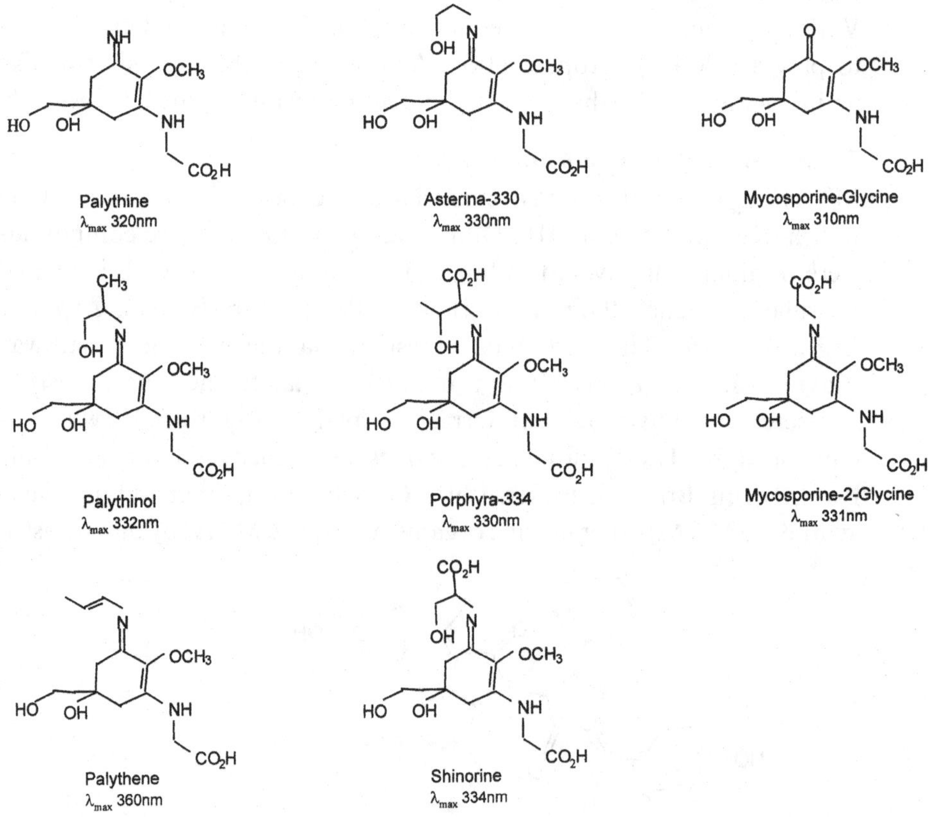

Figure 7.2. The chemical structure of eight of the most common mycosporine-like amino acids. The wavelength of maximum absorbance is indicated below each name.

an action spectra showing that UV radiation is most efficient in eliciting increases in MAAs (Carreto *et al.*, 1990a; Garcia-Pichel, Wingard & Castenholz, 1993).

However, other studies noted that the presence of MAAs was not always sufficient for complete UV protection. Photoinhibition and oxidative stress occurred in a dinoflagellate and a coral despite the presence and accumulation of MAAs (Lesser, 1996a; Lesser & Lewis, 1996). No significant increase in the concentration of MAAs was observed when some sea anemones or corals were transplanted near the surface or acclimatised in the presence of UV light (Stochaj *et al.*, 1994; Banaszak & Trench, 1995b). Furthermore, when other physiological photoprotective and repair mechanisms were inhibited, cyanobacterial cells with high specific contents of MAAs were only slightly more resistant to UV radiation (Garcia-Pichel *et al.*, 1993), indicating that MAAs do not provide complete protection against UV effects. Some algae, particularly diatoms, seem to accumulate very small amounts of MAAs in response to

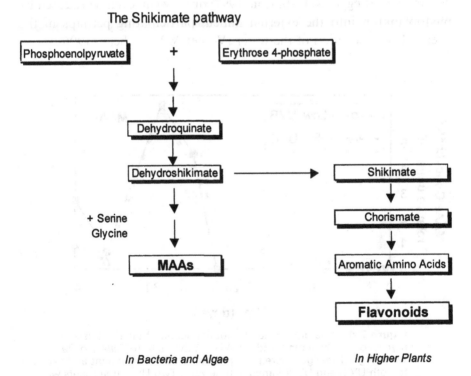

Figure 7.3. Biosynthesis of mycosporine-like amino acids (MAAs) (in bacteria and algae) and of flavonoids (in higher plants) via the shikimate pathway (cf. Favre-Bonvin *et al.*, 1987).

UV exposure, even in Antarctica (Davidson *et al.*, 1994), although accumulation of MAAs is a photoinducible process that can take some time (Carreto, De Marco & Lutz, 1989; Shick, Lesser & Stochaj, 1991; Figure 7.4). Blue light and UV-A wavelengths are most efficient in eliciting the synthesis of these compounds (Carreto *et al.*, 1990b), although UV-B wavelengths can also induce this synthesis in some algae (*Phaeocystis antarctica*, Riegger & Robinson, 1997). Cells grown under high irradiance possess higher concentrations of MAAs than do low light cells (Carreto *et al.*, 1989). Thus previous light history plays an important role in the development of UV screening in photosynthetic organisms.

Cell size is another factor that influences how efficient it is for an organism to synthesise MAAs: a bio-optical model suggests that small cells (<1 μm) cannot physically accumulate enough sunscreen compounds to affect significantly UV penetration inside the cell. Alternatively, accumulating MAAs is potentially a cost-effective strategy for big cells (>10 μm, Garcia-Pichel, 1994). MAAs can sometimes be found outside the cells in the extracellular colonial matrix of some Antarctic prymnesiophytes (Riegger & Robinson, 1997) or they can even be released by phytoplankton into the external milieu, participating perhaps in the overall UV absorption of the water (Vernet & Whitehead, 1996).

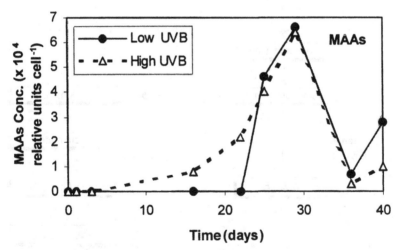

Figure 7.4. The time scale of photoinduction of the synthesis of mycosporine-like amino acids (MAAs) in the diatom *Thalassiosira weissflogii*. The algae were preacclimated under visible light and exposed to both UV-B and UV-A lamps at time zero. Two UV-B treatments were tested: 'Low' UV-B, corresponding to local summer conditions, and 'High' UV-B corresponding to dose rates during the Antarctic ozone hole period (Zudaire, 1999).

Therefore, MAAs may not provide complete protection, but they are frequently part of an overall strategy to diminish the direct and indirect damaging effects of UV radiation (Lesser, 1996a).

7.3.2.2 Other intracellular screening agents

A few other compounds with strong absorbances in the UV range and with potential roles as photoprotectors have been reported (Figure 7.5). These include the lipid-soluble vitamin D and some of its derivatives which absorb maximally between 260 and 281 nm (Holick, Holick & Guillard, 1982), and the water-soluble biopterin glucoside, which absorbs maximally in the UV-A range (320–390 nm) and is found in the marine planktonic cyanobacterium *Oscillatoria* sp. (Matsunaga *et al.*, 1993). In fish eggs and brine shrimps, a metabolite called gadusol also protects against UV radiation. It is chemically related to MAAs and absorbs maximally at 296 nm. Its concentration can reach 4 mg (g dry weight)$^{-1}$ in many fish eggs (Grant, Plack & Thomson, 1980).

7.3.3 Behaviourally mediated UV screening

UV screening can also take the form of inorganic coverings that the cell or

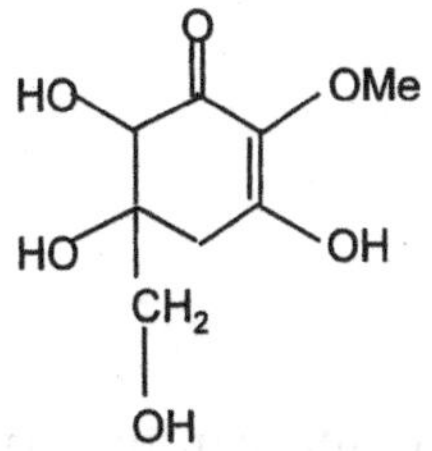

Gadusol

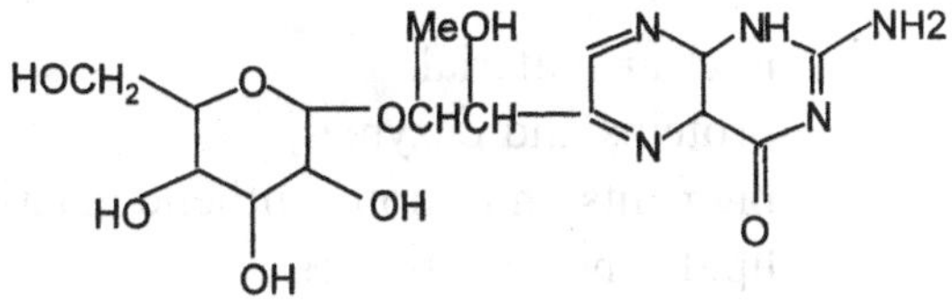

Biopterin Glucoside

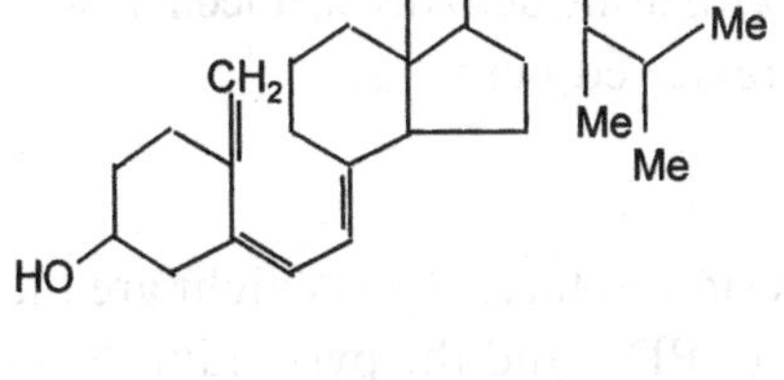

Vitamin D

Figure 7.5. The chemical structure of three intracellular screening agents. Vitamin D and gadusol absorb maximally in the UV-B range, while biopterin glucoside shows maximal absorption in the UV-A range.

the organism attaches to itself. For example, the sea anemone *Anthopleura elegantissima* attaches debris to its column in response to elevated irradiance, likely to shield exposed surfaces (Stochaj *et al.*, 1994). Also, floating oscillatoriacean (cyanobacteria) mats found in Antarctic ponds are encrusted with salt crystals that attenuate solar radiation received (Vincent & Quesada, 1994). Persistent bloom-forming cyanobacteria are protected during exposure to high irradiance by the upper layer of cells at the water surface, which often become bleached (Walsh, 1996). Some intertidal and subtidal seagrass species rely on shading by coexisting seagrass species and epiphytes to reduce the degree of UV exposure (Trocine, Rice & Wells, 1981).

An increase in cell size is frequently observed following UV exposure (cf. Karentz, Cleaver & Mitchell, 1991a). It is not clear whether this increase is a protection strategy (by increasing the pathlength for UV-B through the cell) or a consequence of UV damage to the cell cycle resulting in continued growth without cell division (Behrenfeld, Hardy & Lee, 1992; Buma *et al.*, 1996).

7.4   Repair of UV-induced damages

As seen in preceding chapters, UV radiation has many targets, mainly

> nuclear material,
> proteins and enzymes,
> pigments and photosynthetic reaction centres,
> lipids and membranes.

Most important among those are the damages to nuclear material. However, cells have long developed mechanisms to correct various DNA defects. Repair of damage to other cell constituents generally involves biosynthetic pathways and is mediated by stress proteins. Indirect UV damage caused by reactive oxygen species (ROS) such as superoxide or oxygen peroxide (cf. Chapter 4) are dealt with using detoxifying mechanisms. These various repair mechanisms are reviewed here.

7.4.1   DNA repair

The two most frequent types of DNA lesion induced by UV light are the *cis–syn* cyclobutane pyrimidine dimers (CPDs) and the pyrimidine (6–4) pyrimidone photoproducts (also called (6–4) photoproducts, cf. Chapter 4). Repair of DNA damage can take place by various mechanisms that include photoreactivation (light-induced repair), nucleotide and base excision repair and recombinational repair (Friedberg, 1985; Pfeifer, 1997). In photosynthetic organisms, which possess DNA in both the nucleus and chloroplasts, repair pathways in both compartments are similar, but not identical (Small, 1987).

7.4.1.1 Photoreactivation

Photorepair of DNA involves the direct monomerization of CPDs by the action of the enzyme DNA photolyase in the presence of light (Sutherland, 1981). This enzyme binds to pyrimidine dimers in a light-independent reaction and, upon absorption of a photon of light between 300 and 500 nm, donates an electron to the dimer, initiating an electronic reorganisation that eventually produces two intact pyrimidines (Selby & Sancar, 1990). The photolyase enzyme is found in almost all organisms, from bacteria to humans. It has been isolated and purified from the green alga *Scenedesmus acutus* and found to contain two different flavin chromophores (Eker, Hessels & van de Velde, 1988). The enzyme is generally specific for CPDs, but was recently found also to bind to other DNA damage areas and enhance or suppress excision repair ability (Fox, Feldman & Chu, 1994; Ozer *et al.*, 1995). It has no effect on the other major UV-induced DNA lesion, the (6–4) photoproduct. However, a newly described photoreactivating enzyme, isolated from the fruit fly (*Drosophila melanogaster*), specifically repairs UV-induced (6–4) photo-products (Todo *et al.*, 1993). Plants also appear to contain this second photolyase activity (Taylor, Tobin & Bray, 1997).

Although widely distributed, photoreactivating activity varies widely amongst species, even within a single taxonomic group: great diversity was observed for both DNA damage and repair in 12 species of Antarctic diatoms (Karentz *et al.*, 1991a). Between 1% and 74% of the initial damage induced by UV-B was removed within 6 h. Small cells sustained more damage per unit of DNA. Photoreactivation has also been observed in cyanobacteria (Asato, 1972; O'Brien & Houghton, 1982; Blakefield & Harris, 1994) and in chlorophytes (Eker *et al.*, 1988; Takao *et al.*, 1989).

Previous exposure to photoreactivating light a few hours before UV-B irradiation increases the rate of photoreactivation following the UV-B treatment (Mitchell, Scoggins & Morizot, 1993), suggesting that the activity of these repair enzymes is photoinducible.

7.4.1.2 Nucleotide excision repair

Nucleotide excision repair is also known as dark repair because of the absence of light requirement. The following steps are involved :

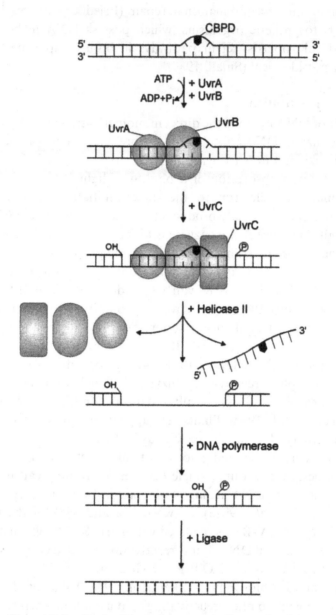

Figure 7.6. Nucleotide excision repair: the mechanism used by *E. coli* (Selby & Sancar, 1990). CBPD, cyclobutane pyrimidine dimer. (Modified from Darnell *et al.*, 1990.)

damage recognition by specific proteins,

assembly of a DNA repair complex (including enzymes that can unwind DNA at the lesion site),

incision of the DNA backbone on either side of the damage and removal of the damaged strand, and

filling of the remaining gap by a DNA polymerase enzyme, followed by attachment to the rest of the DNA with a DNA ligase.

This DNA repair mechanism is present in all organisms. It acts on both the CPDs and the (6–4) photoproducts, although the DNA damage recognition proteins seem to have greater affinity for the (6–4) photoproducts (Pfeifer, 1997). One of the best known mechanisms of excision repair is the one used by the bacterium *Escherichia coli* (Figure 7.6): the damaged site is recognized by the UvrA protein, which brings along the UvrB protein that associates with the damaged strain. After unwinding, the UvrC protein joins in and nucleotide excision is then performed by this excinuclease enzyme (Selby & Sancar, 1990). Final repair is completed by the DNA polymerase and ligase enzymes. Base excision repair is a variant of dark repair using damage-specific DNA glycosylases that remove modified bases from DNA (Lindahl, 1976).

Both photoreactivation and dark repair have been observed in marine planktonic communities: damage sustained by a marine diatom during the day was fully repaired by the next morning when the UV-B treatment was followed by several hours of saturating PAR (400–700 nm) irradiance (Buma *et al.*, 1995), while DNA damage sustained by marine bacterioplankton can be repaired during the night, implicating dark repair pathways (Jeffrey *et al.*, 1996).

7.4.1.3 Other mechanisms of DNA repair

Recombination is the process whereby genetic material can be exchanged by crossing-over groups of genes on homologous chromosomes. Postreplication recombinational repair of DNA involves the insertion of complementary DNA from a sister chromosome into gaps associated with inhibition of DNA replication at or near DNA damage-induced lesions (Miller, 1992). One of the important genes controlling this process in bacteria is the *recA* gene. It codes for a protein that promotes DNA recombination by inserting a section of single-strand DNA into double-stranded DNA (Darnell, Lodish & Baltimore, 1990). Under the influence of UV light, this protein is transformed into an active proteinase that destroys a repressor, leading to the expression of many genes involved in

DNA repair (the SOS response of *E. coli*, Friedberg, Walker & Siede, 1995). Thus, activation of these various UV-induced DNA repair mechanisms is under the control of DNA itself. Signals used for this activation could come either directly from the DNA damage or from free radicals formed by UV-induced membrane damage that elicits changes in membrane-associated proteins or lipids which then activate the UV response (Mount, 1996).

7.4.2   Protein repair

UV can also degrade proteins and enzymes such as those involved in the basic steps of photosynthesis. A 'photoinhibition repair cycle' oversees the biosynthetic recovery of the various components of the photosynthetic reaction centres, including the essential D1 and D2 proteins (Andersson *et al.*, 1992).

Accumulation of damaged proteins in cells usually results in activation of regulatory proteins called heat-shock transcription factors. These factors then bind to heat-shock elements of DNA and increase the rates of transcription of genes encoding the so-called 'heat-shock' or stress proteins (Hightower, 1993). Stress proteins are subdivided into three size classes termed Hsp90, Hsp70 and Hsp60, where the number refers to the relative molecular mass. They play a role in protein folding/unfolding pathways and in the intracellular movement of proteins within the cytoplasm. They are thought to aid in the solubilization of denatured proteins in stressed cells in preparation for either degradation or salvaging via protein-refolding pathways (Hightower, 1993). This function helps to protect protein and protein complexes by restoring damaged proteins to their native forms.

There is evidence that stress proteins are involved in acclimation of mussels and fish to thermal or pollution stresses. Relationships between the amount of stress proteins and stress severity are now common in most organisms, including aquatic species (Sanders, 1993). UV-induced protein damage also results in the subsequent accumulation of stress proteins. This has been recently observed in cyanobacteria and in marine, temperate diatoms (Shibata, Baba & Ochiai, 1991; Döhler, Hoffmann & Stappel, 1995). However, there has generally been little work on the influence of UV on stress proteins in aquatic organisms. Stress proteins specific for light-induced stresses in photosynthetic organisms (early light-induced proteins (ELIPs)) may be involved as they already play a role in photoprotection (Adamska, 1997).

7.4.3 Defences against forms of reactive oxygen

Antioxidation defences comprise three general classes: antioxidant enzymes, lipid-soluble antioxidants residing in cellular membranes (e.g. carotenoids), and water-soluble reductants found in the cytosol (Dunlap & Yamamoto, 1995).

7.4.3.1 Antioxidant enzymes

Univalent reduction of oxygen or oxidation of oxygen-containing compounds such as ferredoxin, flavins, chlorophylls and quinones produces superoxide radicals (Foote, 1976). Continued reduction of these radicals yields hydrogen peroxide and hydroxyl radicals (Fridovich, 1977). Sunlight reacting with oxygen (photosynthetically generated or otherwise present) can produce these active forms of oxygen. A number of enzymes play specific roles in the detoxification of these damaging species of oxygen inside cells. These include: superoxide dismutase (SOD), which removes oxygen radicals but produces hydrogen peroxide in the process; catalase and ascorbate peroxidase, which remove peroxides; and glutathione peroxidase, which acts on lipid hydroperoxides (Fridovich, 1976).

SOD activity is induced by the level of oxidative stress (Dykens & Shick, 1982; Shick et al., 1991) and generally declines with water depth for aquatic organisms (Shick et al., 1995). The specific activity of SOD increases in response to UV-A and UV-B in cultured and freshly isolated zooxanthellae from UV-acclimated sea anemones (Lesser & Shick, 1989). Increasing activity of these antioxidative enzymes in response to light or UV stress is not always sufficient for complete photoprotection (as was observed for MAAs), suggesting that tolerance thresholds can been reached (Lesser, 1996b; Lesser & Lewis, 1996).

7.4.3.2 Carotenoids

Carotenoids play two major roles: light-harvesting and photoprotection (Goodwin, 1980). Their photoprotective role requires a minimum of nine conjugated double bonds in their chemical structure (Mathis & Kleo, 1973). Photoprotection can take place in four basic ways:

1. Deactivating singlet state (excited) oxygen, yielding triplet state (excited) carotenoids and ground state oxygen (Foote, Chang & Denny, 1970; Santus, 1983).
2. Acting as a radical-trapping antioxidant, efficiently removing free radicals (Jialal et al., 1991). β-Carotene is particularly efficient in this role (Burton & Ingold, 1984).
3. Deactivating chlorophyll a triplet (excited) states. These are a

major source of energy for the production of excited, singlet state oxygen (Moore *et al.*, 1982).

4. A last photoprotective role does not involve an antioxidative function: the xanthophyll cycle, characterised by a light-induced rapid increase in the production of the carotenoids zeaxanthin (from violaxanthin) or diatoxanthin (from diadinoxanthin), plays a photoprotective role by promoting heat dissipation of excess light energy accumulated in the light-harvesting centres of chloroplast membranes (Demmig *et al.*, 1987; Demers *et al.*, 1991).

With such potential involvement in photoprotection, it can be expected that the concentration of carotenoids should increase under conditions of excessive light or in response to UV radiation. Accumulation of carotenoids (β-carotene, zeaxanthin, myxoxanthophyll and diadinoxanthin) has been observed for surface populations of chlorophytes, cyanobacteria and red-tide dinoflagellates (Paerl, Tucker & Bland, 1983; Ben-Amotz, Shaish & Avron, 1989; Vernet, Neori & Haxo, 1989). However, this type of response is not universal. Some species of cyanobacteria show no enhancement of carotenoid synthesis under high light (Nultsch & Agel, 1986; Garcia-Pichel & Castenholz, 1991), but the faster degradation of chlorophylls leads to an increased ratio of carotenoids to chlorophylls that can be taken to suggest photoprotection. Accumulation of carotenoids specifically in response to UV light has been reported only for cyanobacteria and chlorophytes (Buckley & Houghton, 1976; Goes *et al.*, 1994).

In the case of the xanthophyll cycle, little information is available on the specific role of this enzyme-mediated pigment transformation with respect to UV exposure. In higher plants, UV-B has been shown to have a detrimental effect on the enzymic conversions involved (Pfündel, Pan & Dilley, 1992), suggesting that the xanthophyll cycle is a UV-B target rather than a photoprotection mechanism. On the other hand, increased activity of this cycle was observed in a study of photoinhibition by visible light supplemented with UV-B, reflecting a greater dissipation of excess light energy through non-photochemical quenching (Bornman & Sundby-Emanuelsson, 1995).

7.4.3.3 Other antioxidants

Apart from the carotenoids, another fat-soluble antioxidant plays an important detoxifying role in the cell. Vitamin E, or α-tocopherol, is an efficient physical quencher and chemical scavenger of photo-generated singlet oxygen (Fryer, 1993). It can also function as a 'chain-breaking' antioxidant by trapping fatty acyl peroxy radicals formed during lipid

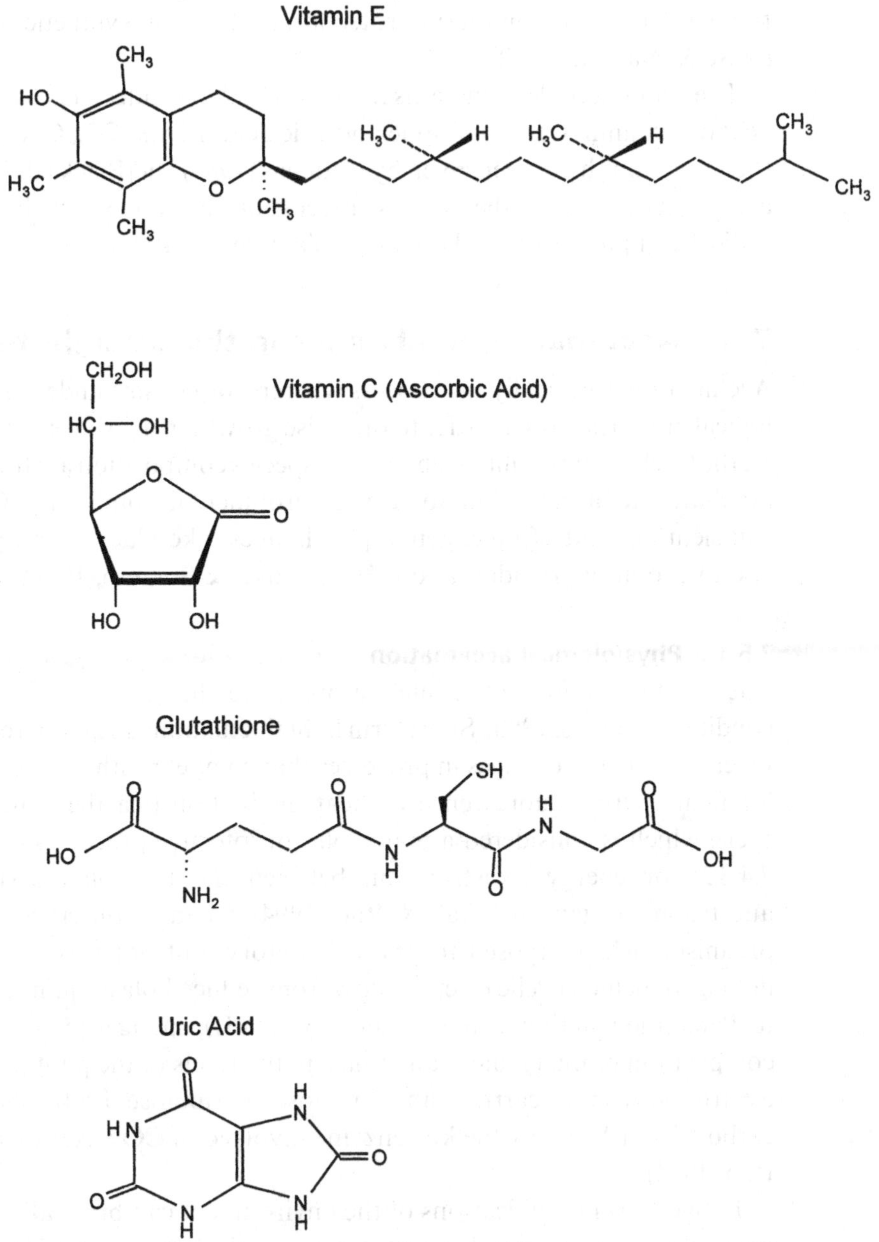

Figure 7.7. The chemical structure of various antioxidants that prevent UV-related damage by reactive oxygen species.

peroxidation and can reduce superoxide to hydrogen peroxide. It appears to be implicated in long-term protection of the photosynthetic pigments (Wise & Naylor, 1987).

The water-soluble reductants include small molecules such as ascorbic acid (i.e. vitamin C), glutathione and uric acid (Figure 7.7). Overviews of their functions have been given by Cadenas (1989) and Black (1987). They also include one of the MAAs, mycosporine-glycine, which exhibits radical-trapping activity (Dunlap & Yamamoto, 1995).

7.5   Acclimation to changes in the UV light field

Acclimation is basically the process whereby organisms undergo physiological modifications in order to optimise growth in a given environment. At the level of communities, changes in species composition are frequently attributed to acclimation to new environmental conditions. Given a sufficient amount of time, genotypic changes take place, thereby giving rise to evolutionary adaptation (Hochachka & Somero, 1973).

7.5.1   Physiological acclimation

There is a continuum of cellular responses to changes in environmental conditions (Harris, 1980). Short term light fluctuations elicit fast, reversible reactions, such as changes in processes that compete with photosynthesis for light energy: fluorescence or heat dissipation (via the xanthophyll cycle, which is considered a major photoprotective process, see Section 7.4.3.2), or energy redistribution between the two photosystems of autotrophic organisms (Hall & Rao, 1994). Changes observed in these organisms when exposed to photoinhibitory (but not necessarily UV) light also include cyclic electron flow from reduced plastoquinones back to P680 (the reaction centre of photosystem II), changes in the size and composition of the pigment antennae, in the levels of the photosynthetic electron transport carriers and in those of ribulose 1,5-bisphosphate carboxylase (Rubisco), the key enzyme involved in CO_2 fixation (Hall & Rao, 1994).

Longer-term modifications of the environment can bring about more permanent physiological changes (i.e. physiological acclimation, Hochachka & Somero, 1973). UV-induced physiological acclimation can involve any of the protection strategies previously discussed in this chapter, notably the synthesis of MAAs and antioxidants. Water depth can be a natural determinant of UV-induced physiological acclimation, particularly for sessile organisms. In corals, both the vertical profile of MAA concentration and acclimation experiments at various depths suggest that MAAs are involved in acclimation (Dunlap & Chalker, 1986; Kinzie, 1993). The involvement of MAAs in acclimation is also observed in some macroalgae and phytoplankton, notably dinoflagellates (Carreto *et al.*, 1989; Wood, 1989). However, in a number of organisms, UV radiation does not seem to regulate the concentration of MAAs (Stochaj *et al.*, 1994; Banaszak & Trench, 1995b). Moreover, as mentioned before, an increase in these UV-absorbing compounds is not always sufficient to counteract the negative effects induced by UV (Garcia-Pichel *et al.*, 1993; Lesser, 1996a).

There are few long term (several days to weeks) studies of UV-induced physiological acclimation. Two such studies with cultured unialgal species demonstrated different results. Whereas the diatom *Phaeodactylum tricornutum* exhibited no signs of UV acclimation after more than five weeks (Behrenfeld *et al.*, 1992), the dinoflagellate *Prorocentrum micans* showed an acclimation response, after three weeks of UV exposure, manifested by an increase in the concentration of MAAs and antioxidants. However, these changes were not sufficient to counter a significant inhibition of photosynthetic and growth rates (Lesser, 1996a).

UV-induced physiological acclimation apparently involves a diverse and complex pattern of biochemical responses that requires more study. MAAs and antioxidants are probably involved, but these may reflect only part of the response. Stress proteins are another possible candidate, but investigations to date in UV-affected marine organisms have been rudimentary. In all cases, however, successful acclimation to UV exposure should decrease the UV sensitivity of the cell or organism. This will affect the biological weighting functions that relate UV levels to UV-induced biological damage and will thus affect model estimates of UV damages (Lesser, 1996a; Cullen & Neale, 1997).

The UV sensitivity of photosynthetic organisms is affected to a large degree by the light history of the cells: acclimation to high PAR irradiance increases UV-B tolerance significantly through photoinduction of screening compounds, increased activity of photorepair enzymes and other physiological changes related to life under high irradiance (Warner & Caldwell, 1983). This explains some of the discrepancies observed between laboratory (generally low PAR light) and field studies noted in higher plant research (Middleton & Teramura, 1994). It may also explain the difference in UV tolerance observed between tropical and polar phytoplankton: tropical surface water species show marked resistance, while Antarctic species, living in a deeply mixed environment, react more like shade-acclimated algae (Helbling *et al.*, 1992). The interaction between photoacclimation and vertical mixing in aquatic environments is likely a key factor governing UV tolerance.

7.5.2   Community changes

Whenever the UV response of natural communities has been examined, changes in species composition have usually been observed. Early studies showed a widely varying UV sensitivity in marine phytoplankton, irrespective of the major algal groups (Calkins & Thordardøttir, 1980; Worrest *et al.*, 1981b). Obvious alterations of the community composition were observed with increased exposure to UV-B radiation over a few weeks (Worrest, Thomson & Van Dyke, 1981a). More recent investigations confirm this trend. Work in Antarctica suggests that diatoms are generally more UV-tolerant than flagellates or the often dominant Antarctic prymnesiophyte, *Phaeocystis* sp., although there is disagreement about the impact of UV-B on this last species (Helbling, Villafañe & Holm-Hansen, 1994; Karentz & Spero, 1995; Davidson, Marchant & de la Mare, 1996). Acclimatisation studies with Antarctic phytoplankton communities also show an increase in diatoms while flagellates decreased after a two week period (Villafañe *et al.*, 1995). Indeed, Antarctic marine diatoms appear capable of surviving high levels of UV-B irradiance, although this does not seem to be relatend to physiological acclimation through screening compounds (Davidson *et al.*, 1994). Benthic diatoms show an even higher level of resistance to UV-B (Peletier, Gieskes & Buma, 1996).

The time scale of the observations is important when determining acclimation responses. A nice example is provided in a study of freshwater periphytic diatom communities, which exhibited different UV responses according to the duration of the experiment: growth was inhibited during the first two to three weeks, but this trend was reversed after three to four weeks. An increase in growth was observed after five weeks of exposure to UV, accompanied by changes in species composition (Bothwell *et al.*, 1993).

7.6   Summary and perspectives

Although UV light, and particularly UV-B (for which no photosensor seems to exist), can cause damage to aquatic organisms, these possess a range of mechanisms to avoid, screen, repair, detoxify and acclimate to this stress. This is not surprising, since life evolved in the sea under much more drastic UV conditions than are encountered nowadays (Rozema *et al.*, 1997). It is thus expected that a number of species will be able to tolerate the increases in UV-B levels associated with a depleting stratospheric ozone layer. However, although physiological changes are observed in response to UV-B increases, these do not often seem sufficient for full protection of cells or organisms. Hence it has been reported that a number of organisms have already reached their UV-B tolerance limits. Whether this is so, or whether the time-frame of acclimation processes needs to be better defined should be investigated more closely. Will the species composition of natural ecosystems be affected in the long term and what will be the effects of species displacement or disappearance due to the UV stress (Vincent & Roy, 1993)? Answering these questions is one of the important tasks that confront us now.

Acknowledgements

The author thanks the proof-reading help of Drs K. Walsh, S. de Mora and W. Jeffrey. Dr Walsh also helped with some of the figures. Financial support for the author was provided by an NSERC operational grant. This work follows from a conference, supported by IAI (Inter-American Institute), on effects of UV radiation on aquatic ecosystems at different latitudes (Ensenada, Mexico, September 1996), and is a contribution to the GREC (Groupe de Recherche en Environnement Côtier) and to the UV research team of INRS-Océanologie.

References

Adamska, I. (1997). ELIPs – light-induced stress proteins. *Physiologia Plantarum*, **100**, 794–805.

Andersson, B., Salter, A. H., Virgin, I., Vass, I. & Styring, S. (1992). Photodamage to photosystem II – primary and secondary events. *Journal of Photochemistry and Photobiology. B. Biology*, **15**, 15–31.

Asato, Y. (1972). Isolation and characterization of ultraviolet light-sensitive mutants of the blue-green alga *Anacystis nidulans*. *Journal of Bacteriology*, **110**, 1058–64.

Banaszak, A. T. & Trench, R. K. (1995a). Effects of ultraviolet (UV) radiation on marine microalgal-invertebrate symbioses. I. Responses of the algal symbionts in culture and *in hospite*. *Journal of Experimental Marine Biology and Ecology*, **194**, 213–32.

Banaszak, A. T. & Trench, R. K. (1995b). Effects of ultraviolet (UV) radiation on marine microalgal-invertebrate symbioses. II. The synthesis of mycosporine-like amino acids in response to exposure to UV in *Anthopleura elegantissima* and *Cassiopeia xamachana*. *Journal of Experimental Marine Biology and Ecology*, **194**, 233–50.

Behrenfeld, M. J., Hardy, J. T. & Lee, II, H. (1992). Chronic effects of ultraviolet-B radiation on growth and cell volume of *Phaeodactylum tricornutum* (Bacillariophyceae). *Journal of Phycology*, **28**, 757–60.

Ben-Amotz, A., Shaish, A. & Avron, M. (1989). Mode of action of the massively accumulated β-carotene of *Dunaliella bardawil* in protecting the alga against damage by excess irradiation. *Plant Physiology*, **91**, 1040–3.

Black, H. S. (1987). Potential involvement of free radical reactions in ultraviolet light-mediated cutaneous damage. *Photochemistry and Photobiology*, **46**, 213–21.

Blakefield, M. K. & Harris, D. O. (1994). Delay of cell differentiation in *Anabaena aequalis* caused by UV-B radiation and the role of photoreactivation and excision repair. *Photochemistry and Photobiology*, **59**, 204–8.

Booth, C. R., Morrow, J. H., Coohill, T. P., Cullen, J. J., Frederick, J. E., Häder, D.-P., Holm-Hansen, O., Jeffrey, W. H., Mitchell, D. L., Neale, P. J., Sobolev, I., Van der Leun, J. & Worrest, R. C. (1997). Impacts of solar UVR on aquatic microorganisms. *Photochemistry and Photobiology*, **65**, 252–69.

Bornman, J. F. & Sundby-Emanuelsson, C. (1995). Response of plants to UV-B radiation: some biochemical and physiological effects. In *Environmental and Plant Metabolism: Flexibility and Acclimation*, ed. N. Smirnoff, pp. 245–62. BIOS Science Publications, Oxford.

Bothwell, M. L., Sherbot, D., Roberge, A. C. & Daley, R. J. (1993). Influence of natural ultraviolet radiation on lotic periphytic diatom community growth, biomass accrual, and species composition: short-term versus long-term effects. *Journal of Phycology*, **29**, 24–35.

Buckley, C. E. & Houghton, J. A. (1976). A study of the effects of near UV radiation on the pigmentation of the blue-green alga *Gloeocapsa alpicola*. *Archives of Microbiology*, **107**, 93–7.

Buma, A. G. J., Van Hannen, E. J., Roza, L., Veldhuis, M. J. W. & Gieskes, W. W. C. (1995). Monitoring ultraviolet-B-induced DNA damage in individual diatom cells by immunofluorescent thymine dimer detection. *Journal of Phycology*, **31**, 314–21.

Buma, A. G. J., Zemmelink, H. J., Sjollema, K. & Gieskes, W. W. C. (1996). UVB radiation modifies protein and photosynthetic pigment content, volume and ultrastructure of marine diatoms. *Marine Ecology Progress Series*, **142**, 47–54.

Burton, G. W. & Ingold, K. U. (1984). β-Carotene: an unusual type of lipid antioxidant. *Science*, **224**, 569–73.

Cadenas, E. (1989). Biochemistry of oxygen toxicity. *Annual Review of Biochemistry*, **58**, 79–110.

Calkins, J. & Thordardøttir, T. (1980). The ecological significance of solar UV radiation on aquatic organisms. *Nature*, **283**, 563–6.

Carreto, J. I., Carignan, M. O., Daleo, G. & De Marco, S. G. (1990a). Occurrence of mycosporine-like amino acids in the red-tide dinoflagellate *Alexandrium excavatum*: UV-photoprotective compounds? *Journal of Plankton Research*, **12**, 909–21.

Carreto, J. I., De Marco, S. G. & Lutz, V. A. (1989). UV-absorbing pigments in the dinoflagellates *Alexandrium excavatum* and *Prorocentrum micans*. Effects of light intensity. In *Red Tides: Biology, Environmental Science, and Toxicology*, ed. T. Okaichi, D. M. Anderson & T. Nemoto, pp. 333–6. Elsevier, Amsterdam.

Carreto, J. I., Lutz, V. A., De Marco, S. G. & Carignan, M. O. (1990b). Fluence and wavelength dependence of mycosporine-like amino acid synthesis in the dinoflagellate *Alexandrium excavatum*. In *Toxic Marine Phytoplankton*, ed. E. Granéli, B. Sundstrom, L. Edler & D. Anderson, pp. 275–9. Elsevier, Amsterdam.

Carroll, A. K. & Shick, J. M. (1996). Dietary accumulation of UV-absorbing mycosporine-like amino acids (MAAs) by the green sea urchin (*Stongylocentrotus droebachiensis*). *Marine Biology*, **124**, 561–9.

Cullen, J. J. & Neale, P. J. (1997). Biological weighting functions for describing the effects of ultraviolet radiation on aquatic systems. In *Effects of Ozone Depletion on Aquatic Ecosystems*, ed. D.-P. Häder, pp. 97–118. R.G. Landes, Austin, TX.

Darnell, J., Lodish, H. & Baltimore, D. (1990). *Molecular Cell Biology*, Scientific American Books Inc., New York.

Davidson, A. T., Bramich, D., Marchant, H. J. & McMinn, A. (1994). Effects of UV-B irradiation on growth and survival of Antarctic marine diatoms. *Marine Biology*, **119**, 507–15.

Davidson, A. T., Marchant, H. J. & de la Mare, W. K. (1996). Natural UVB exposure changes the species composition of Antarctic phytoplankton in mixed culture. *Aquatic Microbial Ecology*, **10**, 299–305.

Demers, S., Roy, S., Gagnon, R. & Vignault, C. (1991). Rapid light-induced changes in cell fluorescence and in xanthophyll-cycle pigments of *Alexandrium excavatum* (Dinophyceae) and *Thalassiosira pseudonana* (Bacillariophyceae): a photo-protection mechanism. *Marine Ecology Progress Series*, **76**, 185–93.

Demmig, B., Winter, K., Krüger, A. & Czygan, F.-C. (1987). Photoinhibition and zeaxanthin formation in intact leaves. *Plant Physiology*, **84**, 218–24.

Döhler, G., Hoffmann, M. & Stappel, U. (1995). Pattern of proteins after heat shock and UV-B radiation of some temperate marine diatoms and the Antarctic *Odontella weissflogii*. *Botanica Acta*, **108**, 93–8.

Dunlap, W. C. & Chalker, B. E. (1986). Identification and quantitation of near-UV absorbing compounds (S-320) in a hermatypic scleractinian. *Coral Reefs*, **5**, 155–9.

Dunlap, W. C., Chalker, B. E. & Oliver, J. K. (1986). Bathymetric adaptations of reef-building corals at Davies Reef, Great Barrier Reef, Australia. III. UV-B absorbing compounds. *Journal of Experimental Marine Biology and Ecology*, **104**, 239–48.

Dunlap, W. C. & Shick, J. M. (1998). Ultraviolet radiation-absorbing mycosporine-like amino acids in coral reef organisms: a biochemical and environmental perspective. *Journal of Phycology*, **34**, 418–30.

Dunlap, W. C. & Yamamoto, Y. (1995). Small-molecule antioxidants in marine organisms: antioxidant activity of mycosporine-glycine. *Comparative Biochemistry and Physiology*, **112B**, 105–14.

Dykens, J. A. & Shick, J. M. (1982). Oxygen production by endosymbiotic algae controls superoxide dismutase activity in their animal host. *Nature*, **297**, 579–80.

Eker, A. P. M., Hessels, J. K. C. & van de Velde, J. (1988). Photoreactivating enzyme from the green alga *Scenedesmus acutus*. Evidence for the presence of two different flavin chromophores. *Biochemistry*, **27**, 1758–65.

Favre-Bonvin, J., Arpin, N. & Brevard, C. (1976). Structure de la mycosporine (P 310). *Canadian Journal of Chemistry*, **54**, 1105–13.

Favre-Bonvin, J., Bernillon, J., Salin, N. & Arpin, N. (1987). Biosynthesis of mycosporines: mycosporine glutaminol in *Trichothecium roseum*. *Phytochemistry*, **26**, 2509–14.

Ferreyra, G. A. (1995). Effets du rayonnement ultraviolet sur le plancton des régions froides tempérées et polaires. PhD thesis, Université du Québec à Rimouski.

Foote, C. S. (1976). Photosensitized oxidation and singlet oxygen: consequences in biological systems. In *Free Radicals in Biology*, ed. W. A. Pryor, pp. 85–133. Academic Press, New York.

Foote, C. S., Chang, Y. C. & Denny, R. W. (1970). Chemistry of singlet oxygen. X. Carotenoid quenching parallels biological protection. *Journal of the American Chemical Society*, **92**, 5216–18.

Fox, M. Y., Feldman, B. J. & Chu, G. (1994). A novel role for DNA photolyase: binding to DNA damaged by drugs is associated with enhanced cytotoxicity in *Saccharomyces cerevisiae*. *Molecular and Cell Biology*, **14**, 8071–7.

Fridovich, I. (1976). Oxygen radicals, hydrogen peroxide, and oxygen toxicity. In *Free Radicals in Biology*, ed. W. A. Pryor, pp. 239–75. Academic Press, New York.

Fridovich, I. (1977). Oxygen is toxic! *BioScience*, **27**, 462–6.

Friedberg, E. C. (1985). *DNA Repair*. W.H. Freeman & Co., New York.

Friedberg, E. C., Walker, G. C. & Siede, W. (1995). *DNA Repair and Mutagenesis*. ASM, Washington, DC.

Fryer, M. J. (1993). Evidence for the photoprotective effects of vitamin E. *Photochemistry and Photobiology*, **58**, 304–12.

Garcia-Pichel, F. (1994). A model for internal self-shading in planktonic organisms and its implications for the usefulness of ultraviolet sunscreens. *Limnology and Oceanography*, **39**, 1704–17.

Garcia-Pichel, F. & Castenholz, R. W. (1991). Characterization and biological implications of scytonemin, a cyanobacterial sheath pigment. *Journal of Phycology*, **27**, 395–409.

Garcia-Pichel, F., Wingard, C. E. & Castenholz, R. W. (1993). Evidence regarding the UV sunscreen role of a mycosporine-like compound in the cyanobacterium *Gloeocapsa* sp. *Applied and Environmental Microbiology*, **59**, 170–6.

Goes, J. I., Handa, N., Taguchi, S. & Hama, T. (1994). Effect of UV-B radiation on the fatty acid composition of the marine phytoplankter *Tetraselmis* sp.: relationship to cellular pigments. *Marine Ecology Progress Series*, **114**, 259–74.

Goodwin, T. W. (1980). *The Biochemistry of the Carotenoids*. Chapman & Hall, London.

Grant, P. T., Plack, P. A. & Thomson, R. H. (1980). Gadusol, a metabolite from fish eggs. *Tetrahedron Letters*, **21**, 4043–4.

Häder, D.-P., Worrest, R. C., Kumar, H. D. & Smith, R. C. (1995). Effects of increased solar ultraviolet radiation on aquatic ecosystems. *Ambio*, **24**, 174–80.

Häder, D.-P. & Worrest, R. C. (1991). Effects of enhanced solar ultraviolet radiation on aquatic ecosystems. *Photochemistry and Photobiology*, **53**, 717–25.

Hall, D. O. & Rao, K. K. (1994). *Photosynthesis*. Cambridge University Press, Cambridge.

Helbling, E. W., Villafañe, V., Ferrario, M. & Holm-Hansen, O. (1992). Impact of natural ultraviolet radiation on rates of photosynthesis and on specific marine phytoplankton species. *Marine Ecology Progress Series*, **80**, 89–100.

Helbling, E. W., Villafañe, V. & Holm-Hansen, O. (1994). Effects of ultraviolet radiation on Antarctic marine phytoplankton photosynthesis with particular attention to the influence of mixing. In *Ultraviolet Radiation in Antarctica: Measurements and Biological Effects*, Antarctic Research Series, ed. C. S. Weiler & P. A. Penhale, pp. 207–28. American Geophysical Union, Washington, DC.

Hessen, D. O. & Soerensen, K. (1990). Photoprotective pigmentation in alpine zooplankton populations. *Aquatic Fenn*, **20**, 165–70.

Hessen, D. O., Van Donk, E. & Andersen, T. (1995). Growth responses, P-uptake and loss of flagellae in *Chlamydomonas reinhardtii* exposed to UV-B. *Journal of Plankton Research*, **17**, 17–27.

Hightower, L. E. (1993). A brief perspective on the heat-shock response and stress proteins. *Marine Environmental Research*, **35**, 79–83.

Hochachka, P. W. & Somero, G. N. (1973). *Strategies of Biochemical Adaptation*. W.B. Saunders Company, Philadelphia.

Holick, M. F., Holick, S. A. & Guillard, R. L. (1982). On the origin and metabolism of vitamin D in the sea. In *Comparative endocrinology of Calcium Regulation*, Proceedings of a satellite symposium of the IXth International Symposium on

Comparative Endocrinology, ed. C. Oguro & P. K. T. Pang, pp. 85–91. Japan Scientific Societies Press, Tokyo.

Jeffrey, W. H., Pledger, R. J., Aas, P., Hager, S., Coffin, R. B., Von Haven, R. & Mitchell, D. L. (1996). Diel and depth profiles of DNA photodamage in bacterioplankton exposed to ambient solar ultraviolet radiation. *Marine Ecology Progress Series*, **137**, 283–91.

Jialal, I., Norkus, E. P., Cristol, L. & Grundy, S. M. (1991). β-Carotene inhibits the oxidative modification of low-density lipoprotein. *Biochimica et Biophysica Acta*, **1086**, 134–8.

Karentz, D., Bothwell, M. L., Coffin, R. B., Hanson, A., Herndl, G. J., Kilham, S. S., Lesser, M. P., Lindell, M., Moeller, R. E., Morris, D. P., Neale, P. J., Sanders, R. W., Weiler, C. S. & Wetzel, R. G. (1994). Impact of UV-B radiation on pelagic freshwater ecosystems: Report of working group on bacteria and phytoplankton. *Ergebnisse der Limnologie (Archive für Hydrobiologie. Beiheft)*, **43**, 31–69.

Karentz, D., Cleaver, J. E. & Mitchell, D. L. (1991a). Cell survival characteristics and molecular responses of Antarctic phytoplankton to ultraviolet-B radiation. *Journal of Phycology*, **27**, 326–41.

Karentz, D., McEuen, F. S., Land, M. C. & Dunlap, W. C. (1991b). Survey of mycosporine-like amino acid compounds in Antarctic marine organisms: potential protection from ultraviolet exposure. *Marine Biology*, **108**, 157–66.

Karentz, D. & Spero, H. J. (1995). Response of a natural Phaeocystis population to ambient fluctuations of UVB radiation caused by Antarctic ozone depletion. *Journal of Plankton Research*, **17**, 1771–89.

Kinzie, III, R. A. (1993). Effects of ambient levels of solar ultraviolet radiation on zooxanthellae and photosynthesis of the reef coral *Montipora verrucosa*. *Marine Biology*, **116**, 319–27.

Leach, C. M. (1965). Ultraviolet-absorbing substances associated with light-induced sporulation in fungi. *Canadian Journal of Botany*, **43**, 185–200.

Lesser, M. P. (1996a). Acclimation of phytoplankton to UV-B radiation: oxidative stress and photoinhibition of photosynthesis are not prevented by UV-absorbing compounds in the dinoflagellate *Prorocentrum micans*. *Marine Ecology Progress Series*, **132**, 287–97.

Lesser, M. P. (1996b). Elevated temperatures and ultraviolet radiation cause oxidative stress and inhibit photosynthesis in symbiotic dinoflagellates. *Limnology and Oceanography*, **41**, 271–83.

Lesser, M. P. & Lewis, S. (1996). Action spectrum for the effects of UV radiation on photosynthesis in the hermatypic coral *Pocillopora damicornis*. *Marine Ecology Progress Series*, **134**, 171–7.

Lesser, M. P. & Shick, J. M. (1989). Effects of irradiance and ultraviolet radiation on photoadaptation in the zooxanthellae of *Aiptasia pallida*: primary production, photoinhibition, and enzymic defenses against oxygen toxicity. *Marine Biology*, **102**, 243–55.

Lindahl, T. (1976). New class of enzymes acting on damaged DNA. *Nature*, **259**, 64–6.

Lowe, C. & Goodman-Lowe, G. (1996). Suntanning in hammerhead sharks. *Nature*, **383**, 677.

Lukas, K. J. & Hoffman, E. J. (1984). New endolithic cyanophytes from the North Atlantic ocean. III. *Hyella pyxis* Lukas & Hoffman sp. nov. *Journal of Phycology*, **20**, 515–20.

Mathis, P. & Kleo, J. (1973). The triplet state of β-carotene and of analog polyenes of different length. *Photochemistry and Photobiology*, **18**, 343–6.

Matsunaga, T., Burgess, J. G., Yamada, N., Komatsu, K., Yoshida, S. & Wachi, Y. (1993). An ultraviolet (UV-A) absorbing biopterin glucoside from the marine planktonic cyanobacterium *Oscillatoria* sp. *Applied Microbiology and Biotechnologies*, **39**, 250–3.

Middleton, E. M. & Teramura, A. H. (1994). Understanding photosynthesis, pigment and growth responses induced by UV-B and UV-A irradiances. *Photochemistry and Photobiology*, **60**, 38–45.

Miller, R. V. (1992). recA. In *Encylopedia of Microbiology*, ed. J. Lederberg, pp. 509–17. Academic Press, Inc., San Diego, CA.

Mitchell, D. L., Scoggins, J. T. & Morizot, D. C. (1993). DNA repair in the variable platyfish (*Xiphophorus variatus*) irradiated *in vivo* with ultraviolet B light. *Photochemistry and Photobiology*, **58**, 455–9.

Moore, A. L., Joy, A., Tom, R., Gust, D., Moore, T. A., Bensasson, R. V. & Land, E. J. (1982). Photoprotection by carotenoids during photosynthesis: motional dependence of intramolecular energy transfer. *Science*, **216**, 982–4.

Mount, D. W. (1996). Reprogramming transcription. *Nature*, **383**, 763–4.

Nägeli, C. & Schwenderer, S. (1877). *Das Mikroskop*. Willhelm Engelmann Verlag, Leipzig.

Neale, P. J., Davis, R. F. & Cullen, J. J. (1998). Interactive effects of ozone depletion and vertical mixing on photosynthesis of Antarctic phytoplankton. *Nature*, **392**, 585–9.

Neale, P. J., Lesser, M. P. & Cullen, J. J. (1994). Effects of ultraviolet radiation on the photosynthesis of phytoplankton in the vicinity of McMurdo station, Antarctica. In *Ultraviolet Radiation in Antartica: Measurements and Biological Effects*, Antartic Research Series, ed. C. S. Weiler & P. A. Penhale, pp. 125–42. American Geophysical Union, Washington, DC.

Nultsch, W. & Agel, G. (1986). Fluence rate and wavelength dependence of photobleaching in the cyanobacterium *Anabaena variabilis*. *Archiv für Microbiologie*, **144**, 268–71.

O'Brien, P. A. & Houghton, J. A. (1982). Photoreactivation and excision repair of UV induced pyrimidine dimers in the unicellular cyanobacterium *Gloeocapsa alpicola* (*Synechocystis* PCC 6308). *Photochemistry and Photobiology*, **35**, 359–64.

Ozer, Z., Reardon, J. T., Hsu, D. S., Malhotra, K. & Sancar, A. (1995). The other function of DNA photolyase: stimulation of excision repair of chemical damage to DNA. *Biochemistry*, **34**, 15886–9.

Paerl, H. W., Tucker, J. & Bland, P. T. (1983). Carotenoid enhancement and its role in maintaining blue-green algal (*Microcystis aeruginosa*) surface blooms. *Limnology and Oceanography*, **28**, 847–57.

Peletier, H., Gieskes, W. W. C. & Buma, A. G. J. (1996). Ultraviolet-B radiation resistance of benthic diatoms isolated from tidal flats in the Dutch Wadden Sea. *Marine Ecology Progress Series*, **135**, 163–8.

Pennington, J. T. & Emlet, R. B. (1986). Ontagenetic and diel migration of a planktonic echinoid larva, *Dendraster excentricus* (Eschscholtz): occurrence, causes, and probable consequences. *Journal of Expermental Marine Biology and Ecology*, **104**, 69–95.

Pfeifer, G. P. (1997). Formation and processing of UV photoproducts: effects of DNA sequence and chromatin environment. *Photochemistry and Photobiology*, **65**, 270–83.

Pfündel, E. E., Pan, R.-S. & Dilley, R. A. (1992). Inhibition of violaxanthin deepoxidation by ultraviolet-B radiation in isolated chloroplasts and intact leaves. *Plant Physiology*, **98**, 1372–80.

Proteau, P. J., Gerwick, W. H., Garcia-Pichel, F. & Castenholz, R. (1993). The structure of scytonemin, an ultraviolet sunscreen pigment from the sheaths of cyanobacteria.

Experientia, **49**, 825–9.

Quiñones, M. A. & Zeiger, E. (1994). A putative role of the xanthophyll, zeaxanthin, in blue light photoreception of corn coleoptiles. *Science*, **264**, 558–61.

Reynolds, C. S., Oliver, R. L. & Walsby, A. E. (1987). Cyanobacterial dominance: the role of buoyancy regulation in dynamic lake environments. *New Zealand Journal of Marine and Freshwater Research*, **21**, 379–90.

Riegger, L. & Robinson, D. (1997). Photoinduction of UV-absorbing compounds in Antarctic diatoms and *Phaeocystis antarctica*. *Marine Ecology Progress Series*, **160**, 13–25.

Rozema, J., van de Staaij, J. & Björn, L. O. (1997). UV-B as an environmental factor in plant life: stress and regulation. *Trends in Ecology and Evolution*, **12**, 22–8.

Rüdiger, W. & López-Figueroa, F. (1992). Photoreceptors in algae. *Photochemistry and Photobiology*, **55**, 949–54.

Sanders, B. M. (1993). Stress proteins in aquatic organisms: an environmental perspective. *Critical Reviews in Toxicology*, **23**, 49–75.

Santus, R. (1983). Mécanismes moléculaires de l'action photoprotectrice des caroténoïdes 'in vivo'. *Océanis*, **9**, 91–102.

Scherer, S., Chen, T. W. & Böger, P. (1988). A new UV-A/B protecting pigment in the terrestrial cyanobacterium *Nostoc commune*. *Plant Physiology*, **88**, 1055–7.

Selby, C. P. & Sancar, A. (1990). Molecular mechanisms of DNA repair inhibition by caffeine. *Proceedings of the National Academy of Sciences, USA*, **87**, 3522–5.

Shibata, H., Baba, K. & Ochiai, H. (1991). Near UV irradiation induces shock proteins in *Anacystis nidulans* R-2; possible role of oxygen. *Plant Cell Physiology*, **32**, 771–6.

Shick, J. M., Dunlap, W. C., Chalker, B. E., Banaszak, A. T. & Rosenzweig, T. K. (1992). Survey of ultraviolet radiation-absorbing mycosporine-like amino acids in organs of coral reef holothuroids. *Marine Ecology Progress Series*, **90**, 139–48.

Shick, J. M., Lesser, M. P., Dunlap, W. C., Stochaj, W. R., Chalker, B. E. & Wu Won, J. (1995). Depth-dependent responses to solar ultraviolet radiation and oxidative stress in the zooxanthellate coral *Acropora microphthalma*. *Marine Biology*, **122**, 41–51.

Shick, J. M., Lesser, M. P. & Stochaj, W. R. (1991). Ultraviolet radiation and photooxidative stress in zooxanthellate anthozoa: the sea anemone *Phyllodiscus semoni* and the octocoral *Clavularia* sp. *Symbiosis*, **10**, 145–73.

Small, D. (1987). Repair systems for nuclear and chloroplast DNA in *Chlamydomonas reinhardtii*. *Mutation Research*, **181**, 31–5.

Smith, K. C. & Macagno, E. R. (1990). UV photoreceptors in the compound eye of *Daphnia magna* (Crustacea, Branchiopoda). A fourth spectral class in single ommatidia. *Journal of Comparative Physiology*, **A166**, 597–606.

Stochaj, W. R., Dunlap, W. C. & Shick, J. M. (1994). Two new UV-absorbing mycosporine-like amino acids from the sea anemone *Anthopleura elegantissima* and the effects of zooxanthellae and spectral irradiance on chemical composition and content. *Marine Biology*, **118**, 149–56.

Sutherland, B. M. (1981). Photoreactivation. *Bioscience*, **31**, 439–44.

Takao, M., Oikawa, A., Eker, A. P. M. & Yasui, A. (1989). Expression of an *Anacystis nidulans* photolyase gene in *Escherichia coli*: functional complementation and modified action spectrum of photoreactivation. *Photochemistry and Photobiology*, **50**, 633–7.

Taylor, R. M., Tobin, A. K. & Bray, C. M. (1997). DNA damage and repair in plants. In *Plants and UV-B: Responses to Environmental Change*, ed. P. Lumsden, pp. 53–76. Cambridge University Press, Cambridge.

Tevini, M., Braun, J. & Fieser, G. (1991). The protective function of rye seedlings against ultraviolet-B-radiation. *Photochemistry and Photobiology*, **53**, 329–33.

Tilzer, M. M. (1973). Diurnal periodicity in the phytoplankton assemblage of a high mountain lake. *Limnology and Oceanography*, **18**, 15–30.

Todo, T., Takemori, H., Ryo, H., Ihara, M., Matsunaga, T., Nikaido, O., Sato, K. & Nomura, T. (1993). A new photoreactivating enzyme that specifically repairs ultraviolet light-induced (6–4) photoproducts. *Nature*, **361**, 371–4.

Trocine, R. P., Rice, R. D. & Wells, G. N. (1981). Inhibition of seagrass by ultraviolet-B radiation. *Plant Physiology*, **68**, 74–81.

Vernet, M., Brody, E. A., Holm-Hansen, O. & Mitchell, B. G. (1994). The response of Antarctic phytoplankton to ultraviolet radiation: absorption, photosynthesis, and taxonomic composition. In *Ultraviolet Radiation in Antartica: Measurements and Biological Effects*, Antartic Research Series, ed. C. S. Weiler & P. A. Penhale, pp. 143–58. American Geophysical Union, Washington, DC.

Vernet, M., Neori, A. & Haxo, F. T. (1989). Spectral properties and photosynthetic action in red-tide populations of *Prorocentrum micans* and *Gonyaulax polyedra*. *Marine Biology*, **103**, 365–71.

Vernet, M. & Whitehead, K. (1996). Release of ultraviolet-absorbing compounds by the red-tide dinoflagellate *Lingulodinium polyedra*. *Marine Biology*, **127**, 35–44.

Villafañe, V. E., Helbling, E. W., Holm-Hansen, O. & Chalker, B. E. (1995). Acclimatization of Antarctic natural phytoplankton assemblages when exposed to solar ultraviolet radiation. *Journal of Plankton Research*, **17**, 2295–306.

Vincent, W. F. & Quesada, A. (1994). Ultraviolet radiation effects on cyanobacteria: implications for Antarctic microbial ecosystems. In *Ultraviolet Radiation in Antartica: Measurements and Biological Effects*, Antartic Research Series, ed. C. S. Weiler & P. A. Penhale, pp. 111–24. American Geophysical Union, Washington, DC.

Vincent, W. F. & Quesada, A. (1997). Microbial niches in the polar environment and the escape from UV radiation in non-marine habitats. In *Antarctic Communities: Species, Structure and Survival*, ed. B. Battaglia, J. Valencia & D. Walton, pp. 388–95. Cambridge University Press, Cambridge.

Vincent, W. F. & Roy, S. (1993). Solar ultraviolet-B radiation and aquatic primary production: damage, protection, and recovery. *Environmental Review*, **1**, 1–12.

Walsh, K. (1996). Biochemical alterations in the cyanobacterium *Microcystis aeruginosa* in response to irradiance and iron. PhD thesis. University of Newcastle, Newcastle, NSW.

Warner, C. W. & Caldwell, M. M. (1983). Influence of photon flux density in the 400–700 nm waveband on inhibition of photosynthesis by UV-B (280–320 nm) irradiation in soybean leaves: separation of indirect and direct effects. *Photochemistry and Photobiology*, **38**, 341–6.

Wise, R. R. & Naylor, A. W. (1987). Chilling-enhanced photooxidation. *Plant Physiology*, **83**, 278–82.

Wood, W. F. (1987). Effect of solar ultra-violet radiation on the kelp *Ecklonia radiata*. *Marine Biology*, **96**, 143–50.

Wood, W. F. (1989). Photoadaptive responses of the tropical red alga *Eucheuma striatum* Schmitz (Gigartinales) to ultra-violet radiation. *Aquatic Botany*, **33**, 41–51.

Worrest, R. C., Thomson, B. E. & Van Dyke, H. (1981a). Impact of UV-B radiation upon estuarine microcosms. *Photochemistry and Photobiology*, **33**, 861–7.

Worrest, R. C., Wolniakowski, K. U., Scott, J. D., Brooker, D. L., Thomson, B. E. & Van Dyke, H. (1981b). Sensitivity of marine phytoplankton to UV-B radiation:

impact upon a model ecosystem. *Photochemistry and Photobiology*, **33**, 223–7.

Xiong, F. S., Komenda, J., Kopecky, J. & Nedbal, L. (1997). Strategies of ultraviolet-B protection in microscopic algae. *Physiologia Plantarum*, **100**, 378–88.

Xiong, F. S., Lederer, F., Lukavsky, J. & Nedbal, L. (1996). Screening of fresh-water algae (Chlorophyta, Chromophyta) for ultraviolet-B sensitivity of the photosynthetic apparatus. *Journal of Plant Physiology*, **148**, 42–8.

Zudaire, L. (1999). Photoprotection et acclimation à long terme au rayonnement ultraviolet (UV). Le cas d'une diatomée marine, *Thalassiosira weissflogii*. M.Sc. thesis, Université du Québec à Rimouski.

8

○ ○

UV radiation effects on heterotrophic bacterioplankton and viruses in marine ecosystems

Wade H. Jeffrey*, Jason P. Kase and Steven W. Wilhelm

8.1 Introduction

Most research examining the effects of UV radiation (UVR) on marine microbial communities has been directed at phytoplankton and primary production. It is now apparent, however, that other microbial trophic levels must also be considered when one is investigating the ecological impact of UVR. The importance of bacterioplankton in oceanic processes has become widely recognised. Bacteria have been found to account for up to 90% of the cellular DNA in oceanic environments (Paul & Carlson, 1984; Paul, Jeffrey & Deflaun, 1985; Coffin et al., 1990) and the role of bacteria in elemental and nutrient cycling has received extensive study (Falkowski & Woodham, 1992). Bacteria have been found to play a vital role in carbon cycling, providing significant amounts of material to higher trophic levels. Various studies have shown that bacteria consume a significant proportion of primary production, although the data are quite variable and depend on location and season (Fuhrman & Azam, 1980; Hansen et al., 1983; Cota et al., 1990; Sullivan et al., 1990).

Viruses are ubiquitous and abundant in marine environments infecting bacteria, phytoplankton and heterotrophic flagellates (for reviews, see Børsheim, 1993; Fuhrman & Suttle, 1993; Bratbak, Thingstad & Heldal, 1994). While viruses are often considered to be pathogens of mammals or higher plants, it has become apparent that the 10^8 to 10^{11} viruses 1^{-1} that are found in marine surface waters, mostly infect marine bacteria and phytoplankton (Suttle, Chan & Cottrell, 1990; Fuhrman & Suttle, 1993). Estimates of the virus-induced mortality of marine bacteria range from a small percentage of production, to rates that imply that viruses are a major mechanism of bacterial mortality in aquatic systems (Proctor & Fuhrman, 1990; Suttle & Chen, 1992; Steward, Smith & Azam, 1996; Wilhelm et al., 1998a). On average, it appears that about 10% to 20% of marine bacteria are lysed by viruses on a daily basis (Suttle, 1994). This significant destruction of host cells by viruses may affect species diversity, nutrient recycling and carbon transfer in pelagic systems (Fuhrman, 1992; Murray & Eldridge, 1994). Despite the information that has been accumulated on viruses, the absolute magnitude of their impact on aquatic systems remains uncertain. Regardless of their ultimate impact, the factors that regulate the abundance of viruses in marine systems will have a direct influence on the role of viruses in marine microbial food webs.

Marine micro-organisms may be affected by UV both directly and indirectly. The close interactions among bacteria, viruses and phytoplankton in the microbial loop (Azam *et al.*, 1983) suggest that, if one group of organisms is directly affected by UVR (e.g. DNA photodamage), this will indirectly affect carbon cycling processes in the other groups. For example, it might be expected that a decrease in phytoplankton production would result in a decline in bacterial production that may be compounded by direct UVR effects on bacterioplankton. Bacteria, in concert with their predators (viruses and eukaryotes), fulfil a crucial nutrient recycling role in the water column. Likewise, it is plausible that some fraction of diminished phytoplankton production attributed to UVR may be the result of reduced bacterial nutrient cycling. Thus, the response of bacterioplankton to UVR may be different in the presence or absence of phytoplankton. UVR may directly impact on viruses, bacteria, phytoplankton or zooplankton via direct DNA damage and reduced rates of production. Direct effects on one trophic group may result in an indirect impact on others. In addition, DOM may be photochemically altered to produce more labile substrates that might increase bacterial production (Miller & Moran, 1997). In contrast, these photochemical reactions may also produce reactive free radicals, which might inhibit growth of any or all of these trophic groups.

This chapter attempts to compile the state of knowledge regarding UVR effects on bacterioplankton and viruses, identify important considerations regarding the effect of ozone depletion on production, and provide a foundation for discussions about potential impact on the planet's biogeochemical cycles.

8.2 Early studies

Early work on the effects of light on bacteria focused on the sensitivity of bacterial cultures to UV-C (200–280 nm). Lamps that emitted wavelengths < 290 nm were used because of their high energy and efficiency in causing significant effects. This work revealed that the biological response of bacteria was nearly identical with the absorption spectrum of DNA and that direct modification of DNA is the primary cause of the lethal and mutagenic effects seen when micro-organisms are exposed to far UVR (Peak & Peak, 1995). While this work formed the basis for much of our mechanistic understanding of DNA damage induction, the ecological relevance of this work is limited by the fact that much of the UVR spectrum used in these experiments (< 290 nm, UV-C) is filtered out completely in the atmosphere, is not part of the solar spectrum reaching the surface of the earth and, thus, is of no consequence in environmental systems (Mitchell, 1995).

Environmental engineers concerned with the persistence of pathogenic and enteric bacteria released with sewage into the marine environment suspected that sunlight may play a role in determining the lifetime of such organisms once released. However, the idea that solar radiation could effect bacteria was dismissed by many researchers on the assumption that only wavelengths < 290 nm were detrimental to bacteria and direct damage to nucleic acids was the only biological effect of UVR (Stanier & Cohen-Bazire, 1957; Krinsky, 1976; Hendricks, 1978). Gameson & Saxon (1967) were the first to demonstrate that sunlight could significantly effect the survival of coliform bacteria. Following this, other studies demonstrated that sunlight could significantly reduce the viability of *Escherichia coli* and other indicator organisms of faecal contamination in the marine environment (Fujioka *et al.*, 1981; Kapuscinski & Mitchell, 1983; Evison, 1988). Viability, reflected by the T_{90} (time required to reduce the viable number by 90%) for indicator bacteria in seawater was often reduced from a few days to only a few hours when the organisms were exposed to sunlight (Fujioka *et al.*, 1981; Solic & Krstulovic, 1992). In addition, inhibition was demonstrated to be due partly due to wavelengths > 370 nm and to occur at depths of up to 30 m (Kapuschinski & Mitchell, 1983; Pommepuy *et al.*, 1992; Solic & Krstulovic, 1992). Taken together, these studies suggest that sunlight is a significant factor controlling the survival and persistence of *E. coli* and other indicator bacteria in the marine environment. Furthermore, these results imply that solar radiation may play equally important roles in the regulation of natural populations of bacterioplankton.

When ozone depletion and possible increases in UV-B became a concern in the mid 1980s attention was focused once again on the effect of solar radiation on micro-organisms in the marine environment. Much of the work was directed at understanding the impact of ozone depletion on the photosynthetic micro-organisms that drive nearly all the carbon fixation in the oceans and the possible ecological implications of increased UV-B on primary production. Compared to reports of UVR effects on phytoplankton, however, studies of the impact and effects of UVR on bacterioplankton are in their relative infancy. The concept of the 'microbial loop' (Azam *et al.*, 1983) introduced the idea that natural communities of heterotrophic bacterioplankton are intimately involved in the flow of carbon through marine ecosystems and help to drive productivity in the marine environment. Recognising that heterotrophic bacterioplankton play an important role in the transfer of biomass between primary and secondary production, a basic understanding of the effects of solar UVR on production in natural assemblages of heterotrophic bacterioplankton must exist before an assessment of the impacts of ozone depletion on carbon cycling can be made. Despite the awareness that significant numbers of bacterioplankton involved in the recycling of carbon are located in the surface waters and, thus, are subjected to solar radiation, relatively few studies have examined the effect of solar radiation on the productivity of these populations. A 1994 bibliography of > 300 papers on the photobiology of aquatic bacteria, microalgae and macroalgae (Karentz *et al.*, 1994) included only 5% devoted to studies on bacteria and none on viruses.

8.3   Solar UVR effects on bacterioplankton production

A seasonal study of the effect of solar radiation on the uptake of amino acids by Chesapeake Bay bacteria determined that the number of bacteria capable of taking up tritiated amino acids (^{3}H-AA) decreased after exposure to solar radiation relative to samples incubated in the dark. The relative number of cells taking up amino acids increased when Mylar (excludes UV-B) and UF-3 (excludes UV-A and UV-B) filters were used during incubations, suggesting that UV-A and UV-B inhibited the uptake viability of bacterioplankton (Bailey *et al.*, 1983). Sieracki & Sieburth (1986) found that sunlight delayed growth in marine bacteria. This delay was decreased when filters were used to exclude UV-B or UV-B and UV-A from samples exposed to sunlight. Because the decrease in growth delay was greatest when UV-A was excluded, they concluded that UV-A was primarily responsible for the lag in growth seen when samples are exposed to solar radiation. The viability, based on colony-forming units (CFU), of a natural bacterial assemblage and two marine isolates from

Antarctic waters decreased significantly when exposed to UVR (Helbling *et al.*, 1995). The inhibition of viability due to UV-A was consistently higher than UV-B in all cases but, interestingly, there were marked differences in the tolerance of the natural assemblage and two isolates to solar radiation, suggesting heterogeneity in sensitivity among species.

Dissimilar response to solar UVR can also be demonstrated by the survival curves of two bacterial isolates, *Vibrio natriegens* and *Pseudomonas stutzeri* (Figure 8.1). Variability between species may be indicative of variable responses seen in natural communities in dissimilar environments (see below). This variability illustrates the difficulty associated with using the response of isolates or even mixed cultures to make conclusions about natural assemblages and points to the need to use methods that emphasise a 'community' approach to determine the effect of solar UVR on bacterioplankton growth and productivity.

The use of radiolabelled precursor molecules such as ^{3}H-thymidine (TdR; Fuhrman & Azam, 1982) and [3]H- or ^{14}C-leucine (Leu) (Chin-Leo & Kirchman, 1988; Simon & Azam, 1989) is a common method for

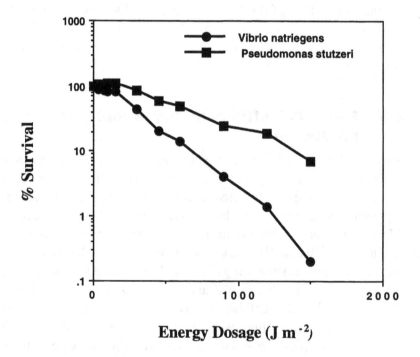

Figure 8.1.   UV-B survival curves of marine bacterial isolates. Cultures were grown to stationary phase, harvested by centrifugation, and resuspended in nutrient-free media. After exposure to the indicated fluence, cells were diluted and plated on nutrient agar plates. All treatments were done in subdued light to minimise photoreactivation.

estimating bacterial activity and production. The method is based on the assumption that exogenously supplied ^{3}H-TdR or ^{3}H-Leu is taken up by actively growing bacteria and incorporated into DNA or protein, respectively. Although these methods are not without their limitations (for a review, see Robarts & Zohary, 1993), they remain widely used and accepted. In most experiments to date, broadband cut-off filters (e.g. Mylar 500D, UF-3 Plexiglas) have been used to exclude selectively UV-B or UV-B + UV-A. The rate of incorporation of radiolabelled substrate can be compared among treatments and relative to a control sample incubated in the dark. Using this method, Herndl, Müller-Niklas & Frick (1993) demonstrated that incorporation of TdR was inhibited relative to a dark control in surface water samples taken from the northern Adriatic Sea after 4 h incubations. Forty-eight per cent of the inhibition was attributed to UV-B, as determined by incubations under Mylar, while approximately 40% was caused by longer wavelengths (UV-A and photosynthetically available radiation (PAR)). Aas et al. (1996) sought to define further the spectral sensitivity of TdR and Leu incorporation in natural bacterial populations. Incubations with TdR and Leu of surface waters from a mesotrophic estuary were performed on six separate occasions. Following exposure for 6.5 to 9 h, significant differences in inhibition of incorporation of TdR and Leu relative to a dark control for treatments in full sunlight and with UV-B excluded were apparent. Inhibition of TdR incorporation by full sunlight was 44%. UV-B contributed to 39% of this total inhibition, followed by UV-A at 37% and PAR at 23%. In contrast, Leu incorporation was inhibited 29% by full sunlight; 83% due to UV-B and 17% due to UV-A. There was a net stimulation of Leu incorporation of 10% (110% of a dark control) when only PAR was included in the treatment. Similar experiments with natural assemblages from surface water samples collected in the northern Adriatic Sea demonstrated approximately 70% inhibition of ^{3}H-TdR and ^{14}C-Leu incorporation against a dark control. Inhibition was not due to UV-B, however, but was almost equally divided between UV-A and PAR (Sommaruga et al., 1997). The minimal contribution of UV-B to the inhibition of incorporation found in this study conflicts with the results of Aas et al. (1996), who found that the contribution of UV-B to total inhibition was nearly equal to or greater than the contribution of UV-A for ^{3}H-TdR and ^{3}H-Leu, respectively.

Several reasons have been suggested for the apparent difference in spectral sensitivity between and within experiments. These reasons generally fall into one of two categories: (a) methodological or

(b) environmental. Foremost among methodological differences is whether exposures are conducted in conjunction with radiolabelled substrate or if incubations are performed following exposure. It would be preferable to allow exposure and uptake to occur together, but inconsistent results have been noted when this procedure is used and may account for the some of the variability in UVR response (Sommaruga *et al.*, 1997). An alternative is to follow exposures by short incubations in the dark. Very little recovery of incorporation has been shown to occur during the first hour in incubations with ^{3}H-TdR and ^{14}C-Leu performed in the dark after exposure to sunlight for 3 h (Kaiser & Herndl, 1997). This method has been suggested to prevent any photodegradation of radiolabelled substrate due to solar exposure (Sommaruga *et al.*, 1997). However, this procedure may not be appropriate when bacterial productivity is particularly low and requires > 1 h incubations to achieve statistically significant incorporation rates. Incubations lasting 3 h have shown recovery (~ 20%) when performed in the dark (Kaiser & Herndl, 1997). This could cause inhibition to be underestimated. Further investigation into the stability of TdR and Leu during exposure to UVR and the dynamics of recovery processes may help to resolve any methodological reasons for differences in apparent sensitivity.

Differences due to environmental factors may not be as easy to resolve but are particularly important in the ecological sense. It has been suggested that differences in inhibition susceptibility may be related to bacterial community composition, nutrient status, dissolved organic matter (DOM), presence and composition of primary producers, level of solar radiation, composition of solar spectrum, and other factors (Aas *et al.*, 1996). Photoadaptation or changes in community structure could influence the sensitivity of the community to solar radiation. Helbling *et al.* (1995) noted a large difference in the reduction of viability between two marine isolates due to UV-B after exposure to incident solar radiation. Filtering whole water samples to reduce grazers and primary producers resulted in generally lower incorporation rates of ^{3}H-TdR and ^{3}H-Leu in the dark and often resulted in greater inhibition in samples exposed to sunlight, suggesting an increased sensitivity to UVR when eukaryotes are reduced (Aas *et al.*, 1996) or an effect caused by changes in dissolved organic carbon availability. In addition, inhibition due to UV-A and UV-B of ^{3}H-TdR and ^{14}C-Leu incorporation against a dark control in cultured freshwater bacterial assemblages was significantly reduced in the presence of a green algae (Sommaruga *et al.*, 1997).

The decreased sensitivity of bacterioplankton to UVR in the presence of primary production may be related to growth rates. Phytoplankton are fairly 'leaky' cells and tend to exude small organic compounds during normal growth that can be readily utilised by bacteria (Sondergaard, Riemann & Jorgenesen, 1985). The utilisation of these algal exudates by bacteria may increase growth rates with a concomitant reduction in UVR sensitivity. The addition of a solution of C, N and P or an algal extract to surface water samples in the Gulf of Mexico was used to stimulate growth prior to solar exposure. Reduced inhibition of ^{3}H-TdR incorporation due to PAR, UV-A and UV-B was observed in the faster growing cells (J. P. Kase, unpublished data). In another experiment, seawater samples from the Yucatán peninsula were amended with casamino acids and incubated in the dark for 12 h to stimulate growth. This resulted in TdR incorporation rates six-fold greater than unamended samples after 12 h. Solar inhibition of TdR incorporation then dropped from 60% in the unamended sample to 10% in the faster growing samples when compared with their respective dark controls (W. H. Jeffrey, unpublished data). These data suggest that faster growing bacterioplankton may be less sensitive to solar UVR than slower growing or starved bacteria, and that the presence of primary producers (Aas et al., 1996; Summaruga et al., 1997) may be an important source of these nutrients. The significance of nutrient availability and growth rate on UVR sensitivity remains to be determined, but this preliminary information warrants a closer look at the environmental factors and their potential impact on bacterioplankton production.

The ecological significance of solar radiation effects on bacterioplankton must be considered with regard to the attenuation of light through this column. As solar UVR penetrates through the water column, it is attenuated by the presence of DOM and chlorophyll (Jerlov, 1976; Smith & Baker, 1978; Kirk et al., 1994). The rate of attenuation is higher for short wavelength UVR (UV-B) and lower for longer wavelengths (UV-A) (Smith & Baker, 1979). In the clearest ocean waters, the depth at 1% surface irradiance for 305 nm can be in the range of 20 m or more (Herndl, 1997; Wilhelm et al., 1998a; W. H. Jeffrey, unpublished data) while in coastal environments penetration of 305 nm can be limited to the upper few metres. Because of the change in quality and quantity of light through the water column, UVR effects at the surface will not apply uniformly through the euphotic zone. Therefore, it is important to assess the effects of changing irradiance regimes, with depth in the water column, on bacterial production.

Loss of viability, measured by the number of CFUs compared to dark incubations, in a natural assemblage and two marine isolates diminished exponentially with depth in Antarctic waters (Helbling *et al.*, 1995). Although the depth at which no significant loss of viability varied, most of the viability lost was due to UV-A. *In situ* measurements of ^{3}H-TdR incorporation in the Gulf of Mexico on three different occasions demonstrated a decrease in inhibition with depth (Aas *et al.*, 1996). In addition, total inhibition extended beyond the depth to which biologically effective UV-B penetrated (as determined by dimer formation in DNA dosimeters; Aas *et al.*, 1996). Depth profiles done in a clear, high mountain lake showed inhibition of TdR and Leu incorporation was primarily due to UV-A at a depth of 2.5 m, and that a net stimulation of TdR incorporation occurred at 8.5 m (Sommaruga *et al.*, 1997). Using neutral density screens to attenuate surface sunlight, Herndl (1993) predicted that UV-B would have a inhibitory effect at > 10 m in the water column. However, neutral density screens attenuate all wavelengths equally and do not simulate the attenuation occurring in the water column and so may bias predictions to shorter wavelengths. This demonstrates the difficulty in trying to understand biological effects that are so intimately related to light attenuation. Currently available data suggest that inhibition due to UV-A extends deeper in the water column than that due to UV-B and may have a significant effect on bacterial production through the water column.

Understanding the effects of solar UVR on bacterial production in the water column is further complicated by mixing. As bacteria are mixed through the water column the instantaneous irradiance is constantly changing. As bacteria are mixed downward the amount of inhibitory UVR diminishes and may allow repair processes to dominate. As they are mixed upward and UVR increases, inhibitory effects may dominate or overwhelm any repair processes. Furthermore, within the UVR spectrum the rate of change in irradiance as depth changes is not uniform; that is, as bacteria are mixed down UV-B diminishes at a faster rate than that in UV-A. Conversely, the rate of increase in UV-B is greater than UV-A as bacteria are mixed upward. This non-uniform change in irradiance may become important in the context of recovery processes that occur with mixing (see below).

8.4   UVR effects on microbial biogeochemical cycles

Solar radiation is a primary driving force for biogeochemical cycles. Radiation in the UV and visible wavelengths is diverted into photochemical and photobiological processes that affect global biogeochemical cycles. Because global biomass production is roughly balanced between aquatic and terrestrial ecosystems and because only 0.5% of surface waters are fresh, marine ecosystems are by far the most important in terms of aquatic carbon cycles. Phytoplankton are a major sink for atmospheric CO_2; therefore factors that affect primary production in oceanic environments (e.g. increased UVR) may contribute to global warming (Bowes, 1993; Melillo *et al.*, 1993; Häder *et al.*, 1995). Biogeochemical cycles can be accelerated by photo-oxidative processes or inhibited by detrimental effects on organisms that carry out these processes. Upper water column carbon cycling is intimately linked to levels of phytoplankton production, heterotrophic production and DOM photo-oxidation, each of which may be directly affected by UVR (Häder *et al.*, 1995). In addition, it has been suggested that UVR controls the activity of bacteria that are responsible for biogeochemical transformations in nitrogen and phosphorus cycles. Hooper & Terry (1974) reported photoinactivation of ammonia oxidation by *Nitrosomonas* sp. and light has been shown to inhibit nitrification in sea surface films (Horrigan, Carlucci & Williams, 1981). More recently, Guerrero & Jones (1996) reported UVR inhibition of both ammonia and nitrite oxidation. In contrast to direct negative effects of UVR, solar radiation may indirectly stimulate bacterial productivity by increasing the bioavailability of recalcitrant DOM via photo-oxidation (Lindell, Granéli & Tranvik, 1995, Miller & Moran, 1997; Moran & Zepp, 1997). Photodegradation of DOM in the upper ocean has other multiple effects on biogeochemical cycles ranging from changes in penetration of UV-B to enhanced formation of dissolved inorganic carbon and carbon monoxide. Sulfur cycling may be affected by changes in UV-B that result in changes in air–sea flux of carbonyl sulfide and dimethylsulfide (Kiene, 1990). Trace metals and bioavailability of trace nutrients, such as iron and manganese, are also sensitive to UVR changes (Zepp, Callaghan & Erickson, 1995). As a result, solar-induced photochemistry plays a very important role in determining organic and inorganic nutrient availability in the oceans and, therefore, solar radiation may be an indirect regulator of growth rates.

8.5   Effects of UVR on nucleic acids

Direct biological effects of UVR result from absorption of specific wavelengths by specific macromolecules and the dissipation of the absorbed energy via photochemical reactions (Mitchell, 1995). Cellular targets of UVR include nucleic acids, proteins, membrane lipids, the cytoskeleton, and photosystem II (Vincent & Roy, 1993; Mitchell, 1995). Non-DNA chromophores (UVR-absorbing compounds) can absorb

UVR to produce photoproducts that react with target macromolecules and produce damage. With an absorption spectrum that extends into UV-B, DNA itself is a chromophore below 320 nm, and direct absorption by DNA of these shorter wavelength photons can produce lesions in DNA (Peak & Peak, 1995). Other cellular components have strong absorption in UV-A and UV-B portions of the solar spectrum and can act as primary chromophores within the cell. UV-absorbing molecules can be involved in photosensitised reactions in which they absorb a photon's energy and become chemically reactive sensitiser molecules capable of reacting with DNA (Peak & Peak, 1995).

Because the UV-A portion of the solar spectrum comprises a substantially large proportion of the total UVR energy in sunlight reaching the biosphere (Mitchell, 1995), it would seem that UV-A should account for most of the damage caused by UVR and thus be responsible for a significant proportion of the damaging effects of sunlight on the environment. However, lesions introduced into DNA are not equally efficient at causing cell death or mutation. Inhibition of DNA and RNA synthesis by DNA photodamage results in many of the lethal effects of UVR. Lesions that block DNA synthesis are lethal, whereas those that allow polymerase bypass are potentially mutagenic and serve to increase the genetic instability of the organism (Brash *et al.*, 1987). The nature of the lesions introduced by UV-A and UV-B can differ and the properties of these lesions are important in determining the biological effect of wavelengths in these regions.

Dimerisations between adjacent pyrimidine bases are the most prevalent photoreactions resulting from the direct action of UVR on DNA (Mitchell, 1995). The two major photoproducts are the cyclobutyl pyrimidine dimer (CPD) and the pyrimidine (6–4) pyrimidinone photo-product ((6–4) photoproduct), which is converted to its valence photoisomer, the Dewar pyrimidinone (Mitchell, 1995). Formation and structural differences of the (6–4)/Dewar isomer photoproducts and CPDs are significant and determine their different molecular and biological effects (Mitchell & Nairn, 1989). Both CPDs and (6–4) photoproducts can inhibit DNA synthesis and gene transcription, but because of structural differences, the (6–4) photoproduct is 300-fold more efficient at blocking the progression of DNA polymerase than is the CPD. Although induced at lower frequencies, the (6–4) photoproduct and Dewar isomer may be responsible for many of the lethal effects of UV-B radiation.

A common pathway for the indirect damage of DNA by UVR begins with the absorption of a photon by a sensitiser molecule. Common sensitisers in the cell include porphyrins, nicotinamide coenzymes (NADH, NADPH), flavins and tRNA (Eisenstark, 1987; Peak & Peak, 1995). The photon energy excites the molecule and puts it in a highly reactive state capable of combining directly with a target molecule. More often in the cell, this excited sensitiser reacts with an intermediary molecule that then reacts with DNA. Molecular oxygen is the primary intermediate substrate and its photo-oxidation leads to the generation of chemical intermediates such as oxygen and hydroxyl radicals (reactive oxygen species), which interact with DNA to form modified bases, strand breaks, alkali-labile sites and DNA protein cross-links (Mitchell, 1995). In addition to strand breaks, reactive oxygen species can cause base modifications that may be lethal or mutagenic (Demple & Harrison, 1994).

Membrane damage in *E. coli* due to near UV light (290–400 nm) can be severe, with a peak in its action spectrum at 334 nm (Klamen & Tuveson, 1982; Kelland, Moss & Davies, 1984). Membrane damage, however, appears to be outweighed by damage to DNA (Eisenstark, 1987). Although membrane damage by UVR may not be a significant lethal effect, it has been shown that UVR-induced membrane damage is exacerbated by the presence of inorganic salts and can alter permeability and uptake rates of substrates (Kubitschek & Doyle, 1981; Klamen & Tuveson, 1982). These changes may result in an increased sensitivity of sublethal responses to UVR reflected in decreased production or growth rate and should be kept in mind when one is measuring the effect of light on a sublethal endpoint.

Owing to the interest in the effect of increased UV-B associated with ozone depletion, the study of molecular effects of solar UVR on natural bacterioplankton has focused on the induction of DNA dimers and the environmental factors controlling it. Jeffrey *et al.* (1996a,b) measured the induction of CPDs over diel cycles, as a function of cell size, and as a function of depth in the Gulf of Mexico. On two consecutive days, water column samples were collected in the morning and evening at depths ranging from the surface to 25 m and filtered onto 0.8 μm and 0.22 μm pore-size filters for subsequent extraction and quantification of CPDs using a radioimmunoassay (for details, see Mitchell, Haipek & Clarkson, 1985). Skies were cloudless on both days, but sea state differed significantly. One day was completely calm, while the other day was marked by 15 knot winds and seas to approximately 1 m. UVR penetration and total flux for both days were not significantly different and potential for DNA damage, as determined from DNA dosimeters (calf thymus solutions of DNA in quartz tubing), was very similar on both days. In general, damage in the dosimeters was greatest near the surface and decreased exponentially with depth, similar to the attenuation of UV-B radiation. On the calm day, damage accumulation in the bacterioplankton in the water column followed a similar pattern and could be seen to accumulate to approximately 10 m. In contrast, when seas were approximately 1 m in height, minimal net DNA damage occurred over the course of the daylight hours. More recently, similar patterns of DNA damage have been in surface waters of the Southern Ocean (Jeffrey *et al.*, 1997). The results suggest that wind-driven surface mixing may have a significant impact on the distribution of DNA damage to micro-organisms in the water column and biological effects of UVR may not necessarily be predicted by patterns of UVR attenuation in the water column.

The cell size-fraction distribution of DNA damage in surface waters has been examined in the Gulf of Mexico (Jeffrey *et al.*, 1996a,b), coral reef waters (Lyons *et al.*, 1998) and in the Southern Ocean (Jeffrey *et al.*, 1997). Surface water samples collected in the late afternoon were passed through a 120 μm Nytex mesh screen and filtered through either a 0.8 μm or 0.2 μm pore-size Supor filter. In addition, the filtrate from the 0.8 μm pore-size filter was then passed through a 0.22 μm pore-size filter. DNA damage is expressed as CPDs per unit DNA (usually per mega base) and so is independent of the amount of DNA present in each size fraction. In the vast majority of samples, DNA damage was greater in the <0.8 μm fraction (presumably heterotrophic bacterioplankton) than in the >0.8 μm fraction, and the amount of damage in the 'whole' water fraction was also often greater than the >0.8 μm fraction but less than the <0.8 μm fraction. At many stations, the amount of damage was approximately twice as great in the smaller size fraction than in the larger fraction. The results confirm that bacterioplankton are often more damaged by UV-B than the phytoplankton fraction and confirm the need to understand further the impact of UV-B on all of the microbial trophic levels.

8.6   DNA damage repair

The previous section demonstrated that significant DNA damage does occur in bacteria in the surfaces of marine waters. Obviously, marine bacteria possess some mechanism(s) to offset the damage incurred by exposure to solar radiation. This is evident by the mere abundance of active bacterioplankton in the surface waters of the ocean. The existence of repair mechanisms in prokaryotes such as *E. coli* is well documented. Available information seems to indicate that DNA damage in bacteria is removed by two major repair pathways: nucleotide excision repair or photoreactivation.

Photoreactivation is a light-dependent process where CPDs are reversed by the binding of a low molecular weight protein (e.g. photolyase, encoded by the *phr* gene in bacteria) to a CPD followed by absorption of photons within the UV-A/visible range of light, reversal of the damage and, finally, release of the enzyme. As this is an ATP-independent process, it occurs at no metabolic cost to the cell. Photoreactivation was originally described as the restoration of infectivity to UVR-irradiated viruses upon exposure to visible light in the presence of bacteria (Dulbecco, 1949, 1950). It has since been resolved that the blue light (380–450 nm) activated enzyme photolyase is capable of reversing dimerisation in UV-B damaged cells, as well as in infecting viruses (Kim & Sancar, 1993; Friedberg, Walker & Siede, 1995). Photolyase has been shown to be ubiquitously distributed throughout the prokaryotic and eukaryotic communities (Kim & Sancar, 1993; Todo *et al.*, 1996), but quantification of its potential impact on natural systems is limited.

The nucleotide excision repair pathway is well studied in bacteria (i.e. the *E. coli uvrABC* excinuclease) (Mitchell & Karentz, 1993; Mitchell, 1995). The *recA* gene encodes a protein, RecA, which is central to the processes of generalised homologous recombination and DNA repair in bacteria (Miller, 1992). In many species, RecA is a regulatory element that acts as a positive effect or as part of global regulatory networks (referred to as the SOS regulon in *E. coli*) that are induced in response to DNA-damaging agents such as UVR (Miller & Kokjohn, 1990; Miller, 1992). The ubiquitous nature and high genetic conservation have made *recA* an ideal candidate for studies of UVR-induced gene expression in the environment (Miller *et al.*, 1999). The genes of the SOS system are either silent or only minimally transcribed until the cell is exposed to UV radiation or other DNA-damaging or DNA-replication-inhibiting stressors (Walker, 1984). Thus, the levels of mRNA for these genes are greatly increased following exposure to these stressors.

Diel patterns of CPD accumulation have been observed in the surface waters of the Gulf of Mexico (Jeffrey *et al.*, 1996a). Surface water was collected every 2 h and filtered onto 0.8 μm and 0.22 μm filters and assayed for CPDs. A lag in CPD accumulation in both fractions as irradiance increased characterised the first several hours of daylight until approximately noon when a rapid accumulation in CPDs was observed that peaked at 16:00 h. This was followed by a 'plateau' or equilibrium in which net accumulation of CPD was zero for 2 h. After 18:00 h (sunset) a rapid decline in CPDs occurred. In a subsequent experiment, surface waters were collected at regular intervals for 48 h in the central Gulf of Mexico at summer solstice and during very calm seas, i.e. maximal UVR conditions (Jeffrey *et al.*, 1996b). Again, bacterioplankton in surface waters were observed to accumulate and repair DNA damage over the course of a 48 h diel sampling. Damage was seen to begin to accumulate by mid-morning and maximal damage was seen to occur in late afternoon each day. Damage was removed (repaired) during the course of the night back to the lowest daily levels just prior to sunrise. While the majority of damage was repaired by sunrise on the second day, there were successively greater net DNA photoproducts on the mornings of the second and third day. This apparent inability to repair the accumulated damage may have been due to the fact that sampling occurred during the summer solstice in calm, highly oligotrophic waters. This maximum UVR fluence may have induced more damage than normal repair mechanisms could accommodate.

Several lines of evidence suggest that the reduction in net DNA damage during the night is the result of repair processes and not due to growth of non-damaged cells. Diel rates of ^{3}H-thymidine incorporation indicate that bacterial growth during the periods of maximal repair could account for only a small percentage of the bacterial population being replaced (Jeffrey *et al.*, 1996b; W. H. Jeffrey, unpublished observations). In addition, a series of diel studies were conducted examining the expression of the DNA damage repair gene *recA* and observed that maximal expression occurs just after sunset (Miller *et al.*, 1999). An example of diel expression of *recA* is presented in Figure 8.2 in which transcription of the gene is seen to peak soon after sunset, presumably to drive nucleotide excision repair during the night.

Although the diel data support the presence of strong dark repair (nucleotide excision repair) processes, there is additional evidence that suggests that photoreactivation is active during daylight hours. These CPD data represent net dimer accumulation during daylight; that is, damage induced minus any dimers removed via repair processes, including photoreactivation. Removal of CPDs by photoreactivation was examined using surface water collected from the northern Gulf of Mexico. Surface water was dispensed into replicate 20 l Nalgene pans and placed in a flowing seawater table to maintain ambient seawater temperature. DNA damage was allowed to accumulate in all samples until 14:00 h, at which time the contents of one pan were filtered for DNA damage. Remaining pans were divided into each of three treatments. One set of pans was allowed to remain uncovered to continue to accumulate DNA damage, while another was covered with Acrylite OP3 (400 nm cut-off) to prevent additional damage but to allow photoreactivation. A third was placed in the dark to allow only excision repair of existing DNA damage. At regular intervals, contents of pans from each treatment were filtered for subsequent CPD determination. Although DNA damage has been observed to be rapidly repaired during the night, less is known about the dynamics of repair processes occurring during sunlight when photoreactivation appears to outperform dark repair mechanisms in bacterioplankton (Figure 8.3).

Kaiser & Herndl (1997) investigated the recovery of natural bacterial assemblages after exposure to artificial UV-B sources and natural radiation. In the laboratory, they exposed $<0.8\,\mu$m filtered seawater to UV-B levels typical for their location (northern Adriatic Sea) for 2 h and then divided samples for 2 h incubations in the dark, PAR, or UV-A only. Exposure to UV-B for 2 h resulted in $\sim 35\%$ decline in bacterial activity. Exposure to UV-A for only 0.5 h following exposure to UV-B resulted in significantly higher recovery than incubations done in the dark after

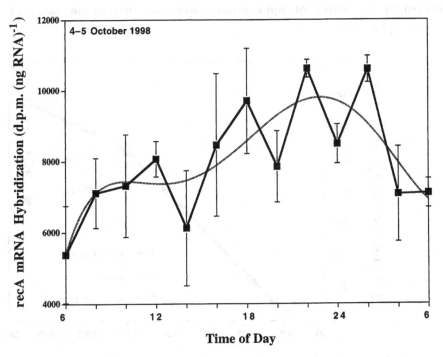

Figure 8.2. Diel pattern of *recA* transcription in the Gerlache Strait, Antarctica. Surface water samples were collected at 4 h intervals and filtered for specific mRNA detection by the method of Jeffrey, Nazaret & Von Haven, (1994). d.p.m., disintegrations per minute.

exposure. Recovery was measured as the change in TdR incorporation after treatment against incorporation prior to exposure. Similar experiments were conducted under natural solar radiation. Samples were incubated at the surface for 3 h around noon and then transferred to the dark, wrapped in Mylar, or left exposed to full sunlight. After 3 h of exposure to full sunlight, TdR incorporation was inhibited ∼40%. Following another 3 h of incubation under the different treatments, samples incubated under Mylar had significantly greater activity than samples incubated in the dark or exposed to full sunlight during one set of experiments. Another set of experiments failed to show a significant difference between light, dark or Mylar treatments. They concluded that UV-A and PAR could be used efficiently to recover from UVR stress, based on the recovery of TdR and Leu incorporation.

It appears that photoreactivation can be a significant DNA damage repair process during solar exposure and that reversal of DNA damage may partially offset the inhibitory effects of UV-B in terms of production. Photoreactivation may be a relevant process to consider in discussions regarding the effect of light attenuation and mixing on rates of bacterioplankton production.

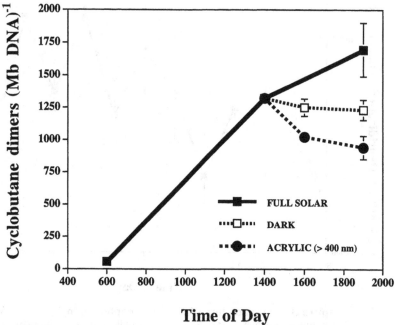

Figure 8.3 DNA damage repair in surface waters from the Gulf of Mexico. Mb, megabase.

There is significant evidence that bacterioplankton possess and utilise nucleotide excision repair and photoreactivation to recover from exposure to UVR, suggesting that UVR lethality may be minimal. Of ecological significance, however, is the energy expenditure used to overcome UVR stress. Energy spent in recovery is unavailable for normal ecological processes mediated by bacteria. These effects may be manifested (and detectable) by changes in growth efficiency. Using oxygen consumption rates, Pakulski *et al.* (1998) have recently reported reduced growth efficiency in bacterioplankton from coral reef environments after exposure to UVR. Reduced growth efficiencies have also been observed in tropical coastal waters by examining respiration of radiolabelled substrates (C. Suttle, W. Jeffrey & H. Maske, unpublished data). These results suggest that in addition to the photoinhibition of growth and potential lethality, the effects of UVR may be subtler and warrant further investigation.

8.7   Influence of UV radiation on viruses – destruction rates of particles and infectivity

It has now been well established that UVR is a major cause of viral destruction in marine surface waters. Among the constituents of a virus most sensitive to UVR are the nucleic acids, where the information that allows the virus to commandeer a host and produce new viral progeny is stored. While UVR can effectively destroy the entire viral particle, it has become apparent that the infectivity of the viral particle is also important, as a virus that is non-infective cannot replicate or destroy host cells (Suttle *et al.*,1993; Wilhelm *et al.*, 1998a,b). To resolve the impact of UVR on viral infectivity, researchers have utilised viral isolates as tracers: viruses for which the hosts are known and in culture. The problem in studying the natural viral community is a continuing inability to culture the majority of marine bacteria and phytoplankton that serve as viral hosts. Model systems using entire natural viral communities for the study of UVR on bacterioplankton therefore remain rare. The use of tracers circumvents this problem as it allows us to ascertain the infectivity of the viruses. This type of study is, however, predicated on the precarious assumption that the viral isolates are representative of the natural community (Garza & Suttle, 1998; Wilhelm *et al.*,1998b).

A variety of researchers have now employed marine viral isolates as models to study the impact of UV radiation on viral infectivity (Suttle & Chen, 1992; Wommack *et al.*, 1996; Noble & Fuhrman, 1997; Garza & Suttle, 1998; Wilhelm *et al.*, 1998a,b). Results from these studies have demonstrated that the UVR-mediated decay rate of viral infectivity in marine surface waters ranges from 0.1 to $1.0\,h^{-1}$ (Table 8.1). Considering the implied turnover rates, these results suggest that many of the virus particles in aquatic systems are uninfective, and that the infectivity of the natural viral community could be destroyed on a daily basis (Suttle *et al.*, 1993).

Other components of viral particles (i.e. the proteinaceous capsids) are also sensitive to UVR. Few studies have directly compared the loss rates of viral particles with the loss rates of viral infectivity (Suttle *et al.*, 1993; Wommack *et al.*, 1996). Studies in the Gulf of Mexico, using a native marine bacteriophage to trace infectivity and epifluorescent microscopy to enumerate total virus particles have demonstrated that rates of decay of viral infectivity consistently exceed rates of decay of viral particles (Wilhelm *et al.*, 1998a). In these studies, viral infectivity decayed two to three-fold faster in nearshore waters and two to four-fold faster in oceanic waters than did viral particle abundance (Table 8.2). However, since the viral abundances in these systems are in steady state (on a day to day basis), this creates a problem. Viral infectivity can be replaced only at a rate comparable to the rate of the production of new particles (or the ocean would fill up with viruses!). This rate should equal the loss rate of viral particles, since the abundance of viruses is relatively constant. Viral infectivity is lost at a greater rate though, implying a discrepancy in the balance of the production and loss of particles in the system. Wilhelm *et al.* (1998a) have attributed this difference to the rate at which infectivity must be restored to particles that must be damaged by UV-B but are not

Table 8.1. Light independent and dependent turnover rates of viruses

Strain	Family	Dark turnover (h)	Light turnover (h)	Reference
PWH3a-P1	Myoviridae	79.2	2.9	Suttle & Chen, 1992
H85/1	Myoviridae	480	12.5	Noble & Fuhrman, 1997
LB1VL-P1b	Podoviridae	40.8	2.2	Suttle & Chen, 1992
H40/1	Styloviridae	480	9.1	Noble & Fuhrman, 1997
CB38	ND	456	9.1	Wommack *et al.*, 1996
CB7	ND	648	16.7	Wommack *et al.*, 1996
Native cyanophage (Gulf of Mexico)	ND	190–706	3.0–3.6	Garza & Suttle, 1998

ND, not determined.

completely destroyed. Their results suggest that DNA repair mechanisms must be restoring infectivity to 40% to 80% of the UV-B-damaged viruses in marine systems in order to maintain a persistent abundance of viruses in marine surface waters. If this were not the case, then the *de novo* production rate of infective viral particles would need to equal the destruction rate of viral infectivity, resulting in a rapid accumulation of non-infective viral capsids.

8.8 Virus repair processes

The fact that the rates of decay of infectivity and particle abundance are imbalanced (see above) implies that the repair of damaged viral DNA by host cell repair mechanisms may be an ecologically important process in marine systems. Laboratory studies have now demonstrated that there are multiple pathways by which organisms can repair UVR-mediated DNA damage; these pathways can be divided into light-dependent and light-independent groups.

Weinbauer *et al.* (1997) presented the first study that attempted to quantify the importance of photoreactivation in natural systems. They started by exposing natural viral communities (water samples where bacterio- and phytoplankton had been removed) to natural sunlight (including damaging UV-B) or dark (control) conditions. After the natural viral communities were damaged by UV-B, natural host populations were reintroduced at ambient concentrations and samples were exposed to either photoreactivating (reduced sunlight with UV-B removed by Mylar filters) or non-photoreactivating (dark) conditions. By comparing the

Table 8.2. Comparison of infectivity and particle decay rates in the Gulf of Mexico. Viral infectivity was determined using the bacteriophage PWH3a-P1. Decay rates for both viral particle abundance and infectivity are integrated over the mixed layer and used to infer the percentage of the damaged population that needs to be repaired in order to maintain a steady-state system

	Decay rate		Inferred repair of infectivity (%)	1% depth (m)		Mixing depth
Station	Infectivity (h^{-1})	Particles (h^{-1})		305 nm	380 nm	
B	0.98	0.22	78	23.0	118.5	22.3
C	0.99	0.46	54	24.4	128.4	26.9
E	0.96	0.59	39	2.1	12.2	10.2
F	0.79	0.26	67	2.0	8.5	28.7

Adapted from Wilhelm *et al.*, 1998a.

differences in the quantity of viruses produced over time between the photoreactivating and non-photoreactivating treatments, they were then able to infer the efficacy of the photoreactivating process. Results from these experiments suggested that in the offshore oligotrophic waters of the Gulf of Mexico, between 21% and 26% of the sunlight-damaged viruses could be repaired by photoreactivation. In nearshore mesotrophic waters, 41% to 52% of the sunlight-damaged viruses could be photorepaired.

In this same study, Weinbauer *et al.* (1997) attempted to quantify the potential increase in production that could occur under photoreactivating conditions. To do this, they exploited the ubiquitous distribution of the bacteriophage PWH3a-P1, which lytically infects the marine bacterium *Vibrio natriegens* strain PWH3a (Suttle & Chen, 1992). By exposing concentrated natural viral communities from the Gulf of Mexico to hosts under photoreactivating and non-photoreactivating conditions, they were able to show that, by numerating viruses under conditions that would allow photorepair to occur, they were able to increase the titre on viruses infecting *V. natriegens* PWH3a by 1.1 to 8.5 times (mean = to 3.0, SD = 2.7) relative to control samples where photoreactivation was not allowed to occur. Moreover, they were able to increase the titre in 67% of the samples they collected during daylight hours by using a protocol that allowed for photoreactivation.

Even less information exists on the role of light-independent DNA repair mechanisms in restoring infectivity to UVR-damaged marine viruses. Kellogg & Paul (1997) have presented a study using caffeine as an inhibitor of excision repair. Their result using a *Vibrio parahaemolyticus* phage–host system suggests that excision repair increases the infectivity of UVR-exposed phages by up to two orders of magnitude. While more information is required, it is apparent that both excision repair and photoreactivation are important processes in maintaining viral populations in UVR-exposed environments.

8.9   DNA damage in the natural viral community

While it has been established that exposure to UV-B can significantly influence the infectivity of marine viral communities, it has become apparent that the ability to quantify the actual amount of DNA damage in the natural viral community is important. Natural viral communities are subject to mixing in the water column, which influences the intensity and period of the light exposure to which they are subjected. In most organisms, the most significant forms of DNA damage appear to be the formation of CPDs and (6–4) photoproducts (Mitchell & Nairn, 1989). Since one distortion in the DNA may be sufficient to stop DNA transcription (Protic-Sabljic & Kraemer, 1985), DNA damage could be a primary cause for sunlight-induced mortality. Since viruses lack pigmentation and have only a thin protein coat, viral DNA might be subjected to severe, sunlight-induced DNA damage.

Information on the concentrations and accumulation rates of sunlight-induced DNA damage in natural microbial communities remains limited. Laboratory studies have examined the accumulation of DNA damage of marine viral isolates and correlated the frequency of cyclobutane pyrimidine dimer formation to the infectivity of the viruses (Figure 8.4). These results have demonstrated that the ability of viruses to persist in environments

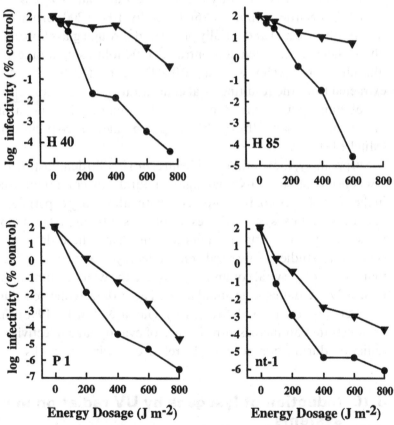

Figure 8.4. Efficacy of photoreactivation in restoring infectivity to UV-C damaged phages. Vibriophages H40/1, H85/1, PWH3a-P1, and nt-1 were exposed to different levels of UV-C and infectivity titred under both photoreactivating (triangles) and non-photoreactivating conditions (circles). (From Wilhelm *et al.*, 1998b.)

where UVR fluences are damaging may be a function of the efficacy of their host photorepair systems. In this study, virus–host systems representing oligotrophic oceanic waters (H40/1 and H85/1) appeared to be more effective than coastal (PWH3a-P1) or estuarine (nt-1) systems at restoring infectivity to viral lysates that had been subjected to significant doses of UV-C radiation. While photoreactivation increased the titre of infective viruses in all the systems, the host repair systems of H40/1 and H85/1 were able to maintain infectivity at several orders of magnitude higher than that in the other systems.

To extrapolate this information to natural systems, several studies have now examined the degree of DNA damage in the viral size fraction in marine surface waters. Ultrafiltration of seawater prefiltered with a 0.2 μm nominal pore-size filter has proved to be an effective means of concentrating the natural viral community (Proctor & Fuhrman, 1990; Suttle *et al.*, 1991; Chen, Suttle & Short, 1996). Typically, 100 to 200 l of marine surface waters are collected and serially prefiltered through glass fibre and 0.2 μm filters. Samples are then concentrated 1000-fold using a tangential flow ultrafiltration system with a 30 000 Da cut-off filter. DNA is then extracted from the resulting viral concentrates, and the concentration of cyclobutane pyrimidine dimers can be quantified using a standard radioimmunoassay with cyclobutane pyrimidine dimer-specific antibodies (Mitchell, 1995).

Surprisingly, the results of Wilhelm *et al.* (1998b) demonstrated a rather low concentration of DNA damage in natural viral communities collected in the Gulf of Mexico. In comparison with laboratory experiments with a native phage–host system, the results of this study suggest that most of the viruses in marine surface waters are infective. These data, in marked contrast to studies using viral tracers deployed at fixed depths, suggest that mixing of the viral community through the water column may greatly reduce DNA damage accumulation. These data (collected during the summer) do concur with studies by Garza & Suttle (1998) that have demonstrated an increased resistance of cyanophage infectivity to solar radiation during summer months relative to winter months.

8.10 Induction of lysogens by UV radiation in marine systems

While it has become apparent that most of the viruses produced in marine systems are the result of lytic infections (Wilcox & Fuhrman, 1994; Weinbauer & Suttle, 1996), there is a also a potential for the production of viruses through the lysogenic pathway. The process of lysogeny involves the maintenance of a viral genome within a host after infection and through the process of host cell replication. Lysogenised cells continue to grow and propagate until some external factor induces the viral genome to excise itself from the host and activate the lytic process, resulting in the destruction of the host cell. While a number of substances and conditions have now been suggested to induce lysogens into the lytic pathway (Jiang & Paul, 1996), the most common naturally occurring inducing agent is exposure to UVR.

Several studies have now looked at the ability of UV light to induce lysogens. Jiang & Paul (1996) utilised UV-C radiation (254 nm) to induce lysogens in samples from Tampa Bay, Florida. Their results, based on increases in the abundance of viruses, suggested that up to 38% of the bacterial population was lysogenised, although exposure to natural sunlight produced no evidence of lysogeny. Similarly, Weinbauer & Suttle (1996), also using a UV-C source, found lower levels of lysogenised marine bacteria, ranging from 0.07% to 1.5%. In consideration of the extreme levels of UV-C required to induce the lysogens (i.e. $3 \times 10^6 \, \mu J \, cm^{-2}$, Weinbauer & Suttle, 1996) it appears that the induction of lysogens by UV-C in marine systems may not be an important source of viral production. It is, however, difficult to resolve the role of UVR in the induction of lysogens as only limited information is available about the cellular mechanisms. Moreover, since these studies examine the increase in viruses produced after UVR exposure, they are further confounded by the detrimental impacts of UV light and the roles of the DNA repair mechanisms described previously.

8.11 Summary

It is now apparent that there is a great potential for UVR to impact microbial carbon cycling and biogeochemical processes. Although variability in effects has been observed, all of the organisms involved may be directly damaged by UV-B and may be indirectly damaged by UV-A. By the nature of food web interactions, any effect on one trophic group will also impact on others. Photochemical reactions may also result in the production of deleterious chemical compounds. In contrast, UV-B and UV-A may photochemically produce more labile substrates that would stimulate bacterial production. UV-A may also serve to provide the energy for photoreactivation. As a result, predicting responses to environmental UVR may be difficult at best. It is likely that the type of response observed will be determined by local environmental conditions, as well as the species composition of the organisms exposed. A better understanding of the environmental conditions influencing UVR response and identification of sensitive (and insensitive) species may enhance the ability to predict when UVR impact may be greatest.

UVR is a natural component of the solar spectrum and micro-organisms have been exposed to it since the beginning of time. There is increasing evidence that the organisms in the aquatic environment have the ability to recover from UVR exposure except under extreme conditions. Perhaps it is now prudent to investigate the mechanisms by which organisms respond to and recover from UVR. Part of this investigation should include the energetic costs to the ecosystem to overcome UVR stress. It is these subtle changes in energetics that may hold the key to determining the ecosystem effects of UVR exposure.

Acknowledgements

This research was supported in part by US Environmental Protection Agency Cooperative Agreement CR-822020-01 through the Gulf Breeze Environmental Research Laboratory, National Science Foundation Office of Polar Programs Grant OPP-9419037, the National Oceanic and Atmospheric Administration's National Undersea Research Center at the University of North Carolina at Wilmington NOAA contract UNCW9204S to W.H.J. We thank Peter Aas, M. Maille Lyons and LeAnna Hutchinson for technical assistance. We are indebted to Curtis Suttle for discussions and ideas on which much of this research was based. S.W.W. was supported by a Natural Science and Engineering Research Council of Canada Visiting Scientists award through Environment Canada, as well as the Office of Naval Research award N00014-92-5-1676, National Science Foundation award OCE-9415602, and Natural Science and Engineering Research Council of Canada grants to Curtis A. Suttle.

References

Aas, P., Lyons, M., Pledger, R., Mitchell, D. L. & Jeffrey, W. H. (1996). Inhibition of bacterial activities by solar radiation in nearshore waters and the Gulf of Mexico. *Aquatic Microbial Ecology*, **11**, 229–38.

Azam, F., Fenchel, T., Field, J. G., Gray, J., Meyer-Reil, L. A. & Thingstad, F. (1983). The ecological role of water column microbes in the sea. *Marine Ecology Progress Series*, **10**, 257–63.

Bailey, C. A., Neihof, R. A. & Tabor, P. S. (1983). Inhibitory effect of solar radiation on amino acid uptake in Chesapeake Bay bacteria. *Applied Environmental Microbiology*, **46**, 44–9.

Børsheim, K. Y. (1993). Native marine bacteriophages. *FEMS Microbiology Ecology*, **102**, 141–59.

Bowes, G. (1993). Facing the inevitable: plants and increasing atmospheric CO_2. *Annual Reviews of Plant Physiology and Plant Molecular Biology*, **44**, 309–32.

Brash, D.E., Seetharam, S., Kraemer, K., Seidman, M. & Bredberg, A. (1987). Photoproduct frequency is not the major determinant of UV base substitution hot spots or cold spots in human cells. *Proceedings of the National Academy of Sciences, USA*, **84**, 3782–6.

Bratbak, G., Thingstad, F. & Heldal, M. (1994). Viruses and the microbial loop. *Microbial Ecology*, **28**, 209–21.

Chen, F., Suttle, C. A. & Short, S. M. (1996). Genetic diversity in marine algal virus communities as revealed by sequence analysis of DNA polymerase genes. *Applied Environmental Microbiology*, **62**, 2869–74.

Chin-Leo, G, & Kirchman, D. L. (1988). Estimating bacterial production in marine waters from the simultaneous incorporation of thymidine and leucine. *Applied Environmental Microbiology*, **54**, 1934–9.

Coffin, R. B., Velinsky, D., Devereux, R., Price, W. A. & Cifuentes, L. (1990). Stable carbon isotope analysis of nucleic acids to trace sources of dissolved substrates used by estuarine bacteria. *Applied Environmental Microbiology*, **56**, 2012–20.

Cota, G. F., Kottmeier, S. T., Robinson, D. H., Smith, W. O. & Sullivan, C. W. (1990). Bacterioplankton in the marginal ice zone of the Weddell Sea, biomass, production and metabolic activities during austral summer. *Deep Sea Research*, **37**, 1145–67.

Demple, B. & Harrison, L. (1994). Repair of oxidative damage to DNA: enzymology and biology. *Annual Reviews of Biochemistry*, **63**, 915–48.

Dulbecco, R. (1949). Reactivation of ultraviolet inactivated bacteriophage by visible light. *Nature*, **163**, 949–50.

Dulbecco, R. (1950). Experiments on photoreactivation of bacteriophages inactivated with ultraviolet radiation. *Journal of Bacteriology*, **59**, 329–47.

Eisenstark, A. (1987). Mutagenic and lethal effects of near-ultraviolet radiation (290–400 nm) on bacteria and phage. *Environmental and Molecular Mutagenesis*, **10**, 317–37.

Evison, L. M. (1988). Comparative studies on the survival of indicator organisms and pathogens in fresh and sea water. *Water Science and Techology*, **20**, 309–15.

Falkowski, P. G. & Woodham, A. D. (1992). *Primary Production and Biogeochemical Cycles in the Sea*. Plenum Press, New York.

Friedberg, E. C., Walker, G. C. & Siede, W. (1995). *DNA Damage and Repair*. ASM Press, Washington, DC.

Fuhrman, J. A. (1992). Bacterioplankton roles in cycling of organic matter: the microbial food web. In *Primary Productivity and Biogeochemical Cycles in the Sea*, ed. P. G. Falkowski & A. D. Woodhead, pp. 361–83. Plenum Press, New York.

Fuhrman, J. A. & Azam, F. (1980). Bacterioplankton secondary production estimates for coastal waters of British Columbia, Antarctica, and California. *Applied Environmental Microbiology*, **39**, 1085–95.

Fuhrman, J. A. & Azam, F. (1982). Thymidine incorporation as a measure of heterotrophic bacterioplankton production in marine surface waters: evaluation and field results. *Marine Biology*, **66**, 109–20.

Fuhrman, J. A. & Suttle, C. A. (1993). Viruses in marine planktonic systems. *Oceanography*, **6**, 51–63.

Fujioka, R. S., Hashimoto, H. H., Siwak, E. B. & Young, R. H. F. (1981). Effect of sunlight on survival of indicator bacteria in seawater. *Applied Environmental Microbiology*, **41**, 690–6.

Gameson, A. L. H. & Saxon, D. J. (1967). Field studies on effect of daylight on mortality of coliform bacteria. *Water Research*, **1**, 279–95.

Garza, D. R. & Suttle, C. A. (1998). The effect of cyanophages on the mortality of *Synechococcus* spp. and selection for UV resistant viral communities. *Microbial Ecology*, **36**, 281–92.

Guerrero, M. A. & Jones, R. D. (1996). Photoinhibition of marine nitrifying bacteria. I. Wavelength-dependent response. *Marine Ecology Progress Series*, **141**, 193–2.

Häder, D. P., Worrest, R. C., Kumar, H. D. & Smith, R. C. (1995). Effects of increased solar ultraviolet radiation on aquatic ecosystems. *Ambio*, **24**, 174–80.

Hansen, R. B., Shafer, D., Ryan, T., Pope, D. & Lowery, H. K. (1983). Bacterioplankton in the Antarctic ocean waters during late austral winter, abundance, frequency of dividing cells, and estimates of production. *Applied Environmental Microbiology*, **45**, 1622–32.

Helbling, E. W., Marguet, E. R., Villafañe, V. E. & Holm-Hansen, O. (1995). Bacterioplankton viability in Antarctic waters as affected by solar ultraviolet radiation. *Marine Ecology Progress Series*, **126**, 293–8.

Hendricks, C. W. (1978). Exceptions to the coliform and the faecal colimar tests. In *Indicators of Viruses in Water and Food*, ed. G. Berg, pp. 99–145. Ann Arbor Science Publishers, Ann Arbor, MI.

Herndl, G. J. (1997). Role of ultraviolet radiation on bacterioplankton activity. In *The Effects of Ozone Depletion on Aquatic Ecosystems*, ed. D. P. Häder, pp. 143–53. Academic Press, San Diego, CA.

Herndl, G. J., Müller-Niklas, G. & Frick, J. (1993). Major role of ultraviolet-B in controlling bacterioplankton growth in the surface layer of the ocean. *Nature*, **361**, 717–19.

Hooper, A. B. & Terry, K. R. (1974). Photoinactivation of ammonia oxidation in *Nitrosomonas*. *Journal of Bacteriology*, **119**, 899–906.

Horrigan, S. G., Carlucci, A. F. & Williams, P. M. (1981). Light inhibition of nitrification in sea-surface films. *Journal of Marine Research*, **39**, 557–65.

Jeffrey, W. H., Aas, P., Lyons, M. M., Coffin, R. B., Pakulski, J. D., EPA, U. S. & Mitchell, D. L. (1997). Ultraviolet radiation induced DNA damage in the microbial community in Antarctic waters. Paper presented at the American Society of Limnology and Oceanography Aquatic Sciences Meeting, Santa Fe, February 10–14. Abstract.

Jeffrey, W. H., Aas, P., Lyons, M. M., Pledger, R., Mitchell, D. L. & Coffin, R. B. (1996a). Ambient solar radiation induced photodamage in marine bacterioplankton. *Photochemistry and Photobiology*, **64**, 419–27.

Jeffrey, W. H., Nazaret, S. & Von Haven, R. (1994). Improved method for the recovery of mRNA from aquatic samples: application to detecting *mer* gene expression. *Applied Environmental Microbiology*, **60**, 1814–21.

Jeffrey, W. H., Pledger, R. J., Aas, P., Hager, S., Coffin, R. B., Von Haven, R. & Mitchell, D. L. (1996b). Diel and depth profiles of DNA photodamage in bacterioplankton exposed to ambient solar radiation. *Marine Ecology Progress Series*, **137**, 283–91.

Jerlov, N. G. (1976). *Marine Optics*, 2nd edn. Elsevier, Amsterdam.

Jiang, S. C. & Paul, J. H. (1996). Occurrence of lysogenic bacteria in marine microbial communities as determined by prophage induction. *Marine Ecology Progress Series*, **142**, 27–48.

Kaiser, E. & Herndl, G. J. (1997). Rapid recovery of marine bacterioplankton activity after inhibition by UV radiation in coastal waters. *Applied Environmental Microbiology*, **63**, 4026–31.

Kapuscinski, R. B. & Mitchell, R. (1983). Sunlight-induced mortality of viruses and *Escherichia coli* in coastal seawater. *Environmental Science and Technology*, **17**, 1–6.

Karentz, D., Bothwell, M. L., Coffin, R. B., Hansen, A., Herndl, G. J., Kilham, S. S., Lesser, M. P., Lindell, M., Moeller, R. E., Morris, D. P., Neale, P. J., Sanders, R. W., Weiler, C. S. & Wetzel, R. G. (1994). Impact of UV-B radiation on pelagic freshwater ecosystems: report of a working group on bacteria and phytoplankton. *Ergebnisse der Limnologie (Archiv für Hydrobiologie. Beiheft)*, **43**, 31–69.

Kelland, L. R., Moss, S. H. & Davies, D. J. G. (1984). Leakage of 86Rb + after ultraviolet irradiation of *E. coli* K12. *Photochemistry and Photobiology*, **39**, 329–35.

Kellogg, C. A. & Paul, J. H. (1997). Multiple repair systems in a marine bacterium reactivate viruses after UV-induced DNA damage. Paper presented at the American Society of Limnology and Oceanography Aquatic Sciences Meeting, Santa Fe, NM. Abstract.

Kiene, R. P. (1990). Dimethyl sulfide production from dimethylsulfonopropionate in coastal seawater samples and bacterial cultures. *Applied Environmental Microbiology*, **56**, 3292–7.

Kim, S. T. & Sancar, A. (1993). Photochemistry, photophysics, and mechanism of pyrimidine dimer repair by DNA photolyase. *Photochemistry and Photobiology*, **57**, 895–904.

Kirk, J. T. O., Hargreaves, B. R., Morris, D. P., Coffin, R., David, B. S., Fredrickson, D., Karentz, D., Lean, D., Lesser, M., Madronich, S., Morrow, J. H., Nelson, N. & Scully, N. (1994). Measurements of UV-radiation in two freshwater lakes: an instrument intercomparison. *Ergebnisse der Limnologie (Archiv für Hydrobiologie. Beiheft)*, **43**, 71–99.

Klamen, D. L. & Tuveson, R. W. (1982). The effect of membrane fatty acids composition on the near-UV (300–400 nm) sensitivity of *Escherichia coli* K1060. *Photochemistry and Photobiology*, **35**, 161–73.

Krinsky, N. I. (1976). Cellular damage initiated by visible light. *Symposium of the Society of General Microbiology*, **26**, 209–30.

Kubitschek, H. E. & Doyle, R. J. (1981). Growth delay induced in *Escherichia coli* by near-ultraviolet radiation: relationship to membrane transport functions. *Photochemistry and Photobiology*, **33**, 695–702.

Lindell, M. W., Granéli, W. & Tranvik, L. J. (1995). Enhanced bacterial growth in response to photochemical transformation of dissolved organic matter. *Limnology and Oceanography*, **40**, 195–9.

Lyons, M. M., Aas, P., Pakulski, J. D., Van Waasbergen, L., Mitchell, D. L., Miller, R. V. & Jeffrey, W. H. (1998). Ultraviolet radiation induced DNA damage in coral reef microbial communities. *Marine Biology*, **130**, 537–43.

Melillo, J. M., McGuire, A. D., Kicklighter, D. W., Moore, B., Vorosmarty, C. J. & Schloss, A. L. (1993). Global climate change and terrestrial net primary production. *Nature*, **363**, 234–40.

Miller, R. V. (1992). RecA. In *Encyclopedia of Microbiology*, vol. 3, ed. S. Luria, pp. 509–17. Academic Press, San Diego, CA.

Miller, R. V., Jeffrey, W., Mitchell, D. & Elasri, M. (1999). Bacterial responses to solar ultraviolet light. *ASM News*, in press.

Miller, R. V. & Kokjohn, T. A. (1990). General microbiology of RecA: environmental and evolutionary significance. *Annual Reviews of Microbiology*, **44**, 365–94.

Miller, W. L. & Moran, M. (1997). Interaction of photochemical and microbial processes in the degradation of refractory dissolved organic matter from a coastal marine environment. *Limnology and Oceanography*, **42**, 1317–24.

Mitchell, D. L. (1995). Ultraviolet radiation damage to DNA. In *Molecular Biology and Biotechnology: A Comprehensive Desk Reference*, ed. R. A. Meyers, pp. 939–43. VCH Publishers, New York.

Mitchell, D. L., Haipek, C. A. & Clarkson, J. M. (1985). (6–4) Photoproducts are removed from the DNA of UV-irradiated mammalian cells more efficiently than cyclobutane pyrimidine dimers. *Mutation Research*, **143**, 109–12.

Mitchell, D. L. & Karentz, D. (1993). The induction and repair of DNA photodamage in the environment. In *Environmental UV Photobiology*, ed. A. R. Young, L. Björn, J. Moan & W. Nultsch, pp. 345–77. Plenum Press, New York.

Mitchell, D. L. & Nairn, R. S. (1989). The biology of the (6–4) photoproduct. *Photochemistry and Photobiology*, **49**, 805–19.

Moran, M. & Zepp, R. G. (1997). Role of photoreactions in the formation of biologically labile compounds from dissolved organic matter. *Limnology and Oceanography*, **42**, 1307–16.

Murray, A. G. & Eldridge, P. M. (1994). Marine viral ecology: incorporation of bacteriophage into the microbial planktonic food web paradigm. *Journal of Plankton Research*, **16**, 627–41.

Noble, R. T. & Fuhrman, J. A. (1997). Virus decay and its causes in coastal waters. *Applied Environmental Microbiology*, **63**, 77–83.

Pakulski, J. D., Aas, P., Jeffrey, W. H., Lyons, M., Von Waasbergen, L., Mitchell, D. & Coffin, R. (1998). Influence of light on bacterioplankton production and respiration in a subtropical reef. *Aquatic Microbial Ecology*, **14**, 137–48.

Paul, J. H. & Carlson, D. 1984. Genetic material in the marine environment: implication for bacterial DNA. *Limnology and Oceanography*, **29**, 1091–7.

Paul, J. H., Jeffrey, W. H. & Deflaun, M. F. (1985). Particulate DNA in subtropical oceanic and estuarine planktonic environments. *Marine Biology*, **90**, 95–101.

Peak, M. J. & Peak, J. G. (1995). Photosensitized reactions of DNA. In *CRC Handbook of Organic Photochemistry and Photobiology*, ed. W. M. Horspool and P. Song, pp. 1318–25. CRC Press, Boca Raton, FL.

Pommepuy, M., Guillaud, J. F., Dupray, E., Derrien, A., Le Guyader, F. & Cormier, M. (1992). Enteric bacteria survival factors. *Water Science and Technology*, **25**, 93–103.

Proctor, L. M. & Fuhrman, J. A. (1990). Viral mortality of marine bacteria and cyanobacteria. *Nature*, **343**, 60–2.

Protic-Sabljic, M. & Kraemer, K. H. (1985). One pyrimidine dimer inactivates expression of a transfected gene in *Xeroderma pigmentosum* cells. *Proceedings of the National Academy of Sciences, USA*, **83**, 6622–6.

Robarts, R. S. & Zohary, T. (1993). Fact or fiction – bacterial growth rates and production as determined by [methyl-^{3}H]thymidine? In *Advances in Microbial Ecology*, ed. J. G. Jones, vol. 13, pp. 371–425. Plenum Press, New York.

Sieracki, M. E. & Sieburth, J. M. (1986). Sunlight-induced growth delay of planktonic marine bacteria in filtered seawater. *Marine Ecology Progress Series*, **33**, 19–27.

Simon, M. & Azam, F. (1989). Protein content and protein synthesis rates of planktonic marine bacteria. *Marine Ecology Progress Series*, **51**, 201–13.

Smith, R. C. & Baker, K. S. (1978). Optical classification of natural waters. *Limnology and Oceanography*, **23**, 260–7.

Smith, R. C. & Baker, K. S. (1979). Penetration of UV-B and biologically effective dose-rates in natural waters. *Photochemistry and Photobiology*, **29**, 311–23.

Solic, M. & Krstulovic, N. (1992). Separate and combined effects of solar radiation, temperature, salinity, and pH on the survival of faecal coliforms in seawater. *Marine Pollution Bulletin*, **24**, 441–16.

Sommaruga, R., Obernosterer, I., Herndl, G. J. & Psenner, R. (1997). Inhibitory effect of solar radiation on thymidine and leucine incorporation by freshwater and marine bacterioplankton. *Applied Environmental Microbiology*, **63**, 4178–84.

Sondergaard, M., Riemann, B. & Jorgenesen, N. O. G. (1985). Extracellular organic carbon (EOC) released by phytoplankton and bacterial production. *Oikos*, **45**, 323–32.

Stanier, R. Y. & Cohen-Bazire, G. (1957). The role of light in the microbial world: some facts and speculations. *Symposium of the Society of General Microbiology*, **7**, 56–89.

Steward, G. F., Smith, D. C. & Azam, F. (1996). Abundance and production of bacteria and viruses in the Bering and Chukchi Seas. *Marine Ecology Progress Series*, **131**, 287–300.

Sullivan, C. W., Cota, G. F., Krempin, D. W. & Smith, W. O. Jr (1990). Distribution and activity of bacterioplankton in the marginal ice zone of the Weddell-Scotia Sea during austral spring. *Marine Ecology Progress Series*, **63**, 239–52.

Suttle, C. A. (1994). The significance of viruses to mortality in aquatic microbial communities. *Microbial Ecology*, **28**, 237–43.

Suttle, C. A., Chan, A. M., Chen, F. & Garza, D. R. (1993). Cyanophages and sunlight: a paradox. In *Trends in Microbial Ecology*, ed. R. Guerro & Pedros-Alio, pp. 303–8. Spanish Society of Microbiology, Barcelona.

Suttle, C. A., Chan, A. M. & Cottrell, M. T. (1990). Infection of phytoplankton by viruses and reduction of primary productivity. *Nature*, **347**, 467–9.

Suttle, C. A., Chan, A. M. & Cottrell, M. T. (1991). Use of ultrafiltration to isolate viruses from seawater which are pathogens of marine phytoplankton. *Applied Environmental Microbiology*, **57**, 721–6.

Suttle, C. A. & Chen, F. (1992). Mechanisms and rates of decay of marine viruses in seawater. *Applied Environmental Microbiology*, **58**, 3721–9.

Todo, T., Ryo, H., Yamamoto, K., Toh, H., Inui, T., Ayaki, H., Nomura, T. & Ikenga, M. (1996). Similarity among Drosophila (6–4) photolyase, a human photolyase homolog, and the DNA photolyase-blue-light photoreceptor family. *Nature*, **272**, 109–12.

Vincent, W. F. & Roy, S. (1993). Solar ultraviolet-B radiation and aquatic primary production: damage, protection, and recovery. *Environmental Review*, **1**, 1–12.

Walker, G. (1984). Mutagenesis and inducible responses to deoxyribonucleic acid damage in *Escherichia coli*. *Microbiological Reviews*, **48**, 60–93.

Weinbauer, M. & Suttle, C. A. (1996). Potential significance of lysogeny to bacteriophage production and bacterial mortality in coastal waters of the Gulf of Mexico. *Applied Environmental Microbiology*, **62**, 4374–80.

Weinbauer, M. G., Wilhelm, S. W., Suttle, C. A. & Garza, D. R. (1997). Photoreactivation compensates for UV damage and restores infectivity to natural marine viral communities. *Applied and Environmental Microbiology*, **63**, 2200–5.

Wilcox, R. M. & Fuhrman, J. A. (1994). Bacterial viruses in coastal seawater: lytic rather than lysogenic production. *Marine Ecology Progress Series*, **114**, 35–45.

Wilhelm, S. W., Weinbauer, M. G., Suttle, C. A. & Jeffrey, W. H. (1998a). The role of sunlight in the removal and repair of viruses in the sea. *Limnology and Oceanography*, **43**, 586–92.

Wilhelm, S. W., Weinbauer, M. G., Suttle, C. A., Pledger, R. J. & Mitchell, D. L. (1998b). Measurements of DNA damage and photoreactivation imply that most viruses in marine surface water are infective. *Aquatic Microbial Ecology*, **14**, 215–22.

Wommack, K. E., Hill, R. T., Muller, T. A. & Colwell, R. R. (1996). Effects of sunlight on bacteriophage viability and structure. *Applied Environmental Microbiology*, **62**, 1336–41.

Zepp, R. G., Callaghan, T. V. & Erickson, D. J. (1995). Effects of increased solar ultraviolet radiation on biogeochemical cycles. *Ambio*, **24**, 181–7.

9

○ ○ ○ ○ ○ ○ ○ ○ ○ ○ ○ ○ ○ ○ ○ ○ ○ ○ ○ ○

Effects of UV radiation on the physiology and ecology of marine phytoplankton

Maria Vernet

9.1   Introduction

Phytoplankton processes govern the synthesis of organic carbon in marine surface waters. The rate of change of phytoplankton concentration (P) in the euphotic zone is a balance between growth (d^{-1}) and losses from cell respiration, carbon excretion, sedimentation and grazing. In addition, advection and diffusion can locally increase or decrease phytoplankton. These processes can be summarised as:

$$dP/dt = P(\text{growth} - \text{respiration} - \text{excretion} - \text{sedimentation} - \text{grazing})$$

Algal growth depends on carbon input by photosynthesis and is influenced by environmental conditions as well as ecological factors. In turn, physiological adaptation and cell metabolism respond to the environmental forcing to maximise survival. Light quality and quantity, nutrient availability and temperature all affect carbon uptake. Thus, phytoplankton are limited to surface waters as they are dependent on visible light (400–700 nm) for photosynthesis. The requirement for solar radiation makes them vulnerable to UV radiation (UVR) (280–400 nm) exposure (Raven, 1991; Gieskes & Buma, 1997), in particular UV-A (320–400 nm), although UV-B (280–320 nm) is more damaging per photon. UV-B, which accounts approximately for 0.01% of the photons absorbed by phytoplankton (Falkowski & Raven, 1997), is damaging to different molecular targets (see Chapters 6 and 7). Net damage at the cellular and population level is a balance between damage and repair (Vincent & Roy, 1993; Cullen & Neale, 1994) as documented for photosynthesis (Cullen *et al.*, 1992; Lesser, Neale & Cullen, 1996; Hazzard, Lesser & Kinzie, 1997) and growth (Quesada, Mouget & Vincent, 1995). Exposure experiments to study UVR effects on phytoplankton have ranged from minutes to months, from molecular to community level, and for orders of magnitude difference in UVR. The relative rate of damage and repair in these disparate conditions is expressed both physiologically and ecologically. The extrapolation of these results to ecological meaningful predictions is still unresolved (Smith *et al.*, 1980; Vincent & Roy, 1993; Cullen & Neale, 1994) but this is a challenge common to all environmental science. In Nature, knowledge of phytoplankton responses to UVR needs to be considered in the broader context of food chain dynamics (Karentz *et al.*, 1994; Vernet & Smith, 1997; and see Chapter 11) and ecosystem research (Caldwell *et al.*, 1995).

The number of studies of UVR effects on organisms, in particular on phytoplankton, has increased exponentially in the last decade. Much has been learned about new mechanisms of damage and the variability in both sensitivity and resistance to UVR. Increased documentation on UVR inhibition of phytoplankton processes has helped in defining the generality of early experiments. Unfortunately, the diversity of experimental approaches has increased exponentially as well. Results from different experiments are difficult to compare, making the search for general patterns as elusive as the lack of data. The following review attempts to address net damage on phytoplankton, whenever possible, from a variety of experimental conditions, field and laboratory experiments, undertaken over a wide range of time and space scales. Effects of UVR on main phytoplankton processes have been grouped to emphasise physiological or ecological aspects, although in Nature a continuum of responses is observed.

9.2   Effects of UV radiation on phytoplankton processes

Experiments to assess UVR damage, defence, recovery and avoidance mechanisms have been carried out in incubation vessels. The size of the vessel varies from several millilitres to hundreds of litres, as in the case of mesocosms (Sakka *et al.*, 1997; Belzile *et al.*, 1998). Most field experiments are based on 4–12 h incubations, either *in situ* or in simulated *in situ* conditions. For *in situ* experiments, phytoplankton are incubated in UVR-transparent containers (quartz bottles or Whirl-pak bags) at the same depth from which they had been sampled (Smith *et al.*, 1992). For simulated *in situ* experiments, the containers are incubated under surface sunlight on a ship's deck or outside a coastal laboratory (see e.g. Helbling *et al.*, 1992). Several limitations are inherent in these approaches, i.e. differences between UVR spectral treatments with comparison to phytoplankton exposed to *in situ* UVR, as well as the effect incubations have on experiments conducted under different conditions or from different environments. During incubations, phytoplankton are exposed to sunlight at fixed depths, which is different from what might happen to a free-floating cell in the water column subject to diffusion and turbulent mixing processes. Shorter exposures are thus considered to be more realistic. Simulated *in situ* incubations have an extra limitation, i.e. lack of reproduction of the spectral balance at depth, for UVR and for visible radiation. Time of incubation is also critical to determine the variable under consideration. Gross photosynthesis (carbon uptake without losses for excretion and/or respiration) is measured during shorter incubations, while net primary production is approached in 24 h experiments. Grazing by microzooplankton and mineral recycling by bacteria during these incubations can also affect overall rates of carbon uptake, particularly for tropical environments, thus influencing latitudinal comparisons.

The effect of UVR on phytoplankton is usually measured by screening part (i.e. − UV-B or − UV-A) or all of UVR (− UVR) from sunlight. In this way, the effect of UV-B is measured as the difference between a certain process measured under total sunlight and that measured without UV-B. Similarly, the influence of UV-A is measured as the difference between UV-A-screened and total sunlight exposures. The results can be expressed as either decreased rate or inhibition of function when UVR is present or increased rates due to screened UVR. Increased UV-B radiation with respect to UV-A (UV-B: UV-A) and photosynthetically available radiation (PAR) (UV-B: PAR) can also be used in the experimental design in order to mimic UV-B during conditions of decreased stratospheric ozone. UV-B radiation is added via appropriate lamps (Björn & Teramura, 1993), similar to laboratory experiments. All these experimental conditions need be taken into consideration when one is evaluating results, extrapolating results to Nature, and in the course of comparing results to achieve generalisations.

9.2.1 Primary production

The overall effect of UVR on phytoplankton is to decrease rates of primary production (Smith, 1989; Holm-Hansen, Lubin & Helbling, 1993; Cullen & Neale, 1994; Prézelin, Boucher & Schofield, 1994; Weiler & Penhale, 1994; Smith & Cullen, 1995; Cullen & Neale, 1997). Inhibition of primary production is measured as inhibition of carbon uptake per unit time and unit volume (i.e. $mg\,C\,m^{-3}\,h^{-1}$) during incubations of phytoplankton with ^{14}C-labelled bicarbonate dissolved in seawater (Steeman-

Nielsen, 1952) or oxygen evolution. If cell concentration is measured by either microscopy or flow cytometry, or other bulk measurements such as chlorophyll *a* (chl *a*) are available, inhibition of carbon uptake can be expressed also per unit chl *a* or per cell. As cellular pigment concentration, cell size and cell division are also affected by exposure to UVR, the units in which damage is expressed can alter the results.

9.2.1.1 Carbon uptake

UV-B and UV-A inhibit both light-limited and light-saturated carbon uptake (Steeman-Nielsen, 1964; Maske, 1984; Cullen *et al.*, 1992; Ekelund, 1994; Lesser, 1996) (Figure 9.1). For example, UV-B inhibits carbon incorporation in field populations in temperate and Antarctic phytoplankton by 25% to 50% of shielded samples (see e.g. Maske, 1984; Holm-Hansen, Villafañe & Helbling, 1997). A strong depth gradient is observed from the surface to depth. Measurable UV-B inhibition of primary production is usually constrained to the upper 20–25 m of the water column (Lorenzen, 1979; Holm-Hansen, Mitchell & Vernet, 1989; Gieskes & Kraay, 1990; Karentz & Lutze, 1990; Smith *et al.*, 1992), as expected from UV-B transmission (see Chapter 2). On average, daily depth-integrated primary production decreases by 6% to 12% (Holm-Hansen *et al.*, 1989; Helbling *et al.*, 1992; Smith *et al.*, 1992) during spring time ozone depletion over

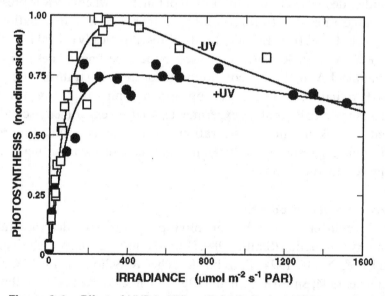

Figure 9.1. Effect of UVR (+UV) on light-limited and light-saturated photosynthesis on Antarctic phytoplankton in 1 h incubation as compared with cultures exposed only to PAR (–UV) PAR, photosynthetically available radiation. (Redrawn from Neale *et al.*, 1992.)

Antarctic waters, although higher water column inhibition has been measured (i.e. 25%; Boucher & Prézelin, 1996b; Figure 9.2). Inhibition of annual primary production was calculated as 2% for the Southern Ocean (Smith *et al.*, 1992). Helbling, Villafañe & Holm-Hansen (1994), based on different assumptions and methodology, calculated the decrease in primary production to be 0.15% for the entire ice-free waters south of the Polar Front. The degree of uncertainty of these measurements increases

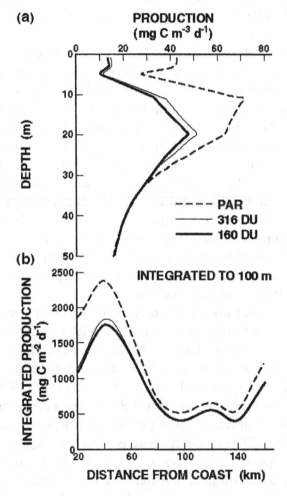

Figure 9.2. Estimates of daily primary production in Antarctic phytoplankton calculated using the BWF of Boucher & Prézelin (1996a). (a) Depth profile of primary production for days with high (thin line) and low (thick line) ozone (316 and 160 Dobson units (DU), respectively), and compared with no UVR (dashed line). (b) Integrated primary production along a transect perpendicular to the coast of the western Antarctic peninsula. Symbols as in (a). PAR, photosynthetically available radiation. (Redrawn from Boucher & Prézelin, 1996b.)

with area- and time-integrated calculations, as they are based on discrete hourly or daily measurements (i.e. depths) taken in a short period of time (i.e. month-long cruise) and space (i.e. several hundred square kilometres as opposed to the whole Southern Ocean). A large database, as can now be compiled from several tens of published studies in Antarctic areas, will improve considerably these estimates and will constrain better the variance of the measurements. In addition, a better knowledge of phytoplankton composition and physiological parameters, in conjunction with UVR inhibition, will provide much needed information on the causes of the observed variability.

The effect of UVR on the inhibition of carbon uptake is not constant across the spectrum (290–390 nm) (see also Chapter 3). In some populations, UV-A can have twice the inhibitory effect relative to UV-B (Holm-Hansen et al., 1997), while other populations are more sensitive to UV-B (Boucher & Prézelin, 1996a). This difference is of importance in assessing the possible effects of decreased stratospheric ozone, which affects only UV-B. The response of phytoplankton is not linear across UVR irradiance levels: the threshold for photosynthetic inhibition of Antarctic coastal phytoplankton has been determined to be $0.5 \, \mathrm{W \, m^{-2}}$ for UV-B and $10 \, \mathrm{W \, m^{-2}}$ for UV-A (Booth et al., 1997). In contrast, an order of magnitude higher sensitivity and no threshold was observed for Arctic phytoplankton sampled from a deep mixed layer (Helbling et al., 1996). The variability has been attributed to differences in phytoplankton composition and to the degree of adaptation to in situ conditions.

UVR inhibition in other marine photosynthetic organisms, such as benthic seaweed, shows a similar decrease under UVR exposure (Häder, Herrman & Santas, 1996). For example, surface blades of *Macrocystis pyrifera* showed a decrease in both light-limited and light-saturated production measured as oxygen evolution (Clendennen et al., 1996).

Experiments on long term exposure to UVR have been performed to study the development of mechanisms of physiological acclimation. The results to date are mixed. The variability suggests influence of temperature and a latitudinal component, with rates of acclimation higher at higher temperatures and at lower latitudes. In Antarctic coastal phytoplankton, the maximum photosynthetic rate $(\mathrm{mg \, C \, (mg \, chl} \, a)^{-1} \, \mathrm{h}^{-1})$ showed a 22% decrease after 9 days (Lesser et al., 1996). The temperate dinoflagellate *Prorocentrum micans* showed a decreased maximum photosynthetic rate and light-limited photosynthesis after 21 days of exposure, similar to results from short term exposure experiments (Lesser, 1996). In this species, the effects were exacerbated at high temperatures (from 20 °C to

31 °C). In contrast, the subtropical diatom *Chaetoceros gracilis* Schutt
was able to counteract UVR effects after 1 day of exposure (Hazzard *et al.*,
1997) (Figure 9.3). The diatom had an initial 12 h (1 solar day) decline in
carbon. After 48 h, the culture had returned to pre-UVR exposure carbon
uptake rates per cell. In summary, the high rate of acclimation observed at
lower latitudes or for tropical isolates suggests that temperature could
contribute to the latitudinal gradient. A possible mechanism might
involve rates of photochemical damage that are temperature independent,
while rates of recovery are driven by enzymatic reactions and are thus
temperature dependent.

The mechanisms involved in the decrease of photosynthetic activity in
Macrocystis pyrifera were associated with decreased size and density of
photosystem II (PS II) reaction centres, and a decreased energy transfer

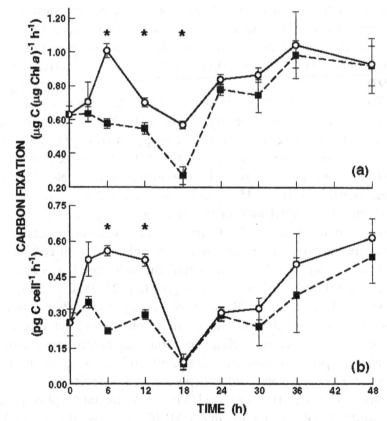

Figure 9.3. Response of light-saturated carbon incorporation in the
subtropical diatom *Chaetoceros gracilis* exposed to ambient UVR. Presented
as the mean of replicate samples for PAR (line) and PAR+UVR (dashed line)
treatments. Asterisk notes samples where treatments were significantly
different from control. (Redrawn from Hazzard *et al.*, 1997. See also p. 278
(*Note*).)

from accessory pigments to chl *a*, as measured by fluorescence excitation spectra (Clendennen *et al.*, 1996). In Antarctic phytoplankton, the lack of complete acclimation after 9 days of incubation showed that Rubisco (ribulose 1,5-biphosphate carboxylase) activity decreased by 22%, similar to the decrease in maximum photosynthetic rates. In contrast, no difference was observed between the treatments with respect to maximum quantum yield of PS II, electron transport, and in the ratio of dark-adapted variable fluorescence to maximum fluorescence (F_v/F_m) (Lesser *et al.*, 1996). In *P. micans*, lack of acclimation correlated with damage in Rubisco and with increased oxidative stress. In addition, cellular chl *a* content decreased and the C:N ratio increased (Lesser, 1996). In *C. gracilis*, the mechanisms of UVR acclimation were similar to those for acclimation to high PAR, including a photoinhibition period followed by acclimation. The initial decline in carbon uptake was related to decreased Rubisco activity and rapid D1 reaction centre protein turnover. After returning to pre-UVR exposure conditions, the turnover rates of protein D1 and chl *a* were higher than pre-UVR acclimation conditions and the carbon uptake per unit chl *a* (but not per cell) was lower (Hazzard *et al.*, 1997).

The comparison of results observed and the mechanisms involved between the previous experiments are to be interpreted with caution owing to the difference in experimental design with respect to UVR + PAR sources and the proportions of UV-B:UV-A:PAR, periodicity of irradiance and other microbes present in the incubation vessels. *Macrocystis pyrifera* was grown at 15 °C and exposed for 2 h per day to UV-B ($10\,\mathrm{W\,m^{-2}}$) and PAR ($400\,\mu\mathrm{mol}$ quanta $\mathrm{m^{-2}\,s^{-1}}$). Similarly, *P. micans* was grown for 21 days at 20 °C under saturating sunlight (Lesser, 1996). Antarctic phytoplankton, however, were collected from coastal waters and were grown at -1.5 °C under sunlight for 9 days (Lesser *et al.*, 1996). The tropical diatom *C. gracilis* was grown at 23–25 °C and exposed to sunlight (Hazzard *et al.*, 1997). *Macrocystis pyrifera*, *P. micans* and *C. gracilis* were unialgal isolates. Antarctic phytoplankton was a mixed population that would include bacteria and microzooplankton, and would be expected to have grazing and mineral recycling. Similarly, UVR treatments varied from limited exposure to UV lamps (*M. pyrifera*) to natural UVR under 41% screened UVR + PAR (Antarctic phytoplankton) to natural UVR under untreated PAR (*C. gracilis*). Although UVR by lamps is considered adequate to understand mechanisms of UV inhibition, differences in UV-B:UV-A:PAR proportions can affect cellular responses and limit the extrapolation of results.

Differences in net damage are due not only to variable recovery rate but also to differences in sensitivity to UVR. Inhibition of primary production in surface phytoplankton was lower than in deep waters (Helbling *et al.*, 1992). Similarly, UVR inhibition was different for phytoplankton populations collected in shallow and in deep mixed layers (Neale, Cullen & Davis, 1998), and for corals (Dunlap, Chalker & Oliver, 1986) and seaweed (Häder *et al.*, 1996) collected at different depths in the same location. For example, specimens of the subtropical brown alga *Padina pavoni* collected from 7 m depth had higher inhibition of the photosynthetic oxygen production than those collected from rock pools, and the inhibition was present for longer periods of time. Genotypic differences affecting sensitivity were considered to be responsible for the response in *P. pavoni* (Herbert, 1990). In the red alga *Porphyra perforata* collected also from the intertidal zone, the observed reduced photosynthetic inhibition was due to lower damage as opposed to higher repair rate.

In conclusion, the variability in UVR inhibition in the field is expressed as a large variance in observed rates of primary production, as well as differences on the effect of UV-A and UV-B to overall inhibition. High inhibition coupled with high recovery, as observed in tropical phytoplankton, or low sensitivity coupled with low recovery rates can result in similar estimates of UVR damage. The worst scenario occurs under conditions of high damage with no or only low recovery. Temperature seems to play a central role in recovery rates and thus influence the balance on net damage. Additionally, genotypic differences might be related to lower sensitivity to UVR. Information on phytoplankton taxonomic composition and estimation of the degree of adaptation not only to UVR but also to other environmental stresses play an important role in explaining variability of carbon uptake measurements in UVR inhibition.

9.2.1.2 Pigmentation

Loss of pigmentation, or pigment bleaching, is considered to be a common response under UVR exposure (see e.g. Gerber & Häder, 1995). Contrary to this general pattern, there are studies that have reported no loss or even an increase in pigmentation. For example, long term exposure of *Selenastrum tricornutum* showed increased cellular chl *a* (Veen, Reuvers & Ronçak, 1997).

Pigment loss can occur after only a few hours of exposure (Sebastian, Scheuerlein & Döhler, 1994; Maske & Latasa, 1997). However, there is a differential sensitivity in the various pigments to UVR. Phycoerythrin concentration in *Cryptomonas maculata* decreased to 50% concentration in 7 h, whereas the decrease in carotenoids took 63 h and chl *a* 132 h (Döhler, 1985). In the Antarctic freshwater cyanobacterium *Phormidium murrayi* exposed to UV-B, cellular chl *a*, phycocyanin (PC), and the ratio PC:chl *a* (PC decreased faster than chl *a*) decreased while the ratio carotenoids:chl *a* increased (Quesada *et al.*, 1995). The interplay of these parameters is such that the ratio PC:chl *a* was a good predictor of population growth rate (Figure 9.4).

Phycobiliproteins seem to be extremely sensitive to UVR. In cyanobacteria from rice paddies, PC decreased steadily in concentration and the intensity of their fluorescence excitation and emission increased with UVR (Sinha *et al.*, 1995b). The quantum yield for the photodestruction of PC (numbers of molecules destroyed by photon absorbed) was four orders of magnitude higher at 295 ± 10 nm than at 647 nm (Lao & Glazer, 1996). At 295 nm, phycobilisomes account for half the absorption of the cells, while the chlorophyll–protein complexes account for the remaining absorbance. Thus, it is expected that PC will have a dominant role in absorbing UV-B photons in natural populations. Phycobilisomes absorb 54 times more radiation than does DNA at 295 nm. Combining the

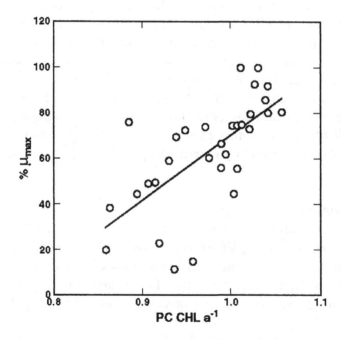

Figure 9.4. Relationship between UV-B-inhibited growth (expressed as % of maximum growth, μ_{max}) and the ratio of phycocyanin:chl *a*. (Redrawn from Quesada *et al.*, 1995. See also p. 278 (*Note*).)

absorption cross-section of the cell and the photodestruction quantum yield for PC, the ratio of damage to PC relative to DNA is approximately 20 (Lao & Glazer, 1996).

Pigment bleaching depends on the damaging wavelength. Although sunlight affected cells of *Cryptomonas maculata* in the green (540 nm) and red (673 nm) portions of the spectrum, UVR bleached cells only in the blue (440, 460 and 500 nm) region of the spectrum (Figure 9.5) and UV-A radiation had no effect on chl *a* or PC concentration, but increased the carotenoid: chl *a* ratio (Häder & Häder, 1991).

For isolates or mixed phytoplankton originating from different latitudes and exposed to solar UV-B, a consistent pattern in pigments is associated to the acclimation responses in carbon uptake (see Section 9.2.1.1). The damage in cyanobacteria from rice paddies was associated with the integrity of the phycobilisome that resulted in decreased efficiency of energy transfer to chl *a* (Sinha *et al.*, 1995b). Antarctic cryptomonads, exposed to natural levels of UV-B + UV-A, showed a decrease in energy transfer from accessory pigments to chl *a*. Increased fluorescence emission of phycoerythrin in field samples was interpreted as pigment uncoupling (Figueroa *et al.*, 1997). Similarly, decreased transfer of energy to chl *a* was observed in the temperate brown alga *Macrocystis pyrifera* (Clendennen *et al.*, 1996) and it was proportional to the time of exposure to UV-B. Cellular pigment concentration did not change. The same lack of change

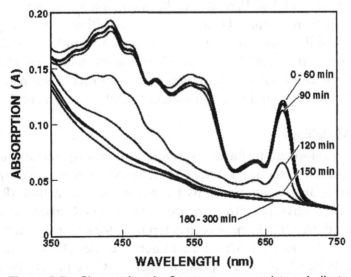

Figure 9.5. Pigment loss in *Cryptomonas maculata* as indicated by loss of absorption at different wavelengths for cultures exposed to sunlight for varying time intervals. (Redrawn from Gerber & Häder, 1995. See also p. 278 *(Note).*)

in cellular concentrations of chl *a* or the ratio carotenoids: chl *a* was observed in the subtropical diatom *Chaetoceros gracilis* Schutt after the cells had acclimated to UV-B (Hazzard *et al.*, 1997). This is different from acclimation to high PAR, where cellular pigment concentrations are known to decrease.

The experimental design of several of the studies on cyanobacteria are not close to environmental UV-B: UV-A: PAR, limiting the comparison with other studies. For example, Sinha *et al.* (1995b) had in their experiment a UV-B: UV-A ratio of 1, well above that of sunlight, while Lao & Glazer (1996) irradiated with UV-B 51 times higher than the average UV-B at 40° zenith angle (but see Quesada *et al.*, 1995).

Although a few patterns seem to be emerging with respect to the effect of UVR on cellular pigmentation, the variability is still too high to have confidence in these generalities. High UVR and short term exposures seem to result in decreased pigmentation. Longer and/or non-lethal exposures show no or positive effect. The damage caused by the different regions of the solar spectrum is related to the absorptive properties of pigments and to the role they play within the photosynthetic apparatus. Thus, UVR can increase carotenoid: chl *a* ratios as a result of increases in photoprotective carotenoids, and change absorption in the blue region of the spectrum. On the other hand, phycobiliproteins absorbing in the orange and red region of the spectrum are consistently bleached by all wavelengths, presumably increasing susceptibility of cyanobacteria and cryptomonads to UVR. Pigment synthesis might also play a role during acclimation to non-lethal UVR conditions resulting in unchanged or increased cellular concentrations in comparison to non-UVR conditions (e.g. *Chaetoceros gracilis*). Non-lethal exposures, while not changing pigment concentration, might decrease energy transfer between auxiliary pigments and chl *a*, and result in an effect similar to pigment decrease.

9.2.1.3 Modelling

The biological weighting functions (BWFs) for photosynthetic inhibition of phytoplankton, necessary to scale UVR to biological effective exposure (see Chapter 3), are determined by the polychromatic method (Rundel, 1983), where phytoplankton are exposed to UVR irradiation with background PAR. BWFs have been determined for Antarctic (Helbling *et al.*, 1992; Lubin *et al.*, 1992; Neale, Lesser & Cullen, 1994; Boucher & Prézelin, 1996a; Neale *et al.*, 1998), temperate (Cullen *et al.*, 1992) and tropical (Behrenfeld *et al.*, 1993a) phytoplankton. In addition to these phytoplankton-specific BWFs, photosynthetic inhibition in natural populations has been explained by biological effective doses obtained with the DNA weighting function (Setlow, 1974; Behrenfeld *et al.*, 1993b) as well as one for the Hill reaction in chloroplasts (Jones & Kok, 1966; Smith *et al.*, 1980). The BWFs differ by the shape of the curve, which results from different relative inhibition at different wavelengths. Differences in the absolute value of the inhibition are also present.

Once a biologically effective exposure is selected, the inhibition of carbon incorporation can be a function of the UVR dose (integrated over time) or of the UVR irradiance (dose rate or fluence rate). The shape of the inhibition to the UVR dose or dose rate has been determined as linear, sigmoidal or hyperbolic (for a review, see Smith & Cullen, 1995). No single response seems to explain photosynthetic inhibition by UVR (but see Behrenfeld *et al.*, 1993a). Inhibition curves in temperate cultures and Antarctic diatoms have been determined as non-linear and dependent on irradiance (or fluence rate) (Cullen *et al.*, 1992; Neale *et al.*, 1994). Using similar experimental design, other Antarctic diatoms have shown a non-linear inhibition response to dose or cumulative exposure (Neale *et al.*, 1998). The latter assumes reciprocity, or no difference between dose $(J\,m^{-2})$ at different irradiances $(W\,m^{-2})$. The effect can also be due to a lack of repair during exposure (Neale *et al.*, 1998). Other studies have shown a linear response of inhibition to weighted dose (Smith *et al.*, 1980; Behrenfeld *et al.*, 1993b). The diversity of BWFs and inhibition responses observed shows the variability in UVR sensitivity by phytoplankton and the different dynamics of net damage or inhibition to be expected in Nature.

Successfully modelling primary production as a function of both PAR and UVR inhibition depends on using an accurate BWF and determining the correct shape of the inhibition response (Cullen *et al.*, 1992; Boucher & Prézelin, 1996b; Neale *et al.*, 1998). Further research in this area continues actively to produce more accurate bases for modelling on larger time and space scales (e.g. Arrigo, 1994; Boucher & Prézelin, 1996b). The main limitation of this approach to date is the difficulty in extrapolating the results of short term photosynthesis (several hours) to time scales of days, weeks and months, relevant to ecological processes (Smith *et al.*, 1980; Smith & Cullen, 1995).

9.2.2 Nutrient metabolism

Several macronutrients (i.e. nitrate, nitrite, orthophosphate, silicic acid, ammonium, urea) as well as micronutrients and metals (i.e. iron, manganese, cadmium, selenium, copper) are needed for plant growth. Special attention is generally given to nitrate and ammonium as nitrogen is considered to be

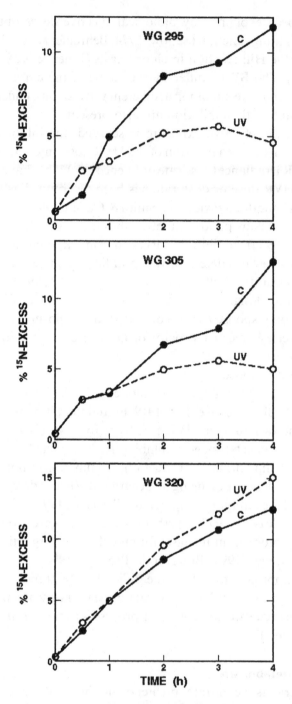

Figure 9.6. Uptake of ^{15}N-ammonium by the tropical diatom *Bellerochea yucatensis* exposed to UV-B (WG 295 and WG 305) and UV-A (WG 320) radiation (dashed lines, UV) with respect to control (solid lines, C). Cultures grown at 22 °C. (Redrawn from Döhler, 1995.)

one of the main limiting factors for phytoplankton growth in the marine environment. Nitrogen metabolism in phytoplankton is inactivated by UV-B radiation (Döhler, 1987). Both nitrate and ammonium uptake are affected (Figure 9.6; Döhler, 1987, 1992; Döhler & Buchmann, 1995), although the presence of ammonium depresses nitrate uptake (Döhler, 1991). Cellular amino acid concentrations also change under UV-B (Döhler, 1985, 1992; Goes *et al.*, 1995b). For example, UVR diminishes synthesis and intracellular accumulation of alanine and valine (Sinha *et al.*, 1995a).

Interspecific variability is observed for UV-B inhibition of nitrate and ammonium uptake. *Phaeocystis pouchetii* has a higher inhibition of nitrate uptake (Döhler, 1992) than do diatoms (Döhler, 1987). Among temperate diatoms, three species showed high sensitivity, while *Lithodesmium variabilis* did not (Döhler, 1987). Higher inhibition of ammonium uptake in comparison to nitrate uptake was observed both in isolates (Döhler, 1991) and in natural populations of Antarctic diatoms (Döhler, 1997).

The mechanisms of UVR inhibition of nitrogen uptake in phytoplankton are different to those of nitrogen assimilation. Decreased uptake by phytoplankton is interpreted as a response to a reduced supply of ATP and NADPH from direct effects of UV-B on the photosynthetic apparatus and pigment bleaching (Döhler, 1992; Döhler & Buchmann, 1995). The assimilation inhibition is interpreted as a UV-B effect on key enzymes in amino acid synthesis inferred from accumulation or decrease of intracellular amino acid concentrations. Thus, accumulation of glutamate and decrease in glutamine is interpreted as inhibition of glutamine synthetase (Döhler, 1985, 1987). Decreased glutamine levels and invariant glutamate pools indicate inhibition of glutamate synthase (Döhler, 1985). Decrease of both glutamate and glutamine may indicate inhibition of glutamate dehydrogenase (Döhler & Buchmann, 1995). These results of nitrogen metabolism under UV-B are similar to metabolic changes observed in phytoplankton under nitrogen limitation (Goes *et al.*, 1995a), suggesting similar mechanisms of nitrogen assimilation of cells under different stress.

In field populations, the action spectra for inhibition of ammonium is half as steep as that of carbon uptake, indicating greater inhibition of ammonium uptake by longer wavelengths (i.e. UV-A) and suggesting a greater effect deeper in the water column (Behrenfeld, Lean & Lee, 1995).

Whereas large areas of the world's ocean are considered to be nitrogen limited, orthophosphate can act as the major limiting nutrient in lakes. The inhibition of UVR on nutrient metabolism is of major importance in these areas, as is determining the combined effects of UVR and nutrient stress. Cullen & Lesser (1991) found that nitrate-limited *Thalassiosira pseudonana* was 8.6 times more sensitive to UV-B than were nitrate-replete cells. In contrast, results from long term exposure experiments are mixed. The marine diatom *Phaeodactylum tricornutum* exposed to UVR showed a lack of growth inhibition due to nitrogen limitation (Behrenfeld, Lee & Small, 1994). The freshwater green alga *Selenastrum capricornutum* grown under UV-B and phosphate limitation showed higher inhibition of photosynthesis and growth than for short term exposure (hours), but a decrease in the inhibition of nutrient limitation (Veen *et al.*, 1997).

Increased UVR inhibition of photosynthesis under short term experiments can be attributed to higher susceptibility of DNA damage in the absence of repair or acclimation of the cell to reduced levels of amino acid and protein, i.e. UV-absorbing compounds (Bothwell *et al.*, 1993). Conversely, the inhibition of phytoplankton growth by UVR was not observed in *P. tricornutum* as 'nutrient limitation exceeded the potential for limitation by UVR' (Behrenfeld *et al.*, 1995). A different mechanism was invoked for *S. capricornutum*, where the lack of growth inhibition under UV-B was not attributed to an insensitivity of the nutrient metabolism to UV-B but rather to a relaxation of nutrient limitation (i.e. an indirect effect) (Veen *et al.*, 1997).

From these different results, it can be hypothesised that UVR inhibits cellular processes for a certain period of time (hours) while the new stress is introduced to the system; this effect can be additive. If cells continue to grow, even at reduced rates, repair of DNA and other cellular processes can acclimatise the population to pre-UVR conditions (days to weeks). Thus, the population is able to return to the initial conditions (i.e. nutrient limitation) and neutralise the effect of the added UV-B stress. Alternatively, in cases where the effect of UVR inhibition is proportionally higher than that of nutrient limitation, the opposite effect, that of reduced nutrient stress under UVR, might be observed.

9.2.3   Respiration

The potential effect of UV-B and UV-A on respiration is of importance, particularly at subsurface depths where lower PAR irradiance, combined with UV-B, and most notably UV-A, could alter net primary production by increasing respiratory losses. Respiration measurements are limited by the presence of heterotrophs in field samples and by confounding oxygen produced via photosynthesis. Thus, respiration experiments are confined to isolates in the dark. The few studies available on the effect of UVR on dark respiration report no changes in rate. For example, experiments with green, red and brown algae show a small effect of UVR on respiration compared to inhibition of photosynthesis (Herrmann *et al.*, 1995). In another study, no increase in dark respiration was observed in the brown alga *Macrocystis pyrifera* (Clendennen *et al.*, 1996) even though photosynthesis decreased by 50% at light-saturated and light-limited rates. The absence of an UVR effect on respiration was also found by Larkum & Wood (1993) for seagrasses, phytoplankton and seaweed (data not shown in their paper).

One study reports stimulation of dark respiration in the green alga *Selenastrum capricornutum* after exposure to UVR. It was measured immediately (minutes) after exposure and correlated with decreased carbon uptake (25% of the control) (Beardall *et al.*, 1997). In the same study, respiration did not increase in the cyanobacterium *Aphanizomenon flos-aquae* after exposure to UVR. This alga had a higher photosynthetic inhibition (33% to 81% of control). Thus, UVR decreased carbon accumulation in the cell through respiration, but only in the species where inhibition of carbon uptake was lower. Is increased respiration a price to pay for lower sensitivity to UVR or for higher repair?

In summary, there is no evidence thus far of an important effect of UVR on phytoplankton respiration. Measurement of respiration under light conditions might change these conclusions. Additional measurements of dark respiration on different taxa and on species with diverse sensitivity to UVR might confirm the findings by Beardall *et al.* (1997). In particular, there is an immediate need to measure respiration rates during periods of acclimation to UVR when higher metabolism might be needed to support repair processes.

9.2.4   Organic matter excretion

The amount of extracellular carbon produced by phytoplankton has been a controversial subject for several decades (Sharp, 1977; Fogg, 1983; Bjørnsen, 1988; Wood *et al.*, 1992). Excretion of carbon by photosynthetic organisms is a widespread process associated with photosynthesis (Mague *et al.*, 1980). On average, phytoplankton excrete 5% to 25% of the carbon incorporated into particulate matter, both in monospecific cultures and natural populations (Fogg, Nalewajko & Watt, 1965; Mague *et al.*, 1980), and the amount excreted is a constant proportion of photosynthetic rates. Additional organic carbon excretion in phytoplankton seems to be associated with physiological imbalance due to events such as nitrogen limitation (Hellebust, 1965; Myklestad & Haug, 1972; Sakshaug *et al.*,

1973; Norrman *et al.*, 1995), in particular under high light conditions (Hellebust, 1965).

In spite of the obvious importance of phytoplankton excretion on phytoplankton physiology and its implications for the dissolved organic carbon pool and as substrate for the microbial loop, no studies have been carried out on the effect of UV-B on excretion. If excretion increases when algae are nutrient stressed, it can be speculated that the effects of UV-B stress would act in a similar way. Additionally, losses of potential UVR-screening compounds, such as amino acids and mycosporine-like amino acids (MAAs) (Vernet & Whitehead, 1996), could add to the cost of recovery.

9.2.5   Swimming and vertical migration

In addition to mixing by physical processes within the mixed layer, phytoplankton are able to control vertical position by active migration through swimming or changes in buoyancy (for a review and definitions, see Nultsch & Häder, 1988). Phytoplankton lack a photoreceptor for UV-B radiation that can cue the cells to avoid damaging radiation. As high UVR is associated with high PAR at or near the surface, both buoyancy and swimming might be expected to help the algae avoid inhibitory PAR irradiance, thus also influencing the UVR environment.

UV-B affects several aspects of cell motility: swimming speed of individual cells and the average speed in a population of flagellates, the gliding of cells living within sheaths, and the buoyancy of diatoms (Booth *et al.*, 1997). For example, exposure to UVR can decrease swimming speed in flagellated algae from $50 \, \mu m \, s^{-1}$ to $20 \, \mu m \, s^{-1}$ in 50 min as opposed to 140 min in controls (Häder, 1986). UV-B can also affect swimming by decreasing the percentage of cells with motility in the population, which results in a decrease in the average swimming speed (Figure 9.7) and alters the ability of cells to control their direction (Figure 9.8). For example, UV-B decreased the percentage of motile cells in a culture of the marine cryptophyte *Cryptomonas maculata* after 20 min of exposure, as opposed to 120 min in the controls (Häder & Häder, 1991). Lower inhibition in motility was expressed also as a delay in the onset of inhibition and by a decrease in the number of non-motile cells at the end of the experiment.

In gliding of cells usually surrounded by a mucilaginous sheath, as in benthic diatoms and cyanobacteria in mats, the gliding velocity, its motility and orientation are affected by UV-B in minutes (Donkor, Damian & Häder, 1993a,b). The recovery is also in minutes; however, it is not complete. There is interspecific variability in the UV-B damage to cell

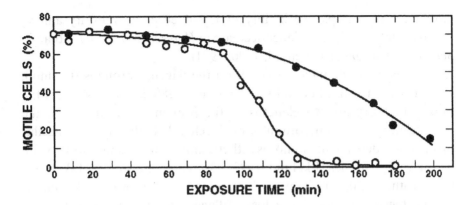

Figure 9.7. Loss of motility by *Cryptomonas maculata* exposed to sunlight with (filled circles) and without (open circles) UV-B. UV-B was excluded by placing a layer of ozone over the experiment. (Redrawn from Häder & Häder, 1991. See also p. 278 (*Note*).)

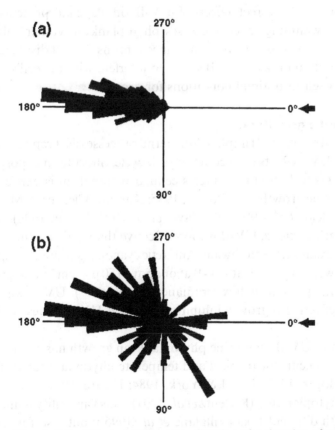

Figure 9.8. Effect of UVR on the direction of swimming of motile cells. (a) Negative phototaxis of *Euglena gracilis* in control experiment. (b) Loss of negative phototaxis after exposure to UVR. The arrow shows the direction of illumination. (Redrawn from Häder, 1986.)

motility. For example, variability of UV-B effect on motility among dinoflagellates showed a higher decrease in *Heterocapsa triquetra* than in *Scripsiella trochoidea* (Ekelund, 1990, 1991).

There is direct damage by UV-B on the flagellar proteins (Ekelund, 1991; Häder, 1993). Maximal sensitivity is at 280–290 nm, which is the range of absorption wavelengths of the proteins associated with the cytoskeleton in the microtubules of flagella. This damage to flagellar proteins is independent of photosynthetic inhibition for some, but not all, dinoflagellate species (Ekelund, 1993). Decreased motility by UV-B can initiate other escape mechanisms. In the case of the freshwater chlorophyte *Chlamydomonas reinhardtii*, a loss of flagellum after exposure to both UV-B and UV-A, resulted in encystment after 9 h of exposure to sunlight and an increase of 30% loss with respect to the controls without UVR (Van Donk & Hessen, 1996).

In summary, the direct effect of UV-B on flagellar proteins and diminishing swimming speed increases phytoplankton vulnerability to UVR by impairing possible avoidance mechanisms. As cell displacement maximises carbon uptake, UV-B seems to interfere with the cell's ability to position itself in optimal conditions for growth.

9.2.6   Specific growth rate

Specific growth rate (d^{-1}) implies 'long term' processes as it represents the synthesis of UVR effects on a complete cell cycle. Short term exposure to UVR (minutes to hours) that affects cellular metabolism is expected to decrease specific growth (e.g. Döhler, 1985; Karentz, Cleaver & Mitchell, 1991; Behrenfeld et al., 1992; Hargraves et al., 1993; Lesser, 1996). As for other cellular functions, UV-B is more effective than UV-A in inhibiting growth. For example, the freshwater Antarctic cyanobactetrium *Phormidium murrayi* showed no growth at UV-B at or above 150 μW cm^{-2} and growth inhibition was present at low irradiance (5 μW cm^{-2}). UV-B was nine times more effective at growth inhibition than was UV-A (Quesada et al., 1995).

The effect of UV-B on marine phytoplankton growth has been shown to be species specific for tropical and temperate phytoplankton (Calkins & Thordardøttir, 1980; Jokiel & York, 1984; Figure 9.9) as well as for Antarctic phytoplankton (Karentz et al., 1991). This variability is manifest not only with different taxa (Villafañe et al., 1995b) but also for species within a taxon (i.e. diatoms; Karentz et al., 1991). Centric (Döhler, 1985, 1995) and pennate (Hargraves et al., 1993) diatoms, as well as dinoflagellates species (Ekelund, 1991), show growth variability under UV-B in the

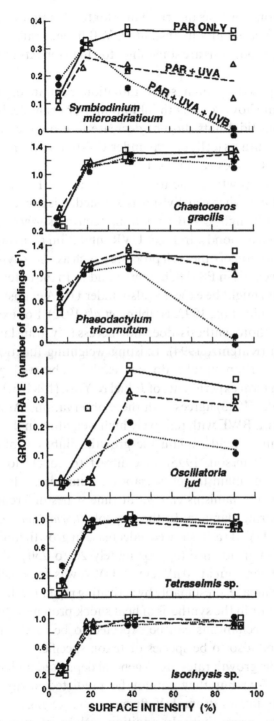

Figure 9.9. Growth rate (in doublings d⁻¹) of several species of phytoplankton exposed to solar UVR. PAR, photosynthetically available radiation. (Redrawn from Jokiel & York, 1984.)

threshold for inhibition as well as the irradiance for total inhibition. As for other parameters, the variability in absolute UV-B fluence rates and the UV-B: UV-A: PAR proportions are sometimes too high for extrapolation to field conditions.

Environments exposed to high solar radiation contain organisms resistant to UVR that show no growth inhibition. For example, benthic diatoms isolated from tidal flats did not decrease in growth compared with UV-B screened cultures (cells grown under UV-B + PAR, no UV-A was added) (Peletier, Gieskes & Buma, 1996).

Direct inhibition of growth is due to UV-B damage of DNA (Buma, Engelen & Gieskes, 1997) when cell division is arrested. It is expected to be maximal in areas with high UV-B radiation, such as surface waters, shallow intertidal zone or ponds. Indirect UVR inhibition of growth can result from photoinhibition of other cell processes, such as photosynthesis, where the target molecule is in PS II (Chapters 1 and 7). In the latter case, a high growth inhibition might be expected also under UV-A exposure (e.g. Cullen *et al.*, 1992; Lubin *et al.*, 1992; Helbling *et al.*, 1992). For example, UV-B inhibition of photosynthesis does not necessarily translate into inhibition of growth (Karentz, 1994). Existing weighting functions for photosynthesis and carbon uptake do not explain observed growth inhibition in the long term experiments of Jokiel & York (1984), while the formulation by Rundel (1983) agrees with these observations. Assuming that reciprocity holds, a BWF with relatively high weighting in the UV-A region of the spectrum would explain the growth inhibition observed (Peterson, Smith & Patterson, 1995), in contrast to conclusions that consider DNA to be the main target (see above, Buma *et al.*, 1997).

Recovery or repair mechanisms can act at time scales different from those used for measuring damage. In the case of *Phormidium murrayi*, increasing the UV-A: UV-B ratio showed a decrease in growth inhibition proportional to UV-A irradiance (approximately 20 to $250\,\mu W\,cm^{-2}$) when UV-B was kept constant at $30\,\mu W\,cm^{-2}$. UV-A was responsible for counteracting UV-B damage, presumably by activating repair mechanisms, either for DNA repair or in the synthesis of heat shock proteins (Quesada *et al.*, 1995). Repair mechanisms can be expected to be as variable as damage processes, and also to be species or taxon specific.

In summary, specific growth rates are in general depressed under UVR (both UV-B and UV-A) but they cannot be predicted always from photosynthetic inhibition or from UV fluence rates owing to repair mechanisms or variable sensitivity. In addition to laboratory cultures, more experiments determining growth rates of individual species in mixed populations are needed in order to improve predictions of specific growth rates under natural UV-B: UV-A: PAR proportions and to elucidate their relationship to other cellular processes.

9.2.7 Cell size

A differential effect of UVR on cell size has been observed for diatom cultures (Karentz *et al.*, 1991), a higher amount of damage being associated with smaller cells. Small cells have a shorter light pathlength with reduced absorption and refraction by cytoplasmic components between the cell membrane and nuclear DNA. As a result, increased UV-B reaches the DNA (Raven, 1991; Garcia-Pichel, 1994; Booth *et al.*, 1997).

Increases in cell size have been observed in cultures under UV-B exposure (Döhler, 1985; Behrenfeld *et al.*, 1992; Veen *et al.*, 1997) and are a consequence of the reduction in specific growth rates. Buma *et al.* (1997) attributed the increased cell size to an arrested cell cycle due to residual DNA damage, as measured by the cellular concentration of thymidine dimer. The increase in cell size is due to carbon uptake in the absence of cell division. This effect is not observed always in association with decreased growth rate, as in the case of ice algae (Nilawati, Greenberg & Smith, 1997).

Coastal waters have, on average, a higher proportion of larger cells than do open waters (Malone, 1980). For example, more than 80% of the nearshore phytoplankton biomass was associated with cells $> 10\,\mu m$ in Terre Adélie, Antarctica, during summer, while, 70 km offshore, cells $> 10\,\mu m$ represented only 30% of the total biomass, and 59% of the cells were between 1 and 10 μm (Fiala & Delille, 1992). Within Antarctic coastal waters, high chl *a* accumulations are dominated by large cells (e.g. $> 20\,\mu m$), while low chl *a* concentrations are dominated by smaller cells (Holm-Hansen & Mitchell, 1991; Bidigare *et al.*, 1996). On the basis of increased inhibition found in smaller cells, presumably owing to their smaller light pathlength, we might hypothesise that oceanic phytoplankton may have a higher sensitivity to UV-B.

9.2.8 Community composition

Damage to organisms exposed to UV-B varies by 100-fold between species (e.g. Ekelund, 1990; Karentz *et al.*, 1991). The differential sensitivity to UV-B suggests a change in species composition due to long term UV-B exposure, with more UV-tolerant species ultimately dominating (Worrest *et al.*, 1981). In general, on the basis of culture studies, diatoms are the most resistant to UVR, followed by prymnesiophytes and other flagellates, such as cryptomonads. Green algae and cyanobacteria are usually considered to be as resistant as diatoms. This differential sensitivity among phyla has been established from several cellular processes, such as nitrogen uptake (Döhler, 1992), radiocarbon uptake (Davidson & Marchant, 1994; Helbling et al., 1994; Vernet et al., 1994; Villafañe et al., 1995a; Beardall et al., 1997; Gieskes & Buma, 1997), specific growth rate (Worrest et al., 1981; Jokiel & York, 1984; Davidson et al., 1994; Karentz, 1994; Villafañe et al., 1995b) and cell abundance in natural populations (Karentz & Spero, 1995). There are exceptions to this trend. For example, Ekelund (1994) reported that the photosynthesis of the diatom *Phaeodactylum tricornutum* and the green flagellate *Euglena gracilis* was more sensitive to UV-B than that of the dinoflagellates *Prorocentrum minimum* and *Heterocapsa triqueta*.

Experiments with mixed populations can be used to test predictions on phytoplankton succession based on differential UVR sensitivity established in the laboratory and from short term field experiments. Davidson, Marchant & de la Mare (1996) found that 2 day UV-B exposures of exponentially growing mixed cultures at 0 °C favoured *Phaeocystis antarctica* over diatoms. They rejected a hypothesis based on previous experiments where they found diatoms were three to five times more resistant than *P. antarctica* (Davidson et al., 1994). Part of the discrepancy can be attributed to differences in experimental design; this latter experiment had $9150 \, J \, m^{-2}$ erythemal UV-B in a 2 day exposure while previous cultures had been exposed to 20% to 650% of average springtime UV-B radiation in the area.

Few studies are available on effects of UVR on communities at longer time scales and those published illustrate the difficulty of extracting generalisations from mixed populations. Although Davidson et al. (1996) did not find differences in their experiments between 2 and 9 day exposures, experiments of several weeks duration indicated community changes due to UV-B. Decreased community diversity was observed in mesocosm experiments (720 litres for two to four weeks) on natural phytoplankton from estuarine waters (Worrest, Van Dyke & Thomson, 1978). These changes were observed after five weeks, with decreased biomass (chl *a* concentration) and decreased autotrophic organisms. A later experiment with diatoms from the same estuary, carried out at four UV-B irradiances and for four weeks, also showed changes in community composition, decreased biomass (ash-free dry weight), chl *a* concentrations, and depressed primary production. It is not clear whether organisms other than diatoms were present in these experiments. Another mesocosm study in the Swedish West Coast with natural phytoplankton showed that diatoms were the most sensitive and were replaced by chrysophytes following UV-B exposure of 0.64 W m^{-2} for 6 h (Wänberg, Selmer & Gustavson, 1996). Cryptophytes, prymnesiophytes and dinoflagellates showed intermediate growth rates. Cell size of the community decreased because diatoms were replaced by smaller flagellates, contrary to reports regarding Antarctic phytoplankton (Figure 9.10; Helbling, Villafañe and Holm-Hansen, 1994; Villafañe et al., 1995a). Conversely, cell size within centric diatoms increased because Chaetoceros spp. were replaced by Rhizosolenia spp. In contrast, the predicted shift from less to more resistance species, e.g. from flagellates to diatoms, was observed in a two week experiment where natural Antarctic populations were exposed to ambient UVR, although similar chl a and particulate carbon accumulation were observed under UVR and UVR + PAR (Villafañe et al., 1995a).

Physiological responses were observed in these long term studies. As a result of the shift from flagellates to diatoms, a decreased sensitivity of photosynthesis was observed in the phytoplankton exposed to UVR and the amount of UV-absorbing compounds (e.g. MAAs) increased as well (Villafañe et al., 1995a). The higher resistance by diatoms, as compared

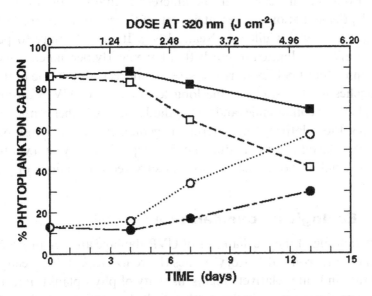

Figure 9.10. Changes in phytoplankton community composition as a function of time and UV treatment. Note decrease of flagellates (circles) and increase in diatoms (squares). Filled symbols correspond to control (PAR only) and open symbols to incubations exposed to UVR + PAR. (Redrawn from Villafañe et al., 1995a.)

with flagellates (in particular the colonial prymnesiophyte *Phaeocystis pouchetii*, Davidson *et al.*, 1994; Karentz & Spero, 1995), seems to be related also to inorganic nitrogen uptake (Döhler, 1992). The shift to larger cell size under UVR is a less consistent change, sometimes but not always present (Davidson *et al.*, 1996; Wänberg *et al.*, 1996).

Information on the effect of enhanced UVR on Antarctic communities on a decadal scale was obtained from changes in diatom composition in sediment cores. The effect of UV-B on natural populations during a 20 year period, coinciding with decreases in ozone, was assessed in laminated Antarctic sediments characterised by high stratification and, due to anoxia, no bioturbation (McMinn, Heijnis & Hodgson, 1994). The changes observed in the relative abundance of dominant diatom species and cell size over this period (McMinn *et al.* 1997) are within the variability observed in the last 600 years. In conclusion, changes due to UV-B, if present, are as yet too small to show above the environmental noise (as pointed out by Smith *et al.* (1992), for primary productivity). The limitations of the above-mentioned study include differential preservation of diatoms in sediments and the unknown relationship between preserved diatom frustules and total diatom abundance in the water column (Bothwell, Karentz & Carpenter, 1995). These limitations are important but not different in scale from the inherent limitations of, for example, fixed-depth incubations of present day estimations of primary productivity (Smith *et al.*, 1980; Cullen & Neale, 1994). Its applicability to pelagic phytoplankton is limited to fjords that have nearly permanent ice cover (McMinn *et al.*, 1994). Nevertheless, sediment records are one of the few approaches for the study of the long term effects of UV-B on marine phytoplankton communities and this method deserves further consideration. It places the relatively short term (in geological times) stress on the Antarctic coastal ecosystem due to ozone depletion (25 years) within the natural variability of a system that has survived for millions of years.

9.3   Ecological considerations

Understanding of the mechanisms of UVR damage and repair processes comes mainly from laboratory studies (Karentz *et al.*, 1991; Veen *et al.*, 1997) that indicate relatively high sensitivity of phytoplankton to UVR. These experiments are challenged by field observations of high phytoplankton concentrations close to the sea surface, under conditions that are predicted to be UVR inhibitory. Even field experiments appear to exaggerate UVR damage due to the constraints imposed by placing the incubation vessels at a constant depth for periods in excess of 1 h. This contradiction between experimental results and field observations suggests that phytoplankton are more resilient to UVR damage than expected or, that the UVR irradiance to which the cells are exposed in the field is such that it favours repair and protection, in particular from UV-B. Present understanding and quantification of UVR damage and repair processes are not close to elucidating the discrepancy. Three approaches are being considered to address the problem: (a) simulation of mixing in incubations with either cultures (Veen *et al.*, 1997) or natural populations (Ferreyra *et al.*, 1994); (b) estimation of UVR damage to populations exposed to different mixing regimes (Jeffrey *et al.*, 1996); and (c) modelling cell movement in the water column, where mixing and vertical movement of the cells are combined with changing UV-B: UV-A: PAR and irradiance with depth, known action spectra and rates of damage and recovery for phytoplankton (Cullen & Neale, 1994; Gieskes & Buma, 1997; Neale, Davis & Cullen, 1998).

Several environmental factors, such as the presence of PAR and the interplay between mixing processes and PAR, are generally considered to decrease UVR inhibition. Several of these factors will be considered in more detail.

9.3.1 Photosynthetically active and UV radiation

The relative sensitivity of phytoplankton to different wavelengths and the effect of irradiance at the various wavelengths are discussed in Chapters 1, 3 and 6 and are best described by an action spectra. There is one such spectrum for each cellular process. The interplay of PAR and UVR has also a role in repair mechanisms (see Chapters 6 and 7). Most of these studies have been carried out for short term processes (< 1 h) and there are few that can be applied to ecological time scales.

There are several studies on UVR inhibition presented in Section 9.2 that describe the effect of PAR. Exposure to high PAR before or during UVR exposure decreased UVR sensitivity in certain organisms (Caldwell & Flint, 1994; Berkelaar, Ormrod & Hale, 1996). For example, decreased damage in growth and motility of dinoflagellates was ascribed to the presence of background PAR (Ekelund, 1991). PAR also affected the level of dark respiration after an illumination period in cells with low UVR inhibition of photosynthesis (Beardall *et al.*, 1997). In field populations, Antarctic phytoplankton dominated by cryptomonads showed higher photosynthetic inhibition when exposed to artificial UV-B + UV-A + PAR at 0 °C for 3 h if collected in the morning than they did at noon (Figueroa *et al.*, 1997), presumably due to the presence of PAR. Future studies should include pre-acclimation to PAR and PAR + UVR separately, in order to discriminate between these two factors, as environmental UVR is associated with higher PAR. As proposed by Hazzard *et al.* (1997), the acclimation to PAR and UVR might be similar, but this hypothesis needs to be tested in the field.

PAR levels also influence pigment recovery after exposure to UVR. For example, chl *a* recovered in cells exposed to high PAR (500 µmol quanta $m^{-2} s^{-1}$) after a 22 h dark period to a higher level than cells exposed to lower PAR (100 or 300 µmol quanta $m^{-2} s^{-1}$; Beardall *et al.*, 1997).

Some populations show higher inhibition by PAR than by UVR, due either to higher sensitivity to PAR or decreased repair rate of UVR damage than that of PAR (Dring, Schmid & Lüning, 1995). For example, phytoplankton in high altitude tropical lakes, when exposed to high solar irradiance owing to the formation of a diel thermocline, were highly inhibited by PAR, although UV-B intensified the effect (Vincent, Neale & Richerson, 1984). Nitrate uptake is more affected by PAR than by UVR, being quite different from the situation with ammonium uptake (Döhler, 1992). PAR had more effect on the recovery of nitrate uptake from PAR and UV-B inhibition than on the recovery of ammonium uptake.

No UVR experiments can provide realistic results without the presence of PAR as it affects estimates of damage. The influence of PAR on repair and recovery processes is one of the main factors to be considered in the effect of mixing on UV photobiology.

9.3.2   Mixing in the upper water column

Mixing of cells in the upper water column within the mixed layer affects the average irradiance to which a cell is exposed during the day (Kullenberg, 1982; Smith & Baker, 1982; Smith, 1989). Several studies have speculated about the possible role of mixing on alleviating UVR inhibition when cells are mixed below the euphotic zone, in particular below the 0.1% incident UV-B (and/or UV-A) (e.g. Bidigare, 1989; Karentz, 1991; Holm-Hansen *et al.*, 1993). The distribution of UVR shows rapid attenuation of UV-B, while UV-A and blue/green irradiances are present at depth (see Chapter 2). It is generally assumed that, if cells are exposed to relatively high PAR with little or no UVR, the repair mechanisms would counteract any damage suffered when in surface waters and the overall water column productivity would not decrease. The overall effect of mixing is a function of the rate of damage and repair of a given process (see e.g. Lesser, Cullen & Neale, 1994), the shape of the damage response function to UVR, and the cumulative damage during the day (or length of incubation). Experiments where UVR intensity was manipulated to resemble mixing in the upper water column showed increased photosynthetic inhibition with respect to fixed controls on sunny days (Helbling *et al.*, 1994). The worsening of the UV effect on photosynthesis by mixing is the result of high inhibition, with a non-linear response to cumulative exposure (Booth *et al.*, 1997; and see chapter 3). In contrast, phytoplankton communities in the Scotia Sea dominated by the diatom *Thalassiosira gravida* showed less photosynthetic inhibition when exposed to variable UVR (Ferreyra *et al.*, 1994). Similar results were observed on cloudy days when the exposure of cells to saturating PAR was interpreted as beneficial to the cells (Helbling *et al.*, 1994). Thus, when photosynthetic inhibition is low, the relationship between response and exposure can be described as linear, and mixing decreases UVR inhibition (Smith & Cullen, 1995; Jeffrey *et al.*, 1996). Although a deep mixed layer can be considered to be beneficial for decreasing UVR damage, it is the ratio between the time the cell is exposed to UVR and the time needed for repair (under blue and UV-A light) that might be more relevant. This ratio will depend on the mixing rate, as well as the relation between the depths of 1% UVB, 1% UV-A and 1% PAR.

9.3.3 Spatial and temporal scales

The effect observed for UV-B exposure during short term (seconds to hours) and long term (days to months) experiments is not similar, as different processes are involved in damage and avoidance mechanisms. Experiments have been done to cover the different time scales in which processes operate. In this way, experiments involved in quenching of oxygen radicals are measured from picoseconds to minutes, studies of DNA repair and photosynthesis span hours to days and experiments for growth span days to months. Ecological and physiological processes are based on the molecular processes presented elsewhere (Chapters 3, 6 and 7). Although separated for presentation in this book, environmental, ecological, physiological and molecular processes operate in a continuum.

Environmental UVR stress in phytoplankton occurs at various temporal and spatial scales. A major source of interannual variability is cloudiness (in minutes to hours), while ozone trends are in decades. At any one location, the temporal variability of atmospheric origin is superimposed over a large seasonal change due to changing sun elevation and day length (Booth *et al.*, 1997; and see Chapter 2). In addition, the most important spatial changes relate to changes in latitude. For aquatic ecosystems, an additional gradient is the vertical distribution of UVR in the water column, function of incident UVR, properties of the air–water interface, and the absorption properties of water, dissolved organic carbon and particulates in the water column, in the absence of very high scattering (Smith & Baker, 1981; Scully & Lean, 1994; Booth et al., 1997).

In general, organisms from high latitudes show higher net damage to UVR (see also Section 9.2.1.1). For example, in laboratory experiments (Döhler, 1996) on two species of the genus Odontella (O. sinensis, a temperate species isolated from the North Sea and grown at 18 °C, and O. weisflogii, an Antarctic species grown at 4 °C) nitrogen uptake was more affected by PAR + UV-B (no UV-A) in the Antarctic than were the temperate species. A similar pattern was described for photosynthetic inhibition in cultures of Antarctic and tropical diatoms (Lesser et al., 1996; Hazzard et al., 1997).

Few field experiments exist to show variability along latitudinal gradients. Notably, Behrenfeld et al. (1992) sampled from 8° N to 60° S on two cruises in the South Pacific, covering equatorial, tropical, temperate

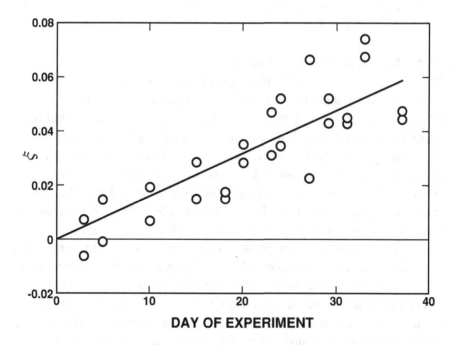

Figure 9.11. Relationship between phytoplankton sensitivity to UV-B (ξ) during the length of the experiments. Sensitivity calculated as specific growth per daily integrated UV-B radiation. (Redrawn from Behrenfeld et al., 1992.)

and subantarctic waters. Their results suggest that photoinhibition of carbon uptake, weighted to DNA action spectrum (Setlow, 1974), is linear with UV-B dose, with no apparent threshold (Figure 9.11). The percentage decrease in carbon uptake per unit of UV-B dose was low and constant from 0 45° S and increased at higher latitudes, with the exception of one sample at the equator that was also as high as the ones in high latitudes. In conclusion, some variability among latitudes was observed in this study in terms of specific sensitivity to UVR. Higher damage in tropical areas might be expected due to higher UVR irradiance.

In another series of studies, Antarctic phytoplankton showed higher photosynthetic inhibition by UVR than did temperate (33° S in the Pacific Ocean) and tropical phytoplankton. On average, under natural UVR on sunny days, spring surface populations in Antarctic coastal waters showed increased carbon incorporation of 50% due to screening of UV-B and an increase of 120% after screening UV-B + UV-A (Holm-Hansen *et al.*, 1993). Temperate surface phytoplankton in Valparaiso Bay, off central Chile, increased carbon incorporation by only 4% (− UV-B) and 14% (− UV-B − UV-A) (Helbling *et al.*, 1993). Tropical phytoplankton from the mixed layer did not show an increase in carbon uptake after screening (Helbling *et al.*, 1992). These results suggest adaptation of phytoplankton at lower latitudes to UV-B and UV-A, either because of a decreased sensitivity or an increased rate of repair mechanisms.

Changes in UVR sensitivity in different seasons in temperate areas can be inferred from the patterns observed between high and low latitudes. It can be hypothesised that winter populations are more sensitive to UVR than those found in the summer. Few studies have been conducted to test this. One such study of seasonal experiments off the coast of Chile by Montecino & Pizarro (1995) (30° and 33° S) showed no UVR sensitivity in carbon uptake for 2–6 h experiments for surface (0–2 m depth) phytoplankton during three seasons. Inhibition of carbon uptake was observed only after 2 h exposure to 305 nm wavelength at $1 \mu W \, cm^{-2} \, nm^{-1}$, suggesting UV-B inhibition of phytoplankton would occur only for the upper 3 m of the water column under severe ozone depletion.

The long term effect of UVR on phytoplankton has been studied in experiments with emphasis on the interaction between food web components (e.g. Bothwell *et al.*, 1993; Bothwell, Sherbot & Pollock, 1994). In these time scales, both biotic and abiotic components are of importance for the overall response of individual components of the food web, such as phytoplankton, to UVR (Vernet & Smith, 1997). There is increasing evidence that effects of several hours to 1–2 days are different from results after several days (see e.g. Villafañe *et al.*, 1995a), and can change again after several weeks (Cabrera, López & Tartarotti, 1997). How long an experiment needs to last to have 'ecological relevance' is a matter of debate. As for terrestrial ecosystems (see e.g. Caldwell & Flint, 1994), responses at the ecosystem level need to consider the whole system, and experiments should last as appropriate for the question addressed. Experiments of several years are needed for the questions on interannual variability to UVR, months for studies of seasonal variability in responses to UVR, and weeks for impacts of a seasonal phenomenon such as the 'ozone hole' or the impact of ice melting. Modelling can be essential in designing these experiments.

9.4   Summary

Several patterns of UVR inhibition have emerged in the last decade, since the discovery of the ozone hole spurred research of UVR effects on Antarctic phytoplankton. The majority of studies, carried out with isolates or with natural populations during 1 to 24 h experiments, have corroborated earlier work from the 1970s. Quantification of the net damage has been documented for several environments, different phytoplankton assemblages and latitudes. Mechanisms of inhibition, and sometimes recovery, have been defined for several cellular processes, such as photosynthesis and growth, with less emphasis on the loss terms, such as respiration and exudation. A few studies, based on longer experiments of days to weeks have shown long term acclimation of isolates and natural mixed populations under non-lethal UV-B radiation, demonstrating different target molecules and mechanisms of damage than at shorter time scales. The species changes observed in these studies are not always predictable from relative inhibition of isolates.

Future studies face an increasing challenge to extrapolate experimental results to natural environments. In spite of a large body of studies available, we are far from understanding how phytoplankton respond to environmental UVR. This limitation is due in part to the lack of realistic irradiances in experimental designs, as in laboratory studies, to a lack of relevant hypotheses of the role of UVR inhibition within existing environmental conditions, and to the difficulty of carrying out the experiments on the appropriate time scales of inhibition and recovery. Unless more care is taken in setting up experiments and describing UVR during exposures, results may be anecdotal and cannot be used on global estimates of UVR effects. It is clear that mechanisms of repair and recovery are fundamental to estimates of net UVR damage and these processes cannot be estimated realistically under irradiances 10 to 50 times higher than environmental UVR. More emphasis should be taken also to reproduce meaningful environmental conditions, representative of the area of interest, in order to assess UVR effects in the context of other limiting factors such as nutrients, temperature and grazing.

There is no unique approach to the study of UVR effects on phytoplankton. Experiments on isolates are indispensable to understand the underlying mechanisms of damage and repair, and are irreplaceable for studies under controlled conditions. Experiments with natural populations provide more realistic phenotypic and genotypic components as well as other necessary properties in studies of irradiation, such as pre-acclimation and physiological conditions. The interaction between these two types of approach will provide the necessary insights to understand, and eventually to predict, UVR effects on phytoplankton.

Acknowledgements

I thank P. A. Matrai, T. Moisan and V. Montecinos for revisions on an earlier version of the chapter. This project was funded by the InteraAmerican Institute for Global Change Research (ISP-2 grant) and by the National Science Foundation grant DPP-9632763.

References

Arrigo, K. R. (1994). Impact of ozone depletion on phytoplankton growth in the Southern Ocean: large-scale spatial and temporal variability. *Marine Ecology Progress Series*, **114**, 1–12.

Beardall, J., Berman, T., Markager, S., Martinez, R. & Montecino, V. (1997). The effects of ultraviolet radiation on respiration and photosynthesis in two species of microalgae. *Canadian Journal of Fisheries and Aquatic Sciences*, **54**, 687–96.

Behrenfeld, M. J., Chapman, J. W., Hardy, J. T. & Lee, H. I. (1993a). Is there a common response to ultraviolet-B radiation by marine phytoplankton? *Marine Ecology Progress Series*, **102**, 59–68.

Behrenfeld, M. J., Hardy, J., Gucinsky, H., Hanneman, A., Li, II, H. & Wones, A. (1993b). Effects of ultraviolet-B radiation on primary production along latitudinal transects in the South Pacific Ocean. *Marine Environmental Research*, **35**, 349–63.

Behrenfeld, M. J., Hardy, J. T. & Lee, III, H. (1992). Chronic effects of Ultraviolet-B radiation on growth and cell volume of *Phaeodactylum tricornutum* (Bacillariophyceae). *Journal of Phycology*, **28**, 757–60.

Behrenfeld, M. J., Lean, D. R. S. & Lee, III, H. (1995). Ultraviolet-B radiation effects on inorganic nitrogen uptake by natural assemblages of oceanic plankton. *Journal of Phycology*, **31**, 25–36.

Behrenfeld, M. J., Lee, III, H. & Small, L. F. (1994). Interactions between nutritional status and long-term responses to ultraviolet-B radiation stress in a marine diatom. *Marine Biology*, **118**, 523–30.

Belzile, C., Demers, S., Lean, D. R. S., Mostajir, B., Roy, S., de Mora, S. J., Bird, D., Gosselin, M., Chanut, J.-P. & Levasseur, M. (1998). An experimental tool for the study of the effects of ultraviolet radiation on planktonic communities: a mesocosm approach. *Environmental Technology*, **19**, 667–82.

Berkelaar, E. J., Ormrod, D. P. & Hale, B. A. (1996). The influence of photosynthetically active radiation on the effects of ultraviolet-B radiation on *Arabidopsis thaliana*. *Photochemistry and Photobiology*, **64**, 110–16.

Bidigare, R. R. (1989). Potential effects of UV-B radiation on marine organisms of the southern ocean: distributions of phytoplankton and krill during austral spring. *Photochemistry and Photobiology*, **50**, 469–77.

Bidigare, R. R., Iriarte, J. L., Kang, S.-H., Karentz, D., Ondrusek, M. E. & Fryxell, G. A. (1996). Phytoplankton: quantitative and qualitative assessments. In *Foundations for Ecological Research West of the Antarctic Peninsula*, ed. R. M. Ross, E. E. Hofmann & L. B. Quetin, pp. 173–98. American Geophysical Union, Washington, DC.

Björn, L. O. & Teramura, A. H. (1993). Simulation of daylight ultraviolet radiation and effects of ozone depletion. In *Environmental UV Photobiology*, ed. C. S. Young, L. O. Björn, J. Moan & W. Nultsch, pp. 41–71. Plenum Press, New York.

Bjørnsen, P. K. (1988). Phytoplankton exudation of organic matter: Why do healthy cells do it? *Limnology and Oceanography*, **33**, 151–4.

Booth, C. R., Morrow, J. H., Coohill, T. P., Cullen, J. J., Frederick, J. E., Häder, D.-P., Holm-Hansen, O., Jeffrey, W.H., Mitchell, D. L., Neale, P. J., Sobolev, I., van der Luen, J. & Worrest, R. C. (1997). Impacts of solar UVR on aquatic microorganisms. *Photobiochemistry and Photobiology*, **65**, 252–69.

Bothwell, M. L., Karentz, D. & Carpenter, E. J. (1995). No UVB effect? *Nature*, **374**, 601.

Bothwell, M. L., Sherbot, D. & Pollock, C. M. (1994). Ecosystem response to solar ultraviolet-B radiation: influence of trophic-level interactions. *Science*, **265**, 97–100.

Bothwell, M. L., Sherbot, D., Roberge, A. C. & Daley, R. J. (1993). Influence of natural ultraviolet radiation on lotic periphytic diatom community growth, biomass accrual, and species composition: short-term versus long-term effects. *Journal of Phycology*, **29**, 24–35.

Boucher, N. P. & Prézelin, B. B. (1996a). An *in situ* biological weighting function for UV inhibition of phytoplankton carbon fixation in the Southern Ocean. *Marine Ecology Progress Series*, **144**, 223–36.

Boucher, N. P. & Prézelin, B. B. (1996b). Spectral modeling of UV inhibition of *in situ* Antarctic primary production using a field-derived biological weighting function. *Photobiochemistry and Photobiology*, **63**, 407–18.

Buma, A. G. J., Engelen, A. H. & Gieskes, W. W. C. (1997). Wavelength-dependent induction of thymine dimers and growth rate reduction in the marine diatom *Cyclotella* sp. exposed to ultraviolet radiation. *Marine Ecology Progress Series*, **153**, 91–7.

Cabrera, S., López, M. & Tartarotti, B. (1997). Phytoplankton and zooplankton response to ultraviolet radiation in a high altitude Andean lake: short- versus long-term effects. *Journal Plankton Research*, **19**, 1565–82.

Caldwell, M. M. & Flint, S. D. (1994). Stratospheric ozone reduction, solar UV-B radiation and terrestrial ecosystems. *Climatic Change*, **28**, 375–94.

Caldwell, M. M., Teramura, A. H., Tevini, M., Bornman, J. F., Björn, L. O. & Kulandaivelu, G. (1995). Effects of increased solar ultraviolet radiation on terrestrial plants. *Ambio*, **24**, 166–73.

Calkins, J. & Thordardøttir, T. (1980). The ecological significance of solar UV radiation on aquatic organisms. *Nature*, **283**, 563–6.

Clendennen, S. K., Zimmerman, R. C., Powers, D. A. & Alberte, R. S. (1996). Photosynthetic response of the giant kelp *Macrocystis pyrifera* (Phaeophyceae) to ultraviolet radiation. *Journal of Phycology*, **32**, 614–20.

Cullen, J. J. & Lesser, M. P. (1991). Inhibition of phytoplankton photosynthesis by ultraviolet radiation as a function of dose and dosage rate: result for a marine diatom. *Marine Biology*, **111**, 183–90.

Cullen, J. J. & Neale, P. J. (1994). Ultraviolet radiation, ozone depletion and marine phytosynthesis. *Photosynthetic Research*, **39**, 303–20.

Cullen, J. J. & Neale, P. J. (1997). Biological weighting functions for describing the effects of ultraviolet radiation on aquatic systems. In *The Effects of Ozone Depletion on Aquatic Ecosystems*, ed. D.-P. Häder, pp. 97–118. R. G. Landes Co., Austin, TX.

Cullen, J. J., Neale, P. J. & Lesser, M. P. (1992). Biological weighting function for the inhibition of phytoplankton photosynthesis by ultraviolet radiation. *Science*, **258**, 646–50.

Davidson, A. T., Bramich, D., Marchant, H. J. & McMinn, A. (1994). Effects of UV-B irradiation on growth and survival of Antarctic marine diatoms. *Marine Biology*, **119**, 507–15.

Davidson, A. T. & Marchant, H. J. (1994). The impact of ultraviolet radiation on *Phaeocystis* and selected species of antarctic marine diatoms. In *Ultraviolet Radiation in Antarctic: Measurements and Biological Effects*, Antartic Research Series, ed. C. S. Weiler & P. A. Penhale, pp. 187–206. American Geophysical Union, Washington, DC.

Davidson, A. T., Marchant, H. J. & de la Mare, W. K. (1996). Natural UVB exposure changes the species composition of Antarctic phytoplankton in mixed culture. *Aquatic Microbial Ecology*, **10**, 299–305.

Döhler, G. (1985). Effect of UV-B radiation (290–320 nm) on the nitrogen metabolism of several marine diatoms. *Journal of Plant Physiology*, **118**, 391–400.

Döhler, G. (1987). Effect of irradiation on nitrogen metabolism in marine diatoms and phytoplankton. *Océanis*, **13**, 487–93.

Döhler, G. (1991). Uptake of ^{15}N-ammonium and ^{15}N-nitrate by Antarctic diatoms: dependence on the daytime and effects of UV-B radiation. *Biochemie und Physiologie der Pflanzen*, **187**, 347–55.

Döhler, G. (1992). Impact of UV-B radiation on uptake of ^{15}N-ammonia and ^{15}N-nitrate by phytoplankton of the Wadden Sea. *Marine Biology*, **112**, 485–9.

Döhler, G. (1995). Impact of UV-A and UV-B irradiance on the patterns of pigments and ^{15}N-ammonium assimilation of the tropical marine diatom *Bellerochea yucatanensis*. *Botanica Marina*, **38**, 513–18.

Döhler, G. (1996). Effect of UV irradiance on utilization of inorganic nitrogen by the Antarctic diatom *Odontella weissflogii* (Janisch) Grünow. *Botanica Acta*, **109**, 35–42.

Döhler, G. (1997). Effect of UV-B radiation on utilization of inorganic nitrogen by Antarctic microalgae. *Photochemistry and Photobiology*, **66**, 831–6.

Döhler, G. & Buchmann, T. (1995). Effects of UV-A and UV-B irradiance on pigments and ^{15}N-ammonium assimilation of the haptophycean *Pavlova*. *Journal of Plant Physiology*, **146**, 29–34.

Donkor, V. A., Damian, H. A. K. & Häder, D.-P. (1993a). Effects of tropical solar radiation on the motility of filamentous cyanobacteria. *FEMS Microbiology and Ecology*, **12**, 143–8.

Donkor, V. A., Damian, H. A. K. & Häder, D.-P. (1993b). Effects of tropical solar radiation on the velocity and photophobic behavior of filamentous gliding cyanobacteria. *Acta Protozoologica*, **32**, 67–72.

Dring, M. J., Schmid, R. & Lüning, K. (1995). Influence of blue light, UV-B radiation and tidal phasing on seaweed photosynthesis in sublittoral coastal ecosystems. In *Photosynthesis: From Light to Biosphere*, ed. P. Mathis, vol. 5, pp. 749–54. Kluwer, Dordrecht.

Dunlap, W. C., Chalker, B. E. & Oliver, J. K. (1986). Bathymetric adaptations of reef-building corals at Davies Reef, Great Barrier Reef, Australia. III. UV-B absorbing compounds. *Journal Experimental Marine Biology and Ecology*, **104**, 239–48.

Ekelund, N. G. A. (1990). Effects of UV-B radiation on growth and motility of four phytoplankton species. *Physiologia Plantarum*, **78**, 590–4.

Ekelund, N. G. A. (1991). The effects of UV-B radiation on dinoflagellates. *Journal of Plant Physiology*, **138**, 274–8.

Ekelund, N. G. A. (1993). The effects of UV-B radiation and humic substances on growth and motility of the flagellate *Euglena gracilis*. *Journal of Plankton Research*, **15**, 715–22.

Ekelund, N. G. A. (1994). Influence of UV-B radiation on photosynthetic light-response curves, absorption spectra and motility of four phytoplankton species. *Physiologia Plantarum*, **91**, 696–702.

Falkowski, P. G. & Raven, J. A. (1997). *Aquatic Photosynthesis*. Blackwell Science, Oxford.

Ferreyra, G. A., Schloss, I. R., Demers, S. & Neale, P. J. (1994). Phytoplankton responses to natural ultraviolet irradiance during early spring in the Weddell–Scotia confluence: an experimental approach. *Antarctic Journal of the United States*, **29**, 268–70.

Fiala, M. & Delille, D. (1992). Variability and interactions of phytoplankton and bacterioplankton in the Antarctic neritic area. *Marine Ecology Progress Series*, **89**, 135–46.

Figueroa, F. G., Blanco, J. M., Jiménez-Gómez, F. & Rodríguez, J. (1997). Effects of ultraviolet radiation on carbon fixation in Antarctic nanophytoflagellates. *Photochemistry and Photobiology*, **66**, 185–9.

Fogg, G. E. (1983). The ecological significance of extracellular products of phytoplankton photosynthesis. *Botanica Marina*, **26**, 3–14.

Fogg, G. E., Nalewajko, C. & Watt, W. D. (1965). Extracellular products of phytoplankton photosynthesis. *Proceedings of the Royal Society, London, Series B*, **162**, 517–34.

Garcia-Pichel, F. (1994). A model for internal self-shading in planktonic organisms and its implications for the usefulness of ultraviolet screens. *Limnology and Oceanography*, **39**, 1704–17.

Gerber, S. & Häder, D.-P. (1995). Effects of enhanced solar irradiation on chlorophyll fluorescence and photosynthetic oxygen production of five species of phytoplankton. *FEMS Microbiology and Ecology*, **16**, 33–42.

Gieskes, W. W. C. & Buma, A. G. J. (1997). UV damage to plant life in a photobiologically dynamic environment: the case of the marine phytoplankton. *Plant Ecology*, **128**, 16–25.

Gieskes, W. W. C. & Kraay, G. W. (1990). Transmission of ultraviolet light in the Weddell Sea: report of the first measurements made in the Antarctic. *Biomass*, **12**, 12–14.

Goes, J. I., Handa, N., Taguchi, S. & Hama, T. (1995a). Changes in the patterns of biosynthesis and composition of amino acids in a marine phytoplankter exposed to ultraviolet-B radiation: nitrogen limitation implicated. *Photochemistry and Photobiology*, **62**, 703–10.

Goes, J. I., Handa, N., Taguchi, S., Hama, T. & Saito, H. (1995b). Impact of UV radiation on the production patterns and composition of dissolved free and combined amino acids in marine phytoplankton. *Journal of Plankton Research*, **17**, 1337–62.

Häder, D.-P. (1986). The effect of enhanced solar UV-B radiation on motile microoganisms. In *Stratospheric Ozone Reduction, Solar Ultraviolet Radiation and Plant Life*, ed. R. C. Worrest & M. M. Caldwell, pp. 547–50. Springer-Verlag, Berlin.

Häder, D.-P. (1993). UV-B effects on phytoplankton. In *Frontiers of Photobiology*, ed. A. Shima, p. 547. Elsevier, Amsterdam.

Häder, D.-P. & Häder, M. (1991). Effects of solar and artificial UV radiation on motility and pigmentation in the marine *Cryptomonas maculata*. *Environmental and Experimental Botany*, **31**, 33–41.

Häder, D.-P., Herrmann, H. & Santas, R. (1996). Effects of solar radiation and solar radiation deprived of UV-B and total UV on photosynthetic oxygen production and pulse amplitude modulated fluorescence in the brown alga *Padina pavonia*. *FEMS Microbiology and Ecology*, **19**, 53–61.

Hargraves, P. E., Zhang, J., Wang, R. & Shimizu, Y. (1993). Growth characteristics of the diatom *Pseudonitzschia pungens* and *P. fraudulenta* exposed to ultraviolet radiation. *Hydrobiologia*, **269/270**, 207–12.

Hazzard, C., Lesser, M. P. & Kinzie, R. A. I. (1997). Effects of ultraviolet radiation on photosynthesis in the subtropical marine diatom, *Chaetoceros gracilis* (Bacillariophyceae). *Journal of Phycology*, **33**, 960–8.

Helbling, E. W., Avaría, S., Letelier, J., Montecino, V., Ramírez, B., Ramos, M., Rojas, W. & Villafañe, V.E. (1993). Respuesta del fitoplancton marino a la radiación ultravioleta en latitudes medias (33° S). *Revista de Biología Marina*, **28**, 219–37.

Helbling, E. W., Eilertsen, H. C., Villafañe, V. E. & Holm-Hansen, O. (1996). Effects of UV radiation on post-bloom phytoplankton populations in Kvalsund, North Norway. *Journal of Photochemistry and Photobiology*, **33**, 255–9.

Helbling, E. W., Villafañe, V. E., Ferrario, M. & Holm-Hansen, O. (1992). Impact of natural ultraviolet radiation on rates of photosynthesis and on specific marine phytoplankton species. *Marine Ecology Progress Series*, **80**, 89–100.

Helbling, E. W., Villafañe, V. E. & Holm-Hansen, O. (1994). Effects of ultraviolet radiation on antarctic marine phytoplankton photosynthesis with particular attention to the influence of mixing. In *Ultraviolet Radiation in Antarctica: Measurements and Biological Effects*, Antarctic Research Series, ed. C. S. Weiler & P. A. Penhale, pp. 207–28. American Geophysical Union, Washington, DC.

Hellebust, J. A. (1965). Excretion of some organic compounds by marine phytoplankton. *Limnology and Oceanography*, **10**, 192–206.

Herbert, S. K. (1990). Photoinhibition resistance in the red alga *Porphyra perforata*. *Plant Physiology*, **92**, 514–19.

Herrmann, H., Ghetti, F., Scheuerlein, R. & Häder, D.-P. (1995). Photosynthetic oxygen and fluorescence measurements in *Ulva laetevirens* affected by solar irradiation. *Journal of Plant Physiology*, **145**, 221–7.

Holm-Hansen, O., Lubin, D. & Helbling, W. E. (1993). Ultraviolet radiation and its effects on organisms in aquatic environments. In *Environmental UV Photobiology*, ed. C. S. Young, L. O. Björn, J. Moan & W. Nultsch, pp. 379–425. Plenum Press, New York.

Holm-Hansen, O. & Mitchell, B. G. (1991). Spatial and temporal distribution of phytoplankton and primary production in the western Bransfield Strait region. *Deep-Sea Research*, **38**, 961–80.

Holm-Hansen, O., Mitchell, B. G. & Vernet, M. (1989). Ultraviolet radiation in Antarctic waters: effect on rates of primary production. *Antarctic Journal of the United States*, **24**, 177–8.

Holm-Hansen, O., Villafañe, V. E. & Helbling, E. W. (1997). Effects of solar ultraviolet radiation on primary production in Antarctic waters. In *Antarctic Communities: Species, Structure and Survival*, ed. B. Battaglia, J. Valencia & D. W. H. Walton, pp. 375–80. Cambridge University Press, Cambridge.

Jeffrey, W. H., Pledger, R. J., Aas, P., Hager, S., Coffin, R. B., Von Haven, R. & Mitchell, D. L. (1996). Diel and depth profiles of DNA photodamage in bacterioplankton exposed to ambient solar ultraviolet radiation. *Marine Ecology Progress Series*, **137**, 283–91.

Jokiel, P. L. & York, R. H. (1984). Importance of ultraviolet radiation in photoinhibition of microalgal growth. *Limnology and Oceanography*, **29**, 192–9.

Jones, L. W. & Kok, B. (1966). Photoinhibition of chloroplast reactions. II. Multiple effects. *Plant Physiology*, **41**, 1044–9.

Karentz, D. (1991). Ecological considerations of Antarctic ozone depletion. *Antarctic Science*, **3**, 3–11.

Karentz, D. (1994). Ultraviolet tolerance mechanisms in Antarctic marine organisms. In *Ultraviolet Radiation in Antarctica: Measurements and Biological Effects*, Antarctic Research Series, ed. C. S. Weiler & P. A. Penhale, pp. 93–110. American Geophysical Union, Washington, DC.

Karentz, D., Bothwell, M. L., Coffin, R. B., Hanson, A. K., Herndl, G. J., Kilham, S. S., Lesser, M. P., Lindell, M., Moeller, R. E., Morris, D. P., Neale, P. J., Sanders, R. W., Weiler, C. S. & Wetzel, R. G. (1994). Impact of UV-B radiation on pelagic freshwater ecosystems: report of working group on bacteria and phytoplankton. *Ergebnisse der Limnologie (Archiv für Hydrobiologie. Beiheft)*, **43**, 31–69.

Karentz, D., Cleaver, J. E. & Mitchell, D. L. (1991). Cell survival characteristics and molecular responses of Antarctic phytoplankton to ultraviolet-B radiation. *Journal of Phycology*, **27**, 326–41.

Karentz, D. & Lutze, L. H. (1990). Evaluation of biologically harmful ultraviolet radiation in Antarctica with a biological dosimeter designed for aquatic environments. *Limnology and Oceanography*, **35**, 549–61.

Karentz, D. & Spero, H. J. (1995). Response of a natural *Phaeocystis* population to ambient fluctuations of UVB radiation caused by Antarctic ozone depletion. *Journal of Plankton Research*, **17**, 1771–89.

Kullenberg, G. (1982). Note on the role of vertical mixing in relation to effects of UV radiation on the marine environment. In *The Role of Solar Ultraviolet Radiation in Marine Ecosystems*, ed. J. Calkins, pp. 283–92. Plenum Press, New York.

Lao, K. & Glazer, A. N. (1996). Ultraviolet-B photodestruction of a light-harvesting complex. *Proceedings of the National Academy of Sciences, USA*, **93**, 5258–63.

Larkum, A. W. D. & Wood, W. F. (1993). The effect of UV-B radiation on photosynthesis and respiration of phytoplankton, benthic macroalgae and seagrass. *Photosynthesis Research*, **36**, 17–23.

Lesser, M. P. (1996). Elevated temperatures and ultraviolet radiation cause oxidative stress and inhibit photosynthesis in symbiotic dinoflagellates. *Limnology and Oceanography*, **41**, 271–83.

Lesser, M. P., Cullen, J. J. & Neale, P. J. (1994). Carbon uptake in a marine diatom during acute exposure to ultraviolet B radation: relative importance of damage and repair. *Journal of Phycology*, **30**, 183–92.

Lesser, M. P., Neale, P. J. & Cullen, J. J. (1996). Acclimation of Antarctic phytoplankton to ultraviolet radiation: ultraviolet-absorbing compounds and carbon fixation. *Molecular Marine Biology and Biotechnology*, **5**, 314–25.

Lorenzen, C. J. (1979). Ultraviolet radiation and phytoplankton photosynthesis. *Limnology and Oceanography*, **24**, 1117–20.

Lubin, D., Mitchell, B. G., Frederick, J. E., Alberts, A. D., Booth, C. R., Lucas, T. & Neuschuler, D. (1992). A contribution toward understanding the biospherical significance of Antarctic ozone depletion. *Journal of Geophysical Research*, **97**, 7817–28.

Mague, T. H., Friberg, E., Hughes, D. J. & Morris, I. (1980). Extracellular release of carbon by marine phytoplankton: a physiological approach. *Limnology and Oceanography*, **25**, 262–79.

Malone, T. C. (1980). Size-fractionated primary productivity of marine phytoplankton. In *Primary Productivity in the Sea*, ed. P. G. Falkowski, pp. 301–19. Plenum Press, New York.

Maske, H. (1984). Daylight ultraviolet radiation and the photoinhibition of phytoplankton carbon uptake. *Journal of Plankton Research*, **6**, 351–7.

Maske, H. & Latasa, M. (1997). Solar ultraviolet radiation dependent decrease of particle light absorption and pigments in lake phytoplankton. *Canadian Journal of Fishery and Aquatic Sciences*, **54**, 697–704.

McMinn, A., Heijnis, H. & Hodgson, D. (1994). Minimal effects of UVB radiation on Antarctic diatoms over the past 20 years. *Nature*, **370**, 547–9.

McMinn, A., Heijnis, H. & Hodgson, D. (1997). Preliminary sediment core evidence against short-term UV-B induced changes in Antarctic coastal diatom communities. In *Antarctic Communities: Species, Structure and Survival*, ed. B. Battaglia, J. Valencia & D. W. H. Walton, pp. 381–7. Cambridge University Press, Cambridge.

Montecino, V. & Pizarro, G. (1995). Phytoplankton acclimation and spectral penetration of UV irradiance off the central Chilean coast. *Marine Ecology Progress Series*, **121**, 261–9.

Myklestad, S. & Haug, A. (1972). Production of carbohydrates by the marine diatom *Chaetoceros affinis* var. *willei* (Gran) Husted. I. Effect of the concentration of nutrients in the culture medium. *Journal of Experimental Marine Biology and Ecology*, **9**, 125–36.

Neale, P. J., Cullen, J. J. & Davis, R. F. (1998). Inhibition of marine photosynthesis by ultraviolet radiation: variable sensitivity of phytoplankton in the Weddell–Scotia Confluence during austral spring. *Limnology and Oceanography*, **43**, 433–48.

Neale, P. J., Davis, R. F. & Cullen, J. J. (1998). Interactive effects of ozone depletion and vertical mixing on photosynthesis of Antarctic phytoplankton. *Nature*, **392**, 585–9.

Neale, P. J., Lesser, M. P. & Cullen, J. J. (1994). Effects of ultraviolet radiation on the photosynthesis of phytoplankton in the vicinity of McMurdo Station, Antarctica. In *Ultraviolet Radiation in Antarctica: Measurements and Biological Effects*, Antartic Research Series, ed. C. S. Weiler & P. A. Penhale, pp. 125–42. American Geophysical Union, Washington, DC.

Neale, P. J., Lesser, M. P., Cullen, J. J. & Goldstone, J. (1992). Detecting UV-induced inhibition of photosynthesis in Antarctic phytoplankton. *Antarctic Journal of the United States*, **27**, 122–4.

Nilawati, J., Greenberg, B. M. & Smith, R. E. H. (1997). Influence of ultraviolet radiation on growth and photosynthesis of two cold ocean diatoms. *Journal of Phycology*, **33**, 215–24.

Norrman, B., Zweifel, L., Hopkinson, C. S. J. & Fry, B. (1995). Production and utilization of dissolved organic carbon during an experimental diatom bloom. *Limnology and Oceanography*, **40**, 898–907.

Nultsch, W. & Häder, D.-P. (1988). Photomovement in motile microorganisms. II. *Photochemistry and Photobiology*, **47**, 837–69.

Peletier, H., Gieskes, W. W. C. & Buma, A. G. J. (1996). Ultraviolet-B radiation resistance of benthic diatoms isolated from tidal flats in the Dutch Wadden Sea. *Marine Ecology Progress Series*, **135**, 163–8.

Peterson, P. J. M., Smith, R. C. & Patterson, K. W. (1995). A biological weighting function for phytoplankton growth inhibition. In *Ultraviolet Radiation and Coral Reefs*, ed. D. Gulko & P. L. Jokiel, Hawaii Institute of Marine Biology Technical Report, pp. 51–60.

Prézelin, B. B., Boucher, N. P. & Schofield, O. (1994). Evaluation of field studies of UVB radiation effects on Antarctic marine primary production. In *North Atlantic Treaty Organization, American Scientific Institute Series*, NATO, pp. 181–94.

Quesada, A., Mouget, J.-L. & Vincent, W. F. (1995). Growth of antarctic cyanobacteria under ultraviolet radiation: UVA counteracts UVB inhibition. *Journal of Phycology*, **31**, 242–8.

Raven, J. A. (1991). Responses of aquatic photosynthetic organisms to increased solar UVB. *Journal of Photochemistry and Photobiology*, **9**, 239–44.

Rundel, R. D. (1983). Action spectra and estimation of biologically effective UV radiation. *Physiologia Plantarum*, **58**, 360–6.

Sakka, A., Gosselin, M., Levasseur, M., Michaud, S., Monfort, P. & Demers, S. (1997). Effects of reduced ultraviolet radiation on aqueous concentrations of dimethyl-sulfoniopROprionate and dimethylsulfide during a microscosm study in the lower St. Lawrence estuary. *Marine Ecology Progress Series*, **149**, 227–38.

Sakshaug, E., Myklestad, S., Krogh, T. & Westin, G. (1973). Production of protein and carbohydrate in the dinoflagellate *Amphidinium carteri*. Some preliminary results. *Norwegian Journal of Botany*, **20**, 211–18.

Scully, N. M. & Lean, D. R. S. (1994). The attenuation of ultraviolet radiation in temperate lakes. *Ergebnisse der Limnologie (Archiv für Hydrobiologie. Beiheft)*, **43**, 135–44.

Sebastian, C., Scheuerlein, R. & Döhler, G. (1994). Effects of solar and artifical ultraviolet radiation on pigment composition and photosynthesis in three *Prorocentrum* strains. *Journal of Experimental Marine Biology and Ecology*, **182**, 251–63.

Setlow, R. B. (1974). The wavelengths in sunlight effective in producing skin cancer: a theoretical analysis. *Proceedings of the National Academy of Sciences, USA*, **71**, 3363–6.

Sharp, J. H. (1977). Excretion of organic matter by marine phytoplankton: do healthy cells do it? *Limnology and Oceanography*, **22**, 381–9.

Sinha, R. P., Kumar, H. D., Kumar, A. & Döhler, G. (1995a). Effects of UV-B irradiation on growth, survival, pigmentation and nitrogen metabolism enzymes in cyanobacteria. *Acta Protozoologica*, **34**, 187–92.

Sinha, R. P., Lebert, M., Kumar, A., Kumar, H. D. & Häder, D.-P. (1995b). Spectroscopic and biochemical analyses of UV effects on phycobiliproteins of *Anabaena* sp. and *Nostoc carmium*. *Botanica Acta*, **108**, 87–92.

Smith, R. C. (1989). Ozone, middle ultraviolet radiation and the aquatic environment. *Photochemistry and Photobiology*, **50**, 459–68.

Smith, R. C. & Baker, K. S. (1980). Stratospheric ozone, middle ultraviolet radiation and carbon-14 measurements of marine productivity. *Science*, **208**, 592–3.

Smith, R. C. & Baker, K. S. (1981). Optical properties of the clearest natural waters (200–800 nm). *Applied Optics*, **20**, 177–84.

Smith, R. C. & Baker, K. S. (1982). Assessmennt of the influence of enhanced UV-B on marine primary productivity. In *The Role of Solar Ultraviolet Radiation in Marine Ecosystems*, ed. J. Calkins, pp. 509–37. Plenum Press, New York.

Smith, R. C., Baker, K. S., Holm-Hansen, O. & Olson, R. (1980). Photoinhibition of photosynthesis in natural waters. *Photochemistry and Photobiology*, **31**, 585–92.

Smith, R. C. & Cullen, J. J. (1995). Effects of UV radiation on phytoplankton. *Review of Geophysics*, **33**, 1211–23.

Smith, R. C., Prézelin, B. B., Baker, K. S., Bidigare, R. R., Boucher, N. P., Coley, T., Karentz, D., MacIntyre, S., Matlick, H. A., Menzies, D., Ondrusek, M. E., Wan, Z. & Waters, K. J. (1992). Ozone depletion: ultraviolet radiation and phytoplankton biology in Antarctic waters. *Science*, **255**, 952–9.

Steeman-Nielsen, E. (1952). The use of radioactive carbon for measuring organic production in the sea. *Journal du Conseil d'Exploration International de la Mer*, **18**, 117–40.

Steeman-Nielsen, E. (1964). On a complication in marine productivity work due to the influence of ultraviolet light. *Journal du Conseil d'Exploration International de la Mer*, **29**, 130–5.

Van Donk, E. & Hessen, D. O. (1996). Loss of flagella in the green alga *Chlamydomonas reinhardtii* due to *in situ* UV-exposure. *Scientia Marina*, **60** (Supplement 1), 107–12.

Veen, A., Reuvers, M. & Ronçak, P. (1997). Effects of acute and chronic UV-B exposure on a green alga: a continous culture study using a computer-controlled dynamic light regime. *Plant Ecology*, **128**, 29–40.

Vernet, M., Brody, E. A., Holm-Hansen, O. & Mitchell, B. G. (1994). The response of Antarctic phytoplankton to ultraviolet light: absorption, photosynthesis, and taxonomic composition. In *Ultraviolet Radiation in Antarctica: Measurements and Biological Effects*, Antartic Research Series, **62**, ed. C. S. Weiler & P. A. Penhale, pp. 143–58. American Geophysical Union, Washington, DC.

Vernet, M. & Smith, R. C. (1997). Effects of ultraviolet radiation on the pelagic Antarctic ecosystem. In *The Effects of Ozone Depletion on Aquatic Ecosystems*, ed. D.-P. Häder, pp. 247–65. R. G. Landes Co., Austin, TX.

Vernet, M. & Whitehead, K. (1996). Release of ultraviolet-absorbing compounds by the red-tide dinoflagellate *Lingulodinium polyedra*. *Marine Biology*, **127**, 35–44.

Villafañe, V. E., Helbling, E. W., Holm-Hansen, O. & Chalker, B. E. (1995a). Acclimatization of Antarctic natural phytoplankton assemblages when exposed to solar ultraviolet radiation. *Journal of Plankton Research*, **17**, 2295–306.

Villafañe, V. E., Helbling, E. W., Holm-Hansen, O. & Diaz, H. F. (1995b). Long-term responses by Antarctic phytoplankton to solar ultraviolet radiation. *Antarctic Journal of the United States*, **30**, 320–3.

Vincent, W. F., Neale, P. J. & Richerson, P. J. (1984). Photoinhibition: algal responses to bright light during diel stratification and mixing in a tropical alpine lake. *Journal of Phycology*, **20**, 201–11.

Vincent, W. F. & Roy, S. (1993). Solar ultraviolet-B radiation and aquatic primary production: damage, protection and recovery. *Environmental Reviews*, **1**, 1–12.

Wängberg, S.-A., Selmer, J.-S. & Gustavson, K. (1996). Effects of UV-B radiation on biomass and composition in marine phytoplankton communities. *Scientia Marina*, **60** (Supplement 1), 81–8.

Weiler, C. S. & Penhale, P. A. (1994). *Ultraviolet Radiation in Antarctica: Measurements and Biological Effects.* American Geophysical Union, Washington, DC.

Wood, A. M., Rai, H., Garnier, J., Kairesalo, T., Gresens, S., Orive, E. & Ravail, B. (1992). Practical approaches to algal excretion. *Marine Microbiological Food Webs*, **6**, 21–38.

Worrest, R. C., Van Dyke, H. & Thomson, B. E. (1978). Impact of enhanced simulated solar ultraviolet radiation upon a marine community. *Photochemistry and Photobiology*, **27**, 471–8.

Worrest, R. C., Wolniakowski, K. U., Scott, J. D., Brooker, D. L., Thomson, B. E. & Van Dyke, H. (1981). Sensitivity of marine phytoplankton to UV-B radiation: impact upon a model ecosystem. *Photochemistry and Photobiology*, **33**, 223–7.

Note added in proof

Figure 9.3: Reproduced from Figure 7A,B, of Hazzard *et al.* (1997), by permission of the *Journal of Phycology*.

Figure 9.4: Reproduced from Figure 6 of Quesada *et al.* (1995) by permission of the *Journal of Phycology*.

Figure 9.5: Reprinted from *FEMS Microbiology and Ecology*, **16**, Gerber, S. & Häder, D.-P., Effects of enhanced solar irradiation on chlorophyll fluorescence and photosynthetic oxygen production of five species of photoplankton, pp. 33–42. Copyright 1995 with permission from Elsevier Science.

Figure 9.7: Reprinted from *Environmental and Experimental Botany*, **31**, Häder, D.-P. & Häder, M., Effects of solar and artificial UV radiation on motility and pigmentation in the marine *Cryptomonas maculata*, pp. 33–41. Copyright 1991 with permission from Elsevier Science.

Figure 9.10: From Villafañe, V. E. *et al.* (1995), Acclimatization of Antarctic natural phytoplankton assemblages when exposed to solar ultraviolet radiation. *Journal of Plankton Research*, **17**, 2295–306, reproduced by permission of Oxford University Press.

10

○ ○ ○ ○ ○ ○ ○ ○ ○ ○ ○ ○ ○ ○ ○ ○ ○ ○ ○ ○

Impact of solar UV radiation on zooplankton and fish

Horacio E. Zagarese* and Craig E. Williamson

10.1 Introduction

Solar UV radiation (UVR) has long been known to be damaging to many aquatic organisms. Reports of sunburn (erythema) in fish have been recognised for many years (Bell & Hoar, 1950; Eisler, 1961) and plankton ecologists have speculated that high UV irradiance could put severe restrictions on the distribution of sensitive species (Brehm, 1938; Thomasson, 1956). The discovery of the 'ozone hole' over Antarctica (Farman, Gardiner & Shanklin, 1985) provided the impetus for much of the recent research on the biological effects of UVR in aquatic environments, but ambient UVR is changing over temperate latitudes as well (Madronich, 1994; Orce & Helbling, 1997).

UVR has damaging effects on a variety of biological molecules and cell components that are common to most living organisms (Young, Moan & Nultsch, 1993; Williamson & Zagarese, 1994; Häder et al., 1995). In many cases, the results obtained with a particular group of organisms are general enough to be extended to other groups. In fact, much of what is now known about the effects of UVR at the molecular and cellular levels comes from studies with bacteria, algae and other unicellular organisms. Comparatively, very few studies have focused on zooplankton or fish. In contrast to the relatively similar effects among different groups of organisms at the cellular and subcellular levels, there are large differences at the levels of the organism, community and ecosystem that prevent straightforward generalisations.

This chapter reviews the effects of UVR on zooplankton and fish. Although this book is concerned mainly with the marine environment, examples are also included from the freshwater literature that complement the findings for marine species and may thus be useful to guide future research.

10.2 Effects and responses to UV radiation in zooplankton and fish

The DNA molecule is often considered to be the primary target for UVB damage. The absorbance maximum of DNA (around 260 nm) is well below the shortest wavelength of solar radiation reaching the earth's surface (around 290 nm). Under natural conditions, the potential for UVR damage to DNA increases exponentially with decreasing wavelength. The photoproducts and mechanisms of molecular damage to DNA are reviewed in Chapter 6.

UVR-induced DNA damage seems to be more serious for embryos, larvae, and cells in culture, in which gene expression is more active (Naganuma, Inoue & Uye, 1997). Similarly, Karanas, Van Dyke & Worrest (1979) found a direct relationship between the age of copepods (*Acartia clausii*) and their ability to tolerate UVB radiation, with the smallest nauplii being the most vulnerable.

In addition to direct effects, UVR may have indirect effects on living organisms through at least two mechanisms: the formation of highly reactive chemical species in the water (see Chapter 6), and the alteration of the food web structure (Bothwell, Sherbot & Pollock, 1994). UVR promotes the formation of free radicals, peroxides, etc., which may react with membranes and other structures. In addition, the differential sensitivity to UVR between organisms may have a profound impact on competitive and predator–prey relationships (see Chapter 11).

10.2.1 Damaging effects

Several lines of investigation have demonstrated the potential for solar UVR damage in zooplankton. Many of the early experiments used artificial UVR sources so the results, even with appropriate weighting functions, are difficult to interpret in terms of the potential for damage from natural solar radiation and related ozone depletion. This is because the biological effectiveness of damaging solar radiation may vary by several orders of magnitude across the wavelengths within the UV region (see Chapter 3). In addition to damage from artificial UVR, there is some evidence that shorter wavelength blue radiation may be damaging to some zooplankton (Hairston, 1979; Ringelberg, Keyser & Flik, 1984; Siebeck & Böhm, 1991, 1994). Experiments with artificial UVR and marine copepods have shown a significant decrease in the survival of adults as well as a reduction in the number of eggs and live nauplii, and deformation of nauplii (Karanas *et al.*, 1979; Karanas, Worrest & Van

Dyke, 1981; Naganuma *et al.*, 1997). For example, Kouwenberg *et al.* (1999b) reported that a significant proportion of *Calanus finmarchicus* nauplii hatched following exposure to artificial UVR (a xenon arc lamp-based solar simulator) exhibited malformations indicative of errors in pattern formation during embryogenesis. Similar lesion patterns have been demonstrated in crab larvae and euphausids exposed to artificial UVR (Damkaer *et al.*, 1980). Sublethal effects of UVB on the marine copepod *Acartia* have been demonstrated in laboratory experiments where virgin copepodids were irradiated, and their survival and reproduction following maturation quantified. Decreases in survival and the production of viable offspring were observed, but the survival of the non-irradiated nauplii produced was not affected (Karanas *et al.*, 1981). Natural solar UVR has been shown to induce mortality in *Daphnia* (Williamson *et al.*, 1994; Zagarese *et al.*, 1994), and freshwater copepods (Luecke & O'Brien, 1981; Williamson *et al.*, 1994; Zagarese, Feldman & Williamson, 1997a).

Because of their transparency and frequent occurrence in shallow waters (Hunter & Taylor, 1982; Crowder & Crawford, 1984), the eggs and larvae of fish are highly vulnerable to natural UVR. Dunbar (1959, cited by Eisler, 1961) observed increased mortality rate in rainbow trout (*Oncorhynchus mykiss*) fingerlings exposed to sunlight. Other examples include yellow perch (*Perca flavescens*) (Williamson *et al.*, 1997), Atlantic cod (*Gadus morhua*) (Béland *et al.*, 1999) and puyen (*Galaxias maculatus*) (Battini *et al.*, submitted) eggs incubated in transparent lakes; as well as Atlantic cod eggs (Kouwenberg *et al.*, 1999), anchovy larvae (Hunter, Taylor & Moser, 1979; Hunter, Kaupp & Taylor, 1981) and sockeye salmon (*Oncorhynchus nerka*) juveniles exposed to artificial UVR (Bell & Hoar, 1950). Hatching failure and larval or juvenile mortality induced by UVR are often associated with more or less severe embryonic malformations, or total mortality of eggs or embryos exposed to high levels of natural solar radiation (Figure 10.1).

Sunburn in fish is characterised by the development of grey focal lesions on the dorsal aspect and extensive erosion of the epidermis, and is usually the most commonly reported UVR-induced damage in juvenile and adult individuals. Evidence of sunburn caused by natural radiation has been reported for juvenile rainbow trout inhabiting outdoor pools (Dunbar, 1959) and at high altitudes in Bolivia (Bullock & Coutts, 1985), juvenile plaice (*Pleuronectes platessa*) from the shallows of the Wadden Sea (Bergham, Bullock & Karakiri, 1993), juvenile paddlefish (*Polyodon spathula*) (Ramos *et al.*, 1994), etc. Similar lesion patterns can be induced by exposure to artificial sources of UVR, e.g. sockeye salmon (Bell &

Hoar, 1950) and Lahontan cutthroat trout (*Onchorhynchus clarki henshawi*) (Blazer *et al.*, 1997).

The damage to the epidermis that develops after sunburn increases the vulnerability to opportunistic infections (e.g. Lahontan cutthroat trout (Little & Fabacher, 1994). Moreover, aside from lesion-related infections, exposure to UVR may depress the immune system and increase vulnerability to other pathogens (Zeeman & Brindley, 1981; Knowles, 1992).

Eye lens damage (cataracts) has been observed in fish confined to shallow depths (i.e. shallow natural waters or hatchery ponds) (Cullen & Monteith-McMaster, 1993). The opacity observed in the lens appears to be mediated by the presence of peroxides. *In vitro* experiments have shown that UVR reduces the ability of dogfish (*Mustelus canis*) lens to protect itself against H_2O_2 insult via catalase (Zigman & Rafferty, 1994).

10.2.2 Recovery from UV radiation damage

Repair mechanisms of nuclear material, proteins, lipids and membranes are reviewed in Chapter 7. In zooplankton and fish, one such repair

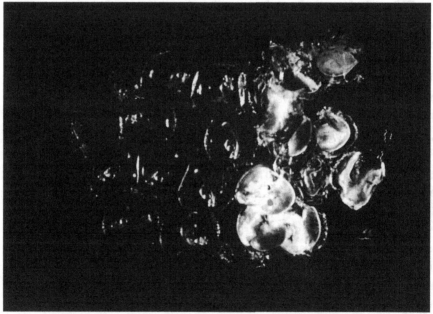

Figure 10.1. Light photomicrograph of yellow perch embryos that were unexposed (dark controls, left) and exposed (in quartz tubes, right) to natural levels of solar radiation at a depth of 0.8 m in UV transparent Lake Giles, north-eastern Pennsylvania, USA (Williamson *et al.*, 1997). The UV-exposed embryos are all dead, as indicated by the white colour, while the unexposed embryos are healthy and developing normally. Egg diameter is 3 mm. (Photo courtesy of Craig Williamson.)

mechanism, namely photoreactivation, has been predominantly studied, and is therefore reviewed briefly here.

It has been known for some time that longer wavelength UVA and visible radiation can stimulate recovery from UVR damage (Kelner, 1949). The underlying process, known as photoreactivation or photorepair, consists of the repair of damaged DNA (cyclobutane pyrimidine dimers, CPDs) mediated by the enzyme photolyase (Sutherland, 1981; Mitchell & Karentz, 1993). Photoreactivation has been demonstrated in both zooplankton and fish, as well as in amphibians and a wide variety of other organisms (Mitchell & Karentz, 1993; Blaustein et al., 1994, 1996; Malloy et al., 1997). Photoreactivation potential is generally estimated by the direct quantification of CPDs (rate of decrease), by photolyase levels, or by differences in survival of UVR-exposed organisms in the presence versus absence of reactivating wavelengths. In the laboratory, substantial photoreactivation is observed even at relatively low light intensities such as that from a single 40 W cool white fluorescent lamp (Damkaer & Dey, 1983; Dey, Damkaer & Heron, 1988; Malloy et al., 1997).

In marine systems, photoreactivation has been documented in a variety of invertebrates including copepods (*Calanus pacificus*) (Dey et al., 1988), shrimp (*Pandalus platyceros*), euphausids (krill, *Euphausia superba*, *Thysanoessa raschii*) (Damkaer & Dey, 1983; Malloy et al., 1997), and opisthobranch slugs (*Elysia tuca, Bursatella leachii, Haminaea antillarum*) (Carlini & Regan, 1995) as well as fish (Kaupp & Hunter, 1981; Regan et al., 1982; Malloy et al., 1997). In freshwater systems, photoreactivation has been less well investigated but has been reported in the cladoceran *Daphnia* (Siebeck, 1978; Siebeck & Böhm, 1991; Zellmer, 1996) and in two out of three species of calanoid copepods tested in Argentina in the genus *Boeckella* (*B. brevicaudata* and *B. gibbosa*, but not *B. gracilipes*) (Zagarese et al., 1997) (Figure 10.2). No evidence of photoreactivation has been found in the one species of diaptomid copepod that has been tested (*Acanthodiaptomus denticornis*) (Ringelberg et al., 1984), but this result cannot be considered definitive due to potential methodological shortcomings. At least some freshwater fish, including the fathead minnow (*Pimephales promelas*) (Applegate & Ley, 1988), the medaka (*Oryzias latipes*) (Funayama, Mitani & Shima, 1993), and the eggs of Atlantic cod (Kouwenberg et al., 1999a) exhibit photoreactivation. Cell cultures derived from goldfish (*Carassius auratus*) have been used extensively to examine photoreactivation (Shima & Setlow, 1984; Yasuhira, Mitani & Shima, 1991, 1992; Ahmed et al., 1993; Funayama et al., 1993; Uchida, Mitani & Shima, 1995). In addition, the platyfish, *Xiphophorus variatus*, is

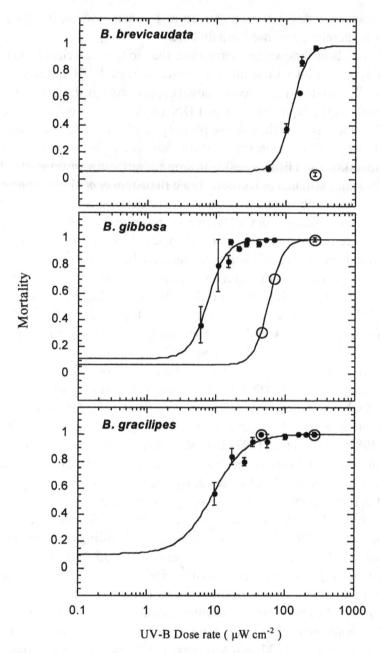

Figure 10.2. Mortality induced by artificial UV-B radiation in three species of *Boeckella* exposed to UVB radiation only (filled symbols) or UV-B and visible radiation (open symbols). Survival is enhanced by the presence of visible radiation in *B. brevicaudata* and *B. gibbosa*, but not in *B. gracilipes*, suggesting that the latter species has little if any potential for photoreactivation. (Redrawn from Zagarese *et al.*, 1997a, with permission of Oxford University Press.)

a popular model organism for studying photoreactivation and Mendelian inheritance of melanoma cancer in humans (Ahmed & Setlow, 1993; Mitchell & Karentz, 1993).

10.2.3 Wavelength dependence of damage and repair

In order to understand the effects of UVR on aquatic organisms we need to know not only the ambient UVR levels, but also the wavelength-specific biological response (see Chapter 3). This relationship expressing the wavelength-dependent response of a given biological (or chemical) process is referred to as an action spectrum or a biological weighting function (BWF) depending on whether it is determined with monochromatic or polychromatic radiation, respectively (Cullen & Neale, 1997; see also Chapter 3). Biological effectiveness (unit of biological damage or response per photon) of solar radiation may vary by several orders of magnitude depending on the wavelength of the radiation, even over the rather narrow UVB range (Setlow, 1974; Caldwell et al., 1986). Shorter wavelengths of UVR are much more damaging. The nature of the BWF will vary for different species depending on a variety of factors, including the extent of photoprotection and photoreactivation.

From an ecosystem perspective, it is important to know the relative tolerance of zooplankton and fish to different wavelengths of UVR. Both UVB and UVA change by orders of magnitude with depth and season within lakes, as well as with changes in the quality and quantity of dissolved organic carbon (DOC) among lakes (Morris et al., 1995; Morris & Hargreaves, 1997). Natural and anthropogenic disturbances such as acid precipitation, climate change, and changes in land use and hydrology may all influence DOC concentrations and hence the exposure of zooplankton and fish to UVR in freshwater ecosystems (Schindler et al., 1996; Williamson et al., 1996; Yan et al., 1996). Stratospheric ozone depletion on the other hand will lead to increases in UVB but not UVA radiation.

Substantial progress has been made in the understanding of the wavelength-specific effects of UVR in phytoplankton through the development of BWFs (Cullen, Neale & Lesser, 1992; Cullen & Neale, 1997). The spectral responses of higher trophic levels, however, are poorly understood. As a consequence, empirical action spectra derived from data on other species or cell cultures are often used (Hunter, Kaupp & Taylor, 1981; Malloy et al., 1997). While mortality is the primary response variable that has been examined for zooplankton and fish, sublethal effects include depression of reproduction, alteration of the sex ratio, induction of deformities, and accumulation of the products of photodamage such as CPDs (Williamson *et al.*, 1994; Chalker-Scott, 1995; Malloy *et al.*, 1997; Naganuma *et al.*, 1997).

Few studies exist that compare the sensitivity to different wavelengths of UVR of zooplankton and fish. Direct *in situ* comparisons of the meroplanktonic larvae of the predatory insect *Chaoborus* and larval sunfish (*Lepomis* spp.) have demonstrated that the larvae of the insect are much less tolerant than those of the fish (Williamson *et al.*, 1999). When incubated in the surface waters of clearwater lakes, *Chaoborus* perishes after less than one day of exposure to ambient solar radiation, while the fish larvae may survive for several days. In these experiments, the fish larvae showed a statistically significant short wavelength UVB effect, while the insect larvae did not, suggesting that longer wavelength UVB or even UVA was responsible for the damage in the *Chaoborus*, in spite of the fact that these wavelength are known to be ineffective in producing alterations in the DNA molecule (Setlow, 1974).

Different species of zooplankton may also vary in their tolerance to different wavelengths of UVR. For example, Williamson *et al.* (1994) found that the cladocerans *Daphnia* and *Diaphanosoma* were sensitive to both UVA and UVB, while the mortality of the copepod *Diaptomus* was increased by UVB but not by UVA, and the rotifer *Keratella* was unaffected by either waveband. Strong differences among species of the same genus of a South American calanoid copepod have also been described (Zagarese *et al.*, 1997a), and there also appear to be differences between populations of the same species. For example, *Boeckella gracilipes* was found to be very sensitive to solar UVR in lakes in Nahuel Huapi National Park (41y S 71° W Argentina) – nearly 80% of the individuals died after three days at a depth of 0.5 m in a clear lake (Zagarese *et al.*, 1997a). In contrast, Cabrera, Lopez & Tartarotti (1997) exposed plankton mesocosms to high UVR levels in Laguna Negra (33° S, 70° W; Chile) at 2700 m above sea level for periods of 30 to 48 days, and found that whereas *Chydorus sphaericus* and *Lepadella ovalis* populations were strongly inhibited by UVB radiation, *B. gracilipes* was virtually unaffected.

Differences in vulnerability to UVR between fish species have also been reported. For example, Blazer *et al.* (1997) found the Lahontan cutthroat trout were more sensitive than razorback suckers (*Xyrauchen texanus*). In cutthroat trout, the lesions were characterised by an initial sloughing of the mucus cells followed by necrosis and oedema of the epidermis, with eventual sloughing and sometimes secondary infections. Razorback suckers also lost the superficial mucus cell layer, but no oedema or lifting and sloughing of the epidermis was observed.

Photoreactivation is also a wavelength-selective process, for which action spectra are known to be related to the absorption properties of the photolyase or related chromophores. These action spectra may peak either in the UVA range around 380–390 nm, or in the visible spectrum closer to 440 nm (Siebeck & Böhm, 1991, 1994; Sancar, 1994). The peak sensitivity for two species of *Daphnia* (*D. galeata* and *D. pulex*) to reactivating radiation has been determined to be between 410–450 nm (Siebeck & Böhm, 1991, 1994). However, these latter studies were action spectra determined with monochromatic radiation, so they do not account for the importance of the potential time dependence and interactions between wavelengths. In this respect, polychromatic action spectra or BWFs are much more useful (Cullen & Neale, 1997).

Information on the UVR tolerance of protozoa is also very sparse. UVR is known to inhibit cell motility in photosynthetic flagellates (Häder & Worrest, 1997), and to alter cell shape, motility and bacterivory in one heterotrophic nanoflagellate, *Bodo saltans* (Sommaruga, Oberleiter & Psenner, 1996). In ciliates, several studies have demonstrated inhibition of motility, and the few available action spectra indicate an exponential decrease in survival with decreasing wavelengths through the UVB and down into the UVC wavelength range (Marangoni, Martini & Colombetti, 1997). Some ciliates (*Euplotes aediculatus* and *Paramecium aurelia*) can detect and avoid UVR with wavelengths less than 350 nm (Barcelo & Calkins, 1979).

10.2.4 Reciprocity with and without repair mechanisms

One of the most fundamental assumptions for constructing accurate dose-dependent BWFs is the principle of reciprocity (Kouwenberg et al., 1999a). Under the assumption of reciprocity, the damage is only a function of the dose, which is customarily assumed to be proportional to the irradiance integrated over the exposure time. It is almost certain that reciprocity will fail when one considers a very broad range of exposure conditions (i.e. same dose administered over a very short or a very long exposure period). Reciprocally, within a narrow range of exposure conditions the chances of meeting reciprocity are maximal. Thus, it is important to specify the range of conditions for which reciprocity holds (or fails). From an environmental perspective, any statement about reciprocity is only meaningful within a range of conditions similar to that experienced in Nature by the organisms under study.

For understanding the dependence of reciprocity on the damage and repair rates, it is useful to consider the following formal definitions:

> *Damage rate* ($\dot{D}$) is the number of units per individual and unit of time that are altered by UVR (e.g. the number of pyrimidine dimers produced in a cell per second due to UV radiation).
> *Damage* (D) is the amount of damage produced after a period of UVR exposure:

$$D = \int \dot{D} \partial t \tag{10.1}$$

Here, only the case where the damage rate ($\dot{D}$) is a function of the irradiance is considered. Under constant irradiance conditions, the damage rate will also be constant, but if irradiance changes over time (as it does under natural conditions), the consequent fluctuations in $\dot{D}$ will be described by

$$\ddot{D} = \frac{\dot{D}}{\partial t} = \frac{\partial^2 D}{\partial t^2} \tag{10.2}$$

where $\ddot{D}$ is the second derivative of D over time.

> *Measured effect* (E) is the variable that is actually measured in experiments. E may be a more or less gradual variable, such as the degree of sunburn in fish, or an all-or-nothing variable that can have only two states, such as 'live' or 'dead'. Most UVR effects considered by zooplankton and fish ecologist are cumulative in nature, i.e. the death of an individual occurs after a certain amount of damage has accumulated.
> *The principle of reciprocity* or *Bunsen–Roscoe principle*. It can be shown that the principle of reciprocity is satisfied only when the damage rate is a linear function of dose rate[1] (assumed to be proportional to irradiance, *I*):
> If

$$\dot{D} = kI \tag{10.3}$$

> then

$$D = \int \dot{D} \partial t = k \int I \partial t \tag{10.4}$$

---

[1] Although for the principle of reciprocity to hold the damage rate must be a linear function of dose rate, the relationship between the measured effect and damage does not need to be linear (see Chapter 3).

where k is a constant.

If repair processes are operating, the net damage (D_{Net}) can be defined as the difference between the damage produced and the amount of such damage that has been repaired (R):

$$D_{Net} = D - R \tag{10.5}$$

Likewise, the rate of net damage can be written as:

$$\frac{\partial D_{Net}}{\partial t} = \frac{\partial D}{\partial t} - \frac{\partial R}{\partial t} \tag{10.6}$$

As already mentioned, for the principle of reciprocity to hold, the damage rate has to be a linear function of irradiance. Similarly, when repair mechanisms are operating, the rate of net damage should be a linear function of irradiance:

$$\frac{\partial D_{Net}}{\partial t} = k'I \tag{10.7}$$

where k' is a constant. But, since

$$\dot{D} = kI \tag{10.3}$$

it follows that for the principle of reciprocity to hold, it should be true that

$$\frac{\partial R}{\partial t} = k'I - kI = (k' - k)I \tag{10.8}$$

In practice, the above condition (10.8) is seldom met. Moreover, there are no theoretical reasons why the rate of repair should be a linear function of irradiance. As a matter of fact, the presence of repair mechanisms is one of the most commonly referred causes for the failure of reciprocity. There have been reports of failure of reciprocity in larval anchovy (*Engraulis mordax*) (Hunter *et al.*, 1979; Hunter *et al.*, 1981), shrimp (*Pandalus platyceros*) larvae and adult euphasiids (*Thysanoessa raschii*) (Damkaer, Day & Heron, 1981; Damkaer & Dey, 1983), and the cladoceran *Ceriodaphnia dubia* (Zagarese *et al.*, 1998b), all of which are capable of photoreactivation. Reciprocity has been found to hold in the copepod *Boeckella gracilipes* that lacks photoreactivation (Zagarese *et al.*, 1998a), and in the eggs of Atlantic cod, which are capable of photorecovery (Kouwenberg *et al.*, 1999a).

Another important consequence of the presence of repair mechanisms is that, even under constant irradiance conditions, the net rate of damage may not be constant. In mathematical terms:

$$\frac{\partial^2 D_{Net}}{\partial t^2} = \frac{\partial^2 D}{\partial t^2} - \frac{\partial^2 R}{\partial t^2} \neq 0 \tag{10.9}$$

The above inequality may result from the fact that repair mechanisms are induced after a period of exposure to UVR (Buma *et al.*, 1996). Although, under the constant irradiance conditions of laboratory experiments, a balance between damage and repair is eventually established (Cullen & Lesser, 1991; Buma *et al.*, 1996), such a steady state may be delayed or even prevented under the fluctuating irradiance of natural conditions.

10.2.5 Indirect effects

The UV-B radiation has been shown to reduce the levels of ω-3 fatty acid eicosapentaenoic acid (EPA, 20:5ω3) and docosahexaenoic acid (DHA, 22:6ω3) (Wang & Chai, 1994) and other polyunsaturated fatty acids (PUFAs) (Goes *et al.*, 1994) in microalgae. Although fatty acids constitute a small portion of the total organic matter synthesised by marine algae, they are an important source of energy for marine trophic food chains. Their importance is qualitative, since certain phytoplankton fatty acids are essential for higher trophic levels. A dietary deficiency of PUFAs in phytoplankton can limit the growth of several herbivores (Watanabe, Kitajima & Fujita, 1983). In the marine environment, the impact of UVB radiation on the biochemical status of phytoplankton could have serious consequences on grazers and higher trophic levels (Goes *et al.*, 1994; Arts & Rai, 1997; Hessen, De Lange & Van Donk, 1997, and references therein). On the other hand, the nutritional status of herbivores can alter their tolerance to UVR as well as their capacity for recovering from UVR damage (Zellmer, 1996).

The interaction of UV-B, oxygen and certain organic compounds can also result in the production of reactive oxidants that are highly toxic to many forms of aquatic life (Scully *et al.*, 1995; Scully, McQueen & Lean, 1996). Hydrogen peroxide and hydroxyl radicals can be generated photochemically in natural aquatic environments. These toxic photochemical products have a broad range of destructive effects on cellular macromolecules (Palenik, Price & Morel, 1991; Vincent & Roy, 1993, and references therein).

10.2.6 Photoprotective compounds

Three major types of photoprotective compounds occur in planktonic crustaceans: carotenoid pigments, cuticular melanin, and mycosporine-like amino acids (MAAs). Carotenoid pigments have been identified from a variety of planktonic organisms, but large differences in concentration exist between different zooplankton groups. For example, the concentrations reported in calanoid copepods (typical levels $\sim 10\,\mu g\,(mg\ dry\ weight)^{-1}$) are nearly ten times the corresponding levels in *Daphnia* and *Bosmina* (Hairston, 1979; Hessen & Sørensen, 1990). Hairston (1976) was the first to demonstrate experimentally that carotenoid pigments are photoprotective to calanoid copepods. Melanin is common in several *Daphnia* species and other cladoceran crustaceans. Melanic clones have been shown to be more resistant to natural and artificial UVR (Hebert & Emery, 1990). Although both carotenoids and melanins have been demonstrated to reduce photodamage, the actual mechanisms involved appear to be different. Unlike melanin, whose main role is radiation screening, the photoprotective properties of carotenoids may be more associated with antioxidant mechanisms such as quenching of triplet oxygen and inhibition of free radical reactions (Palozza & Krinsky, 1992; Hessen, 1994).

UVR-absorbing compounds, the MAAs, were first discovered in fungi by Leach (1965), and in marine organisms by Shibata (1969). Since then, they have been described and characterised in bacteria, dinoflagellates, macroalgae, anthozoans and other organisms (Karentz, McEuen & Dunlap, 1991; Teai *et al.*, 1997, and references therein). Thirteen different MAAs have been chemically identified to date (1997). These compounds have strong absorption maxima in the range 310–360 nm, which correspond to biologically harmful wavelengths of UVR. In addition to their sunscreen properties, MAAs may also serve as antioxidants (Dunlap & Yamamoto, 1995). Carroll & Shick (1996) were the first to document the bioaccumulation of MAAs from the diet in the sea urchin *Strongylocentrotus droebachiensis*. MAAs concentrated in the ovaries and passed to maturing eggs released into the water column may provide protection from damaging solar UVR.

Pigments may be distributed over the entire body (i.e. carotenoids in copepods), more or less concentrated in the most exposed areas (i.e. cuticular melanin in *Daphnia*), or aggregated in complex structures (i.e. chromatophores, scales). The presence of chromatophores protects the larvae of marine decapods from UVR. Darkly pigmented species survived significantly better than lightly pigmented larvae when exposed to sunlight (Morgan & Christy, 1996). Direct solar UV-B radiation has been shown to induce melanosome dispersion in the dermis of Atlantic salmon (*Salmo salar*) (MacArdle & Bullock, 1987). Another protective pigment in fish is the silver, guanine-based reflective covering of the scales.

The efficiency of radiation-blocking compounds changes with the size of the organism. For example, Garcia-Pichel (1996) concluded that the efficient use of sunscreen pigments can be accomplished only by organisms larger than 100 μm. The size range of zooplankton thickness spans from a few micrometres to millimetres or even centimetres. Thus, the efficiency of sunscreen pigments is likely to differ among species. How the efficiency of antioxidant pigments such as carotenoids changes with the size of the organisms remains unknown. In freshwater copepods, it has been observed that the amount of carotenoids per unit of weight increases as the size of the individual decreases. That is, nauplii normally have a higher concentration of carotenoids than do adults. This has been interpreted in terms of a trade-off between photoprotection and vulnerability to visual predators (Hairston, 1979), but differences in pigment efficiency due to differences in size deserve to be considered in the future.

10.2.7 UV radiation vulnerability in relation to habitat

Zooplankton populations from highly exposed habitats (shallow, transparent waters, high elevation lakes, tropics versus higher latitudes) are generally much more tolerant to UVR than those populations from better protected environments. Such a higher tolerance is sometimes due to the presence of photoprotective compounds (Luecke & O'Brien, 1981; Hebert & Emery, 1990), to more efficient recovery mechanisms (Siebeck, 1978), or to a combination of both mechanisms (Zagarese et al., 1997a). Regardless of the specific mechanism behind the increased tolerance, this trend has been interpreted as a strong indication of the importance of UVR in determining the geographical and spatial distribution of planktonic organisms.

Calanoid copepods are typically more darkly pigmented at higher elevation (Brehm, 1938; Byron, 1982). Moreover, even within the same water body, the most pigmented individuals tend to occur at shallower depths than do their pale relatives. For example, Hairston (1980) observed that the carotenoid content of *Diaptomus sicilis* and *D. nevadensis* was inversely correlated with daytime vertical distribution in two lakes. However, no such a correlation has been found in *Daphnia* (De Meester &

Beenaerts, 1993), suggesting that the relationship between pigmentation and depth is probably more complex and likely to be affected by other environmental factors, such as temperature, food and predator distributions. In fact, carotenoid content in zooplankton has long been known to be influenced by diet in addition to radiation intensity (Green, 1957; Herring, 1968).

10.2.8 UV detection

For free swimming organisms, the first response option to increased ambient UVR is to use the external media (water and the substances in it) as a filter to reduce their exposure to UVR (Zagarese & Williamson, 1994). This can be achieved by either permanently migrating out of the surface layers of the water column or by undergoing a diel vertical migration wherein they avoid high UVR irradiance in the surface waters during the day (see below). Thus, the ability to detect and respond to UVR could potentially be of great importance to migratory zooplankton and ichthyoplankton.

UV-A vision has been demonstrated in rainbow trout (Beaudet, Browman & Hawryshyn, 1993), red sea bream (*Pagrus major*) (Kawamura, Miyagi & Anraku, 1997) and roach (*Rutilus rutilus*) (Downing, Djamgoz & Bowmaker, 1986) among other species. Yellow perch have retinal cone photoreceptors with a visual pigment that has a maximum absorption peak at 400 nm, but this ability to detect UVR is lost during early life history stages (Loew *et al.*, 1993; Wahl *et al.*, 1993). A similar loss of UVA photosensitivity during juvenile development has been demonstrated in rainbow trout (Browman & Hawryshyn, 1992). The capacity of young fishes to detect UVA indicates that they are usually exposed to substantial UVA levels in Nature. It has been suggested that sensitivity to polarised UVA can enhance the contrast of an underwater target (Hawryshyn, 1992), thus helping in the detection of prey. Larval and juvenile fish have been reported to lose the ability to detect UV light during ontogenetic development. Interestingly, in the species studied so far, the loss of UVA perception coincides with a diet shift from planktonic feeders to benthic feeders (Beaudet *et al.*, 1991; Loew *et al.*, 1993; Wahl *et al.*, 1993). Several groups of invertebrates, including *Daphnia* (Smith & Macagno, 1990; Hessen, 1994), and even deep sea decapod crustaceans (Frank & Widder, 1994), have also been found to react to UVA. There is presently no evidence for the presence of UVB sensitive channels in the visual system of any organism (H. Browman, personal communication).

An important group of planktivores, notably fish and a few invertebrates, locate their prey by sight, and are collectively referred to as visual predators. Pigments such as carotenoids and melanin increase the conspicuousness of prey, making them more susceptible to visual predators (Hairston, 1979; Luecke & O'Brien, 1981; Hobaek & Wolf, 1991). Although MAAs are highly transparent to visible light, their high absorbance in the near UVR range could potentially favour the detection of prey by predators capable of UVA detection. As far as we know, this possibility has not been explored.

10.3 The pelagic environment

10.3.1 The structure of pelagic habitats

In both marine and freshwater systems, the water column can be viewed as a habitat gradient along which multiple factors vary including temperature, radiation, nutrients, food quantity and quality, and predators. UVR is just one of many selective forces acting on aquatic organisms along this gradient, so the nature of the response of aquatic ecosystems to UVR changes will not be straightforward, but rather will be a function of the interactive effects with these other variables (Williamson, 1996).

In addition to the strong vertical gradient, large numbers of discrete structures can be found in the pelagic environment (Kingsford, 1993). Single structures in the pelagic environment, as well as aggregates (e.g. phytoplankton and marine snow) and flotsam can provide shelter and substratum on which ichthyoplankton may feed. Fish representing 72 families have been found associated with structures in the pelagic environment as larvae or pelagic juveniles (Figure 10.3). Particles of organic material and zooplankton aggregate on marine snow[2]. Copepods such as *Oncaea* spp. use marine snow as a substratum on which to feed. The concentration of biotic structures is often intensified by oceanographic features, such as fronts. It has been proposed that structures may provide fish with an object with which they can orient, shelter, and enhance visibility of prey, or which they can use as a source of food, or a surface on which to prey. There is presently no evidence that such structures may also serve as protection against UVR, but their occurrence should be taken into account as they can influence the distribution of planktonic organisms as well as provide shading.

---

[2] Marine snow refers to agglomerated particles of organic material that are macroscopic in size, fragile and usually white in colour.

10.3.2 UV radiation attenuation and water mixing

The physical processes controlling UVR penetration in natural waters (absorption and scattering) are reasonably well understood (Chapter 3; Kirk, 1994). Within a homogeneous water column, the attenuation of monochromatic UVR roughly obeys the Lambert–Beer law; that is, a constant proportion of photons is absorbed by each metre of water. In general, UVR attenuates less rapidly in high elevation lakes and open oceans as compared with most other lakes and marine coastal areas (Figure 10.4). The irradiance received by sessile organisms such as attached algae or invertebrates may be estimated with relative accuracy from the irradiance at the surface and the extinction coefficient of the water (see Figure 10.4). Moreover, the use of biological dosimeters (Karentz & Lutze, 1990; Regan *et al.*, 1992; Kirk *et al.*, 1994) and '*in situ*' incubations of planktonic organisms (Williamson *et al.*, 1994; Zagarese *et al.*, 1994) have been helpful in understanding the attenuation of biologically effective radiation with depth.

Yet, the estimation of the irradiance conditions experienced by free swimming organisms, such as the plankton, is complicated by the fact that they are exposed to fluctuating radiation levels (Helbling, Holm-Hansen

Figure 10.3. The occurrence of pelagic structures can influence the distribution of planktonic organisms as well as provide shading. In the picture, large numbers of *Trachurus* spp. are associated with *Desmonema chierchianum* (Cyaneidae) off the northeastern coast of New Zealand. (From Kingsford, 1993. Reprinted with the permission of the *Bulletin of Marine Science*.)

& Villafañe, 1994; Jeffrey *et al.*, 1996). In addition to their own vertical displacements, many organisms may be passively cycled in the water column by physical mixing. Wind stress is usually responsible for turbulent mixing in surface waters (Imboden & Wüest, 1995). During calm weather the ambient fluid motion for individual plankton may be laminar, but stormy conditions produce extremely high turbulence near the surface. The intensity of turbulence diminishes with depth (Kiørboe & Saiz, 1995; Yamazaki & Squires, 1996). Thus, depending on wind speed, current shear and stratification, the time scales for cycling of neutrally buoyant particles by turbulent eddies and mixing may vary from about 0.5 h to hundreds of hours for vertical displacements of the order of 10 m (Denman & Gargett, 1983).

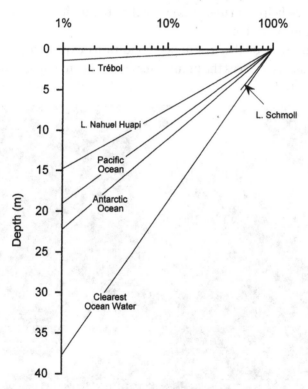

Figure 10.4. Depth–irradiance plots for different fresh and marine waters. The point at which each line transects the *y*-axis is the 1% attenuation depth. Data sources: forest Lakes Trébol and Nahuel Huapi, and mountain Lake Schmoll in southwestern Argentina, Morris *et al.* (1995); Pacific Ocean off the central Chilean coast, Montecino & Pizarro (1995); Antarctic Ocean near Elephant Island, Helbling *et al.* (1994); clearest ocean water, Smith & Baker (1979).

Mixing causes planktonic organisms to be exposed to fluctuating radiation levels that may vary over a range of several orders of magnitude. Under the assumptions of reciprocity (Bunsen–Roscoe principle), the organisms average out the fluctuations in dose rate and consequently the effects can be reasonably modelled as a function of dose. However, reciprocity can hardly be claimed to be the rule in aquatic ecosystems. In most cases, the activity of recovery mechanisms results in the failure to meet reciprocity. Understanding the effects of mixing requires knowledge of the vertical circulation within the water column as well as the kinetics of UVR damage and repair (see Chapters 3, 6 and 7).

A few studies have explored the effects of fluctuating radiation levels on the survival of zooplankton species (which is perhaps the most direct application of the reciprocity principle in the pelagic environment). Zagarese and co-workers incubated two zooplankton crustaceans (one capable of photorecovery and the other not) either at fixed depths or under simulated mixing (Zagarese et al., 1998a,b). To simulate the vertical movement produced by mixing, the organisms were placed in quartz tubes attached to wheels of different diameters. The wheels rotated in the vertical plane, their upper edge being tangential to the lake surface. Additional sets of tubes were suspended at fixed depths. The main goal of these experiments was to assess the extent to which the results from incubations at fixed depths could be extrapolated to vertically moving zooplankton. In order to accomplish this goal, a logistic model was fitted to the survival data from the incubations at fixed depths and subsequently the parameterised model was used to estimate the expected survival for the organisms that were incubated by rotating in the wheels. The incubations at fixed depths accurately predicted the survival of moving individuals for the species without photorecovery (Zagarese et al., 1998a), but failed to do so for the species capable of photorecovery (Zagarese et al., 1998b). This is consistent with the common perception that repair mechanisms are often responsible for the failure of the reciprocity law. In any event, the lack of consideration of vertical mixing processes may produce a distorted picture of effects of UVR in pelagic environments (see also Neale, Davis & Cullen, 1998).

Depending on factors such as weather conditions, basin depth, and size of the organisms, the motion can be either independent of the flow or entirely dictated by the water flow (Yamazaki & Squires, 1996). Stronger swimmers may be able to avoid the most exposed and turbulent surface layers, while organisms with more limited swimming capacities may be forced to cycle near the surface. As swimming speed tends to increase with body size (Peters *et al.*, 1994), large organisms have, in principle, an advantage over small organisms (Kiørboe & Sabatini, 1995). Another factor to consider is the energy required for swimming against the water flow. For many organisms, this may represent a significant proportion of the total energy expenditure (Peters *et al.*, 1994). Thus, for small organisms that regularly occur in the upper mixed layer, it may be more effective to develop protective or repair strategies against high UVR levels instead of actively swimming away from the surface. There have been only a few attempts to assess the cost of maintaining mechanisms to cope with high UVR levels. Zagarese *et al.* (1997b) incubated the copepod *Boeckella gibbosa* in the ultraoligotrophic Lake Toncek. Despite their use of different filter combinations to produce three radiation treatments – (a) visible radiation only, (b) visible + UVA radiation, and (c) visible + UVA + UVB radiation – no differences in the performance of *B. gibbosa* between treatments were observed after 24 days. Enough UVR reached the incubation depth (1 m) to produce about 70% mortality in the species, *B. gracilipes*, that lacked photoreactivation after 3 days. In contrast, *B. gibbosa* tolerated the high UVR levels for over 24 days by virtue of its photorecovery capacity (Zagarese *et al.*, 1997a). Because the damage produced in *B. gibbosa* in the UVR-exposed treatments did not accumulate, it was assumed that photoreactivation had continuously removed it. On the other hand, no damage was produced in UVR-protected treatments, and consequently the individuals did not have to make use of their capacity for photorecovery. Thus, the lack of differences between the three treatments suggested that the cost of repairing the damage was low and had negligible impact on the growth and fecundity of *B. gibbosa*.

10.4 Diel patterns of vertical distribution

Damaging solar radiation has long been recognised as a potentially important proximate and ultimate selective force in determining the vertical distribution and migration patterns of zooplankton (Huntsman, 1924). Recently, however, with the demonstration that changes in vertical distribution and migration patterns can be directly induced by predators (Dawidowicz, Pijanowska & Ciechomski, 1990; Leibold, 1990; Neill, 1990), the radiation damage hypothesis has been all but eclipsed. Accumulation of evidence for the damaging effects of natural solar radiation in the surface waters of lakes and oceans suggests that the predation and radiation damage hypotheses need to be re-examined. This is particularly important because both hypotheses predict the same vertical migration patterns (down during the day, up at night) and are thus difficult to separate in field survey data. In one of the few instances where attempts were made to separate these two hypotheses experimentally, the authors concluded that there was no evidence of induction of vertical migration by UVB in *Acartia hudsonica* (Bollens & Frost, 1990). However, the low level of replication and lack of treatments for potentially damaging UVA radiation in this study renders it less than conclusive.

The size selective nature of both visual predators (with their preference for larger zooplankton prey) and tactile predators (which prefer small zooplankton prey) suggests that small zooplankton may be the most vulnerable to the conflicting selective pressures from predation and damaging solar radiation (Williamson, 1995). For example, during the summer solstice in a high UVR system with visual zooplanktivores, larger predatory zooplankton will migrate down during the day, and smaller zooplankton will tend to exhibit a reverse migration into the surface waters by day to avoid these invertebrate predators. During the summer solstice this is likely to cause a 'solar bottleneck' period during which small zooplankton are 'squeezed' between damaging high UVR in the surface waters and predatory zooplankton in the deeper waters (Williamson, 1995). The prediction that arises from this situation is that smaller zooplankton will tend to be more tolerant of high UVR than larger zooplankton owing to their exposure to higher UVR in the surface waters.

10.5 Statistical issues

Two basic approaches have been used in controlled experiments: 'dose response at a set end-point' and 'time to event'. In the former type of experiment, the investigator controls the duration of exposure and varies the dose. The effect is measured after a prescribed period of time. The results are then reported either directly or as a computed measure of susceptibility, such as the LD_{50}. The aim of the 'time to event' approach is to model the duration of exposure before an individual's death. The cumulative proportion of exposed individuals dying is recorded at a series of time intervals. It is beyond the scope of this chapter to present a discussion of these different methods, but the interested reader can find a complete description and several useful examples in Scheiner & Gurevitch (1993) and Newman (1995).

Because of the smaller size of bacteria and phytoplankton, the variables usually considered (e.g. carbon uptake) represent an average of thousands of individuals. This translates into low between-replicate variance as compared to a typical experiment with zooplankton and fish. Therefore the power of tests performed on the results of experiments with zooplankton and ichthyoplankton is lower, and consequently the ability to detect UVR effects is also lower. Conversely, differences in performance among individuals are more easily seen when one is working with zooplankton and ichthyoplankton. It must be noted that these differences are simply the result of differences in the methods used by the investigators. In fact, large differences among individual cells have been detected in algae using immunofluorescent thymine dimer detection (Buma *et al.*, 1995).

Another important difference between studies performed with unicellular organisms and studies performed with animals relates to the nature of the variable being measured. In studies with algae for example, it is common to focus on some instantaneous measure of cell performance, such as photoinhibition of photosynthesis. Such variables are more probably related to irradiance (or dose rate) than to dose (Cullen & Lesser, 1991; Lesser, Cullen & Neale, 1994). On the other hand, animal researchers have been concerned mainly with variables that are cumulative in nature, such as sunburn, death of the organisms, failure to hatch, etc. Such variables are more directly related to dose, but in cases where reciprocity is violated, it is also important to know the time scale of exposure (or conversely the irradiance during exposure). In cases where the irradiance fluctuates during the experimental period, it should be necessary to know the irradiance at any time.

10.6 Concluding remarks

Not too long ago, the UV component of solar radiation was considered to be inconsequential to the aquatic biota. The last decade has witnessed a great deal of progress in the understanding of the biological effects of UVR and, as a result, UVR is now considered to be a significant and pervasive selective force in aquatic ecosystems. Perhaps the new challenge for future UVR research is to incorporate present knowledge of UVR effects into a broader ecological context. We cannot overemphasise the importance of having good estimates of UVR attenuation, but even the most accurate K_d estimates are of limited value if we ignore the vertical distribution of the organisms and their typical mixing patterns, the wavelength specific kinetics of UVR damage and repair, the potential for acclimation and/or interaction between food chain components, etc. Several molecular techniques now allow the detection of direct UVR damage in a variety of organisms. In future, they could be used in combination with models of water circulation, to produce a more realistic picture of the effects of UVR in the pelagic environment. There is also an urgent need to link the damage at the molecular and cellular level to ecologically relevant variables such as growth, fecundity or mortality. BWFs have proved to be a useful approach to translate raw radiation measurements into some biologically meaningful magnitude. There is, however, a need to develop models that allow us to evaluate the environmental consequences of different radiation scenarios; that is, an ecological weighting function.

Acknowledgements

We thank Chris Osburn, Barbara Tartarotti, Walter Helbling, Virginia Villafañe and two anonymous reviewers for critical reading of the manuscript, and Howard I. Browman for making available copies of his work in press. Figure 10.3 was generously provided by M. J. Kingsford. This work was supported by Universidad Nacional del Comahue (Proyecto B922); Consejo Nacional de Investigaciones Científicas y Técnicas de Argentina (PEI no. 0195/97); Agencia Nacional de Promoción Científica y Tecnológica de Argentina (PICT no. 01-00002-00066), International Foundation for Science (grant H/2325-2), and USA National Science Foundation (grants DEB 9509042 and DEB 9740356).

References

Ahmed, F. E. & Setlow, R. B. (1993). Ultraviolet radiation-induced DNA damage and its photorepair in the skin of the platyfish *Xiphophorus*. *Cancer Research*, **53**, 2249–55.

Ahmed, F. E., Setlow, R. B., Grist, E. & Setlow, N. (1993). DNA damage, photorepair, and survival in fish and human cells exposed to UV radiation. *Environmental and Molecular Mutagenesis*, **22**, 18–25.

Applegate, L. & Ley, R. (1988). Ultraviolet radiation-induced lethality and repair of pyrimidine dimers in fish embryos. *Mutation Research*, **198**, 85–92.

Arts, M. T., & Rai, H. (1997). Effects of enhanced ultraviolet-B radiation on the production of lipid, polysaccharide and protein in three freshwater algal species. *Freshwater Biology*, **38**, 597–610.

Barcelo, J. A. & Calkins, J. (1979). Positioning of aquatic microorganisms in response to visible light and simulated solar UV-B irradiation. *Photochemistry and Photobiology*, **29**, 75–83.

Battini, M., Rocco, V., Lozada, M. Tartarotti, B. & Zagarese, H. E. Effects of ultraviolet radiation on the eggs of landlocked *Galaxias maculatus* (Galaxiidae, pisces) in Northwestern Patagonia. *Freshwater Biology*, submitted.

Beaudet, L., Browman, H. I. & Hawryshyn, C. W. (1991). Ontogenetic loss of U.V.

photosensitivity in rainbow trout, determined using optic nerve compound action potential recording. *Society for Neurosciences Abstracts*, **17**, 299.

Beaudet, L., Browman, H. I. & Hawryshyn, C. W. (1993). Optic nerve response and retinal structure in rainbow trout of different sizes. *Vision Research*, **33**, 1739–46.

Béland, F., Browman, H. I., Alonso Rodriguez, C. & St-Pierre, J.-F. (1999) The effect of solar ultraviolet radiation on the eggs and larvae of Atlantic cod (*Gadus morhua*). *Canadian Journal of Fisheries and Aquatic Science*, in press.

Bell, G. M. & Hoar, W. S. (1950). Some effects of ultraviolet radiation on sockeye salmon eggs and alevins. *Canadian Journal of Research*, **28**, 35–43.

Bergham, R., Bullock, A. M. & Karakiri, M. (1993). Effects of solar radiation on the population dynamics of juvenile flatfish in the shallows of Wadden Sea. *Journal of Fish Biology*, **42**, 329–45.

Blaustein, A. R., Hoffman, P. D., Hokit, D. G., Kiesecker, J. M. W., S. C. & Hays, J. B. (1994). UV repair and resistance to solar UV-B radiation in amphibian eggs: a link to population declines? *Proceedings of the National Academy of Sciences, USA*, **91**, 1791–5.

Blaustein, A. R., Hoffman, P. D., Kiesecker, J. M. & Hays, J. B. (1996). DNA repair activity and resistance to solar UV-B radiation in eggs of the red-legged frog. *Conservation Biology*, **10**, 1398–402.

Blazer, V. S., Fabacher, D. L., Little, E. E., Ewing, M. S. & Kocan, K. M. (1997). Effects of ultraviolet-B radiation on fish: histologic comparison of a UVB-sensitive and a UVB-tolerant species. *Journal of Aquatic Animal Health*, **9**, 132–43.

Bollens, S. M. & Frost, B. W. (1990). UV light and vertical distribution of the marine planktonic copepod *Acartia hudsonica* Pinhey. *Journal of Experimental Marine Biology and Ecology*, **137**, 89–93.

Bothwell, M. L., Sherbot, D. M. J. & Pollock, C. M. (1994). Ecosystem responses to solar ultraviolet-B radiation: influence of trophic-level interactions. *Science*, **265**, 97–100.

Brehm, V. (1938). Die Rotfärbung von Hochgebirgsorganismen. *Biological Reviews*, **13**, 307–18.

Browman, H. I. & Hawryshyn, C. W. (1992). Thyroxine induces a precocial loss of ultraviolet photosensitivity in rainbow trout (*Oncorhynchus mykiss*, Teleostei). *Vision Research*, **32**, 2303–12.

Bullock, A. M. & Coutts, R. (1985). The impact of solar ultraviolet radiation upon the skin of rainbow trout, *Salmo gairdneri* Richardson, farmed at high altitude in Bolivia. *Journal of Fish Diseases*, **8**, 263–72.

Buma, A. G. J., Van Hannen, E. J., Roza, L., Veldhuis, M. J. W. & Gieskes, W. W. C. (1995). Monitoring ultraviolet-B-induced DNA damage in individual diatom cells by immunofluorescent thymine dimer detection. *Journal of Phycology*, **31**, 314–21.

Buma, A. G. J., Van Hannen, E. J. Veldhuis M. J. W. & Gieskes, W. W. C. (1996). UV-B induces DNA damage and DNA synthesis delay in the marine diatom *Cyclotella* sp. *Scientia Marina*, **60** (Supplement 1), 101–6.

Byron, E. R. (1982). The adaptive significance of calanoid copepod pigmentation: a comparative and experimental analysis. *Ecology*, **63**, 1871–86.

Cabrera, S., López, M. & Tartarotti, B. (1997). Phytoplankton and zooplankton response to ultraviolet radiation in a high altitude Andean lake: short- versus long-term effects. *Journal of Plankton Research*, **19**, 1565–82.

Caldwell, M. M., Camp, L. B., Warner, C. W. & Flint, S. D. (1986). Action spectra and their key role in assessing biological consequences of solar UV-B radiation change. In *Stratospheric Ozone Reduction, Solar Ultraviolet Radiation and Plant Life*, ed. R. C. Worrest & M. M. Caldwell, pp. 87–111. Springer-Verlag, New York.

Carlini, D. B. & Regan, J. D. (1995). Photolyase activities of *Elysa tuca, Bursatella leachii,* and *Haminaea antillarum* (Mollusca: Opisthobranchia). *Journal of Experimental Marine Biology and Ecology*, **189**, 219–32.

Carroll, A. K. & Shick, J. M. (1996). Dietary accumulation of UV-absorbing mycosporine-like amino acids (MAAs) by the green sea urchin (*Strongylocentrotus droebachiensis*). *Marine Biology*, **124**, 561–9.

Chalker-Scott, L. (1995).Survival and sex ratios of the intertidal copepod *Tigriopus californicus* following ultraviolet-B (290–320) radiation exposure. *Marine Biology*, **123**, 799–804.

Crowder, L. B. & Crawford, H. L. (1984). Ecological shifts in resource use by bloaters in Lake Michigan. *Transactions of the American Fisheries Society*, **113**, 694–700.

Cullen, A. P. & Monteith-McMaster, C. A. (1993). Damage to the rainbow trout (*Onchorhynchus mykiss*) lens following acute dose of UV-B. *Current Eye Research*, **12**, 97.

Cullen, J. J. & Lesser, M. P. (1991). Inhibition of photosynthesis by ultraviolet radiation as a function of dose and dosage rate: results for a marine diatom. *Marine Biology (Berlin)*, **111**, 183–90.

Cullen, J. J. & Neale, P. J. (1997). Biological weighting functions for describing the effects of ultraviolet radiation on aquatic systems. In *The Effects of Ozone Depletion on Aquatic Ecosystems*, ed. D.-P. Häder, pp. 97–118. Academic Press, San Diego, CA.

Cullen, J. J., Neale, P. J. & Lesser, M. P. (1992). Biological weighting function for the inhibition of phytoplankton photosynthesis by ultraviolet radiation. *Science*, **258**, 646–50.

Damkaer, D. M. & Dey, D. B. (1983). UV damage and photoreactivation potentials of larval shrimp, *Pandalus platyceros*, and adult euphausiids, *Thysanoesa raschii*. *Oecologia* (Berlin), **60**, 169–75.

Damkaer, D. M., Dey, D. B. & Heron, G. A. (1981). Dose/dose-rate responses of shrimp larvae to UV-B radiation. *Oecologia*, **48**, 178–82.

Damkaer, D. M., Dey, D. B., Heron, G. A. & Prentice, E. F. (1980). Effects of UV-B radiation on near-surface zooplankton of Puget Sound. *Oecologia*, **44**, 149–58.

Dawidowicz, P., Pijanowska, J. & Ciechomski, K. (1990). Vertical migration of *Chaoborus* larvae is induced by the presence of fish. *Limnology and Oceanography*, **35**, 1631–7.

De Meester, L. & Beenaerts, N. (1993). Heritable variation in carotenoid content in *Daphnia magna. Limnology and Oceanography*, **38**, 1193–9.

Denman, K. L. & Gargett, A. E. (1983). Time and space scales of vertical mixing and advection of phytoplankton in the upper ocean. *Limnology and Oceanography*, **28**, 801–15.

Dey, D. B., Damkaer, D. M. & Heron, G. A. (1988). UV-B dose/dose-rate responses of seasonally abundant copepods of Puget Sound. *Oecologia*, **76**, 321–9.

Downing, J. E. G., Djamgoz, M. B. A. & Bowmaker, J. K. (1986). Photoreceptors of a cyprinid fish, the roach: morphological and spectral characteristics. *Journal of Comparative Physiology A*, **159**, 859–68.

Dunbar, C. (1959). Sunburn in fingerling rainbow trout. *Progressive Fish Culturist*, **21**, 74.

Dunlap, W. C. & Yamamoto, Y. (1995). Small-molecule antioxidants in marine organisms: antioxidant activity of mycosporine-glycine. *Comparative Biochemistry and Physiology*, **112B**, 105–14.

Eisler, R. (1961). Effects of visible radiation on salmonid embryos and larvae. *Growth*, **25**, 281–346.

Farman, J. C., Gardiner, B. G. & Shanklin, J. D. (1985). Large losses of total ozone in the Antarctica reveal seasonal ClO$_x$/NO$_x$ interaction. *Nature*, **315**, 207–10.

Frank, T. M. & Widder, E. A. (1994). Comparative study of behavioral sensitivity thresholds to near-UV and blue-green light in deep-sea crustaceans. *Marine Biology*, **121**, 229–35.

Funayama, T., Mitani, H. & Shima, A. (1993). Ultraviolet-induced DNA damage and its photorepair in tail fin cells of the medaka, *Oryzias latipes*. *Photochemistry and Photobiology*, **58**, 380–5.

Garcia-Pichel, F. (1996). The absorption of ultraviolet radiation by microalgae: simple optics and photobiological implications. *Scientia Marina*, **60**, 73–9.

Goes, J. I., Handa, N., Taguchi, S. & Hama, T. (1994). Effect of UV-B radiation on the fatty acid composition of the marine phytoplankton *Tetraselmis* sp.: relationship to cellular pigments. *Marine Ecology Progress Series*, **114**, 259–74.

Green, J. (1957). Carotenoids in *Daphnia*. *Proceedings of the Royal Society, London, Series B*, **147**, 292–401.

Häder, D.-P. & Worrest, R. C. (1997). Consequences of the effects of increased solar ultraviolet radiation on aquatic ecosystems. In *The Effects of Ozone Depletion on Aquatic Ecosystems*, ed. D.-P. Häder, pp. 11–30. R. G. Landes Co., Austin, TX.

Häder, D.-P., Worrest, R. C., Kumar, H. D. & Smith, R. C. (1995). Effects of increased solar ultraviolet radiation on aquatic ecosystems. *AMBIO*, **24**, 174–80.

Hairston, N. G. (1976). Photoprotection by carotenoid pigments in the copepod *Diaptomus nevadensis*. *Proceedings of the National Academy of Sciences, USA*, **73**, 971–4.

Hairston, N. G. (1979). The adaptive significance of color polymorphism in two species of *Diaptomus* (Copepoda). *Limnology and Oceanography*, **24**, 15–37.

Hairston, N. G. (1980). The vertical distribution of diaptomid copepods in relation to body pigmentation. In *Evolution and Ecology of Zooplankton Communities*, ed. W. C. Kerfoot, pp. 98–110. University of New England Press, Hanover.

Hawryshyn, C. W. (1992). Polarization vision in fish. *American Scientist*, **80**, 164–75.

Hebert, P. D. N. & Emery, C. J. (1990). The adaptative significance of cuticular pigmentation in *Daphnia*. *Functional Ecology*, **4**, 703–10.

Helbling, E. W., Holm-Hansen, O. & Villafañe, V. (1994). Effects of ultraviolet radiation on Antarctic marine phytoplankton photosynthesis with particular attention to the influence of mixing. In *Ultraviolet Radiation in Antarctica: Measurements and Biological Effects*, Antartic Research Series, **62**, ed. C. S. Weiler & P. A. Penhale, pp. 207–27. American Geophysical Union, Washington, DC.

Herring, P. J. (1968). The carotenoid pigments of *Daphnia magna* Strauss. II. Aspects of pigmentary metabolism. *Comparative Biochemistry and Physiology*, **24**, 205–21.

Hessen, D. O. (1994). *Daphnia* responses to UV-B light. *Ergebnisse der Limnologie (Archiv für Hydrobiologie. Beihefte)*, **43**, 185–95.

Hessen, D. O., De Lange, H. J. & Van Donk, E. (1997). UV-induced changes in phytoplankton cells and its effects on grazers. *Freshwater Biology*, **38**, 513–24.

Hessen, D. O. & Sørensen, K. (1990). Photoprotective pigmentation in alpine zooplankton populations. *Aqua Fennica*, **20**, 165–70.

Hobaek, A. & Wolf, H. G. (1991). Ecological genetics of Norwegian *Daphnia*. II. Distribution of *Daphnia longispina* genotypes in relation to short-wave radiation and water colour. *Hydrobiologia*, **225**, 229–43.

Hunter, J. R., Kaupp, S. E. & Taylor, J. H. (1981). Effects of solar and artificial ultraviolet-b radiation on larval northern anchovy, *Engraulis mordax*. *Photochemistry and Photobiology*, **34**, 477–86.

Hunter, J. R. & Taylor, J. H. (1982). Assessment of the effects of UV radiation on marine fish larvae. In *The Role of Ultraviolet Radiation in Marine Ecosystems*, ed. J. Calkins, pp. 459–97. NATO Conference Series IV.

Hunter, J. R., Taylor, J. H. & Moser, H. G. (1979). Effect of ultraviolet irradiation on eggs and larvae of the northern anchovy, *Engraulis mordax*, and the pacific mackerel, *Scomber japonicus*, during the embryonic stage. *Photochemistry and Photobiology*, **29**, 325–38.

Huntsman, A. G. (1924). Limiting factors for marine animals. I. The lethal effect of sunlight. *Canadian Biology*, **2**, 83–8.

Imboden, D. M. & Wüest, A. (1995). Mixing mechanisms in lakes. In *Physics and Chemistry of Lakes*, ed. A. Lerman, D. M. Imboden & J. R. Gat, pp. 83–138. Springer-Verlag, Berlin.

Jeffrey, W. H., Pledger, R. J., Aas, P., Hager, S., Coffin, R. B., Vonhaven, R. & Mitchell, D. L. (1996). Diel and depth profiles of DNA photodamage in bacterioplankton exposed to ambient solar ultraviolet radiation. *Marine Ecology Progress Series*, **137**, 283–91.

Karanas, J. J., Van Dyke, H. & Worrest, R. C. (1979). Midultraviolet (UV-B) sensitivity of *Acartia clausii* Giesbrecht (Copepoda). *Limnology and Oceanography*, **24**, 1104–16.

Karanas, J. J., Worrest, R. C. & Van Dyke, H. (1981). Impact of UV radiation on the fecundity of the copepod *Acartia clausii*. *Marine Biology* (Berlin), **65**, 125–33.

Karentz, D. & Lutze, L. H. (1990). Evaluation of biologically harmful ultraviolet radiation in Antarctica with a biological dosimeter designed for aquatic environments. *Limnology and Oceanography*, **35**, 549–61.

Karentz, D., McEuen, F. S. & Dunlap, W. C. (1991). Survey of mycosporine-like amino acid compounds in Antarctic marine organisms: potential protection from ultraviolet exposure. *Marine Biology* (Berlin), **108**, 157–66.

Kaupp, S. & Hunter, J. (1981). Photorepair in larval anchovy, *Engraulis mordax*. *Photochemistry and Photobiology*, **33**, 253–6.

Kawamura, G., Miyagi, M. & Anraku, K. (1997). Retinomotor movement of all spectral cone types of red sea bream *Pagrus major* in response to monochromatic stimuli and UV sensitivity. *Fisheries Science*, **63**, 233–5.

Kelner, A. (1949). Effect of visible light on the recovery of *Streptomyces griseus* conidia from ultraviolet irradiation injury. *Proceedings of the National Academy of Sciences, USA*, **35**, 73–9.

Kingsford, M. J. (1993). Biotic and abiotic structure in the pelagic environment: importance to small fishes. *Bulletin of Marine Science*, **53**, 393–415.

Kiørboe, T. & Sabatini, M. (1995). Scaling of fecundity, growth and development in marine planktonic copepods. *Marine Ecology Progress Series*, **120**, 285–98.

Kiørboe, T. & Saiz, E. (1995). Planktivorous feeding in calm and turbulent environments, with emphasis on copepods. *Marine Ecology Progress Series*, **122**, 135–45.

Kirk, J. T. O. (1994). Optics of UV-B radiation in natural waters. *Ergebnisse der Limnologie (Archiv für Hydrobiologie. Beiheft)*, **43**, 1–16.

Kirk, J. T. O., Hargreaves, B. R., Morris, D. P., Coffin, R. B., David, B., Frederickson, D., Karentz, D., Lean, D. R. S., Lesser, M. P., Madronich, S., Morrow, J. H., Nelson, N. B. & Scully, N. M. (1994). Measurements of UV-B radiation in two freshwater lakes: an instrument intercomparison. *Ergebnisse der Limnologie (Archiv für Hydrobiologie. Beiheft)*, **43**, 71–99.

Knowles, J. F. (1992). The effect of chronic radiation on the humoral immune response of rainbow trout *Onchorhynchus mykiss* Walbaum. *International Journal of Radiation Research*, **62**, 239–48.

Kouwenberg, J. H. M., Browman, H. I. Cullen, J. J. Davis, R. F. St-Pierre, J.-F. & Runge, J. A. (1999a). Biological weighting of ultraviolet (280–400 nm) induced mortality in marine zooplankton and fish. I. Atlantic cod (*Gadus morhua*) eggs. *Marine Biology*, in press.

Kouwenberg, J. H. M., Browman, H. I. Runge, J. A. Cullen, J. J. Davis, R. F. & St-Pierre, J.-F. (1999b). Biological weighting of ultraviolet (280–400 nm) induced mortality in marine zooplankton and fish. II. *Calanus finmarchicus* eggs. *Marine Biology*, in press.

Leach, C. M. (1965). Ultraviolet absorbing substances associated with light-induced sporulation in fungi. *Canadian Journal of Botany*, **43**, 185–200.

Leibold, M. A. (1990). Resources and predators can affect the vertical distributions of zooplankton. *Limnology and Oceanography*, **35**, 938–44.

Lesser, M. P., Cullen, J. J. & Neale, P. J. (1994). Carbon uptake in a marine diatom during acute exposure to ultraviolet-b radiation – relative importance of damage and repair. *Journal of Phycology*, **30**, 183–92.

Little, E. E. & Fabacher, D. L. (1994). Comparative sensitivity of rainbow trout and two threatened salmonids, Apache trout and Lahontan cutthroat trout, to ultraviolet-B radiation. *Ergebnisse der Limnologie (Archiv für Hydrobiologie. Beiheft)*, **43**, 217–26.

Loew, E. R., McFarland, W. N., Mills, E. L. & Hunter, D. (1993). A chromatic action spectrum for planktonic predation by juvenile yellow perch, *Perca flavescens*. *Canadian Journal of Zoology*, **71**, 384–6.

Luecke, C. & O'Brien, J. O. (1981). Phototoxicity and fish predation: selective factors in color morphs in *Heterocope*. *Limnology and Oceanography*, **26**, 454–60.

MacArdle, J. & Bullock, C. (1987). Solar ultraviolet radiation as a causal factor of 'summer syndrome' in cage-reared Atlantic salmon, *Salmo salar* L.: a clinical and histopathological study. *Journal of Fish Diseases*, **10**, 255–64.

Madronich, S. (1994). Increases in biologically damaging UV-B radiation due to stratospheric ozone reductions: a brief review. *Ergebnisse der Limnologie (Archiv für Hydrobiologie. Beiheft)*, **43**, 17–30.

Malloy, K. D., Holman, M. A., Mitchell, D. & Detrich III, H. W. (1997). Solar UVB-induced DNA damage and photoenzymatic DNA repair in Antarctic zooplankton *Proceedings of the National Academy of Sciences, USA*, **94**, 1258–63.

Marangoni, R., Martini, B. & Colombetti, G. (1997). Effects of UV-B on ciliates. In *The Effects of Ozone Depletion on Aquatic Ecosystems*, ed. D.-P. Häder, pp. 229–46. Academic Press, San Diego, CA.

Mitchell, D. L. & Karentz, D. (1993). The induction and repair of DNA photodamage in the environment. In *Environmental UV Photobiology*, ed. A. R. Young, L. O. Björn, J. Moan & W. Nultsch, pp. 345–77. Plenum Press, New York.

Montecino, V. & Pizarro, G. (1995). Phytoplankton acclimation and spectral penetration of UV irradiance off the central Chilean coast. *Marine Ecology Progress Series*, **121**, 261–9.

Morgan, S. G. & Christy, J. H. (1996). Survival of marine larvae under the countervailing selective pressures of photodamage and predation. *Limnology and Oceanography*, **41**, 498–504.

Morris, D. P. & Hargreaves, B. R. (1997). The role of photochemical degradation of dissolved organic carbon in regulating the UV transparency of three lakes on the Pocono Plateau. *Limnology and Oceanography*, **42**, 239–49.

Morris, D. P., Zagarese, H. E., Williamson, C. E., Balseiro, E. G., Hargreaves, B. R., Modenutti, B., Moeller, R. & Queimaliños, C. (1995). The attenuation of solar

UV radiation in lakes and the role of disolved organic carbon. *Limnology and Oceanography*, **40**, 1381–91.

Naganuma, T., Inoue, T. & Uye, S. (1997). Photoreactivation of UV-induced damage to embryos of a planktonic copepod. *Journal of Plankton Research*, **19**, 783–7.

Neale, P. J., Davis, R. F. & Cullen, J. J. (1998). Interactive effects of ozone depletion and vertical mixing on photosynthesis of Antarctic phytoplankton. *Nature*, **392**, 585–9.

Neill, W. E. (1990). Induced vertical migration in copepods as a defense against invertebrate predation. *Nature*, **345**, 524–6.

Newman, M. C. (1995). *Quantitative Methods in Aquatic Ecotoxicology*. Lewis Publishers, Boca Raton, FL.

Orce, V. L. & Helbling, E. W. (1997). Latitudinal UVR-PAR measurements in Argentina: extent of the 'ozone hole'. *Global and Planetary Change*, **15**, 113–21.

Palenik, B., Price, N. M. & Morel, F. M. M. (1991). Potential effects of UV-B radiation on the chemical environment of marine organisms: a review. *Environmental Pollution*, **70**, 117–30.

Palozza, P. & Krinsky, N. I. (1992). Astaxanthin and canthaxanthin are potent antioxidants in a membrane model. *Archives of Biochemistry and Biophysics*, **297**, 291–5.

Peters, R. H., Demers, E., Koelle, M. & MacEnzie, B. R. (1994). The allometry of swimming speed and predation. *Internationale Vereinnigung für Theoretische und Angewandte Limology. Verhandlungen*, **25**, 2316–23.

Ramos, K. T., Fries, L. T., Berkhouse, C. S. & Fries, J. N. (1994). Apparent sunburn of juvenile paddlefish. *Progressive Fish Culturist*, **56**, 214–16.

Regan, J. D., Carrier, W. L., Gusinski, H., Olla, B. L., Yoshida, H., Fujimura, R. K. & Wicklund, R. I. (1992). DNA as a solar dosimeter in the ocean. *Photochemistry and Photobiology*, **56**, 35–42.

Regan, J. D., Carrier, W. L., Samet, C. & Olla, B. L. (1982). Photoreactivation in two closely related marine fishes having different longevities. *Mechanisms of Ageing and Development*, **18**, 59–66.

Ringelberg, J., Keyser, A. L. & Flik, B. J. G. (1984). The mortality effect of ultraviolet radiation in a red morph of *Acanthodiaptomus denticornis* (Crustacea: Copepoda) and its possible ecological relevance. *Hydrobiologia*, **112**, 217–22.

Sancar, A. (1994). Structure and function of DNA photolyase. *Biochemistry*, **33**, 2–9.

Scheiner, S. M. & Gurevitch, J. (1993). *Design and Analysis of Ecological Experiments*. Chapman & Hall, New York.

Schindler, D. W., Curtis, P. J., Parker, B. P. & Stainton, M. P. (1996). Consequences of climate warming and lake acidification for UV-B penetration in North American boreal lakes. *Nature*, **379**, 705–8.

Scully, N. M., Lean, D. R. S., McQueen, D. J. & Cooper, W. J. (1995). Photochemical formation of hydrogen peroxide in lakes: effects of dissolved organic carbon and ultraviolet radiation. *Canadian Journal of Fisheries and Aquatic Sciences*, **52**, 2675–81.

Scully, N. M., McQueen, D. J. & Lean, D. R. S. (1996). Hydrogen peroxide formation: the interaction of ultraviolet radiation and dissolved organic carbon in lake waters along a 43–75° N gradient. *Limnology and Oceanography*, **41**, 540–8.

Setlow, R. B. (1974). The wavelengths in sunlight effective in producing skin cancer: a theoretical analysis. *Proceedings of the National Academy of Sciences, USA*, **71**, 3363–6.

Shibata, K. (1969). Pigments and a UV-absorbing substance in corals and blue-green algae living in the Great Barrier Reef. *Plant Cell Physiology*, **10**, 325–35.

Shima, A. & Setlow, R. B. (1984). Survival and pyrimidine dimers in cultured fish cells exposed to concurrent sun lamp ultraviolet and photoreactivating radiations. *Photochemistry and Photobiology*, **39**, 49–56.

Siebeck, O. (1978). Ultraviolet tolerance of planktonic crustaceans. *Internationale Vereinnigung für Theoretische und Angewandte Limology. Verhandlungen*, **20**, 2469–73.

Siebeck, O. & Böhm, U. (1991). UV-B effects on aquatic animals. *Internationale Vereinnigung für Theoretische und Angewandte Limology. Verhandlungen*, **24**, 2773–7.

Siebeck, O. & Böhm, U. (1994). Challenges for an appraisal of UV-B effects upon planktonic crustaceans under natural conditions with a non-migrating (*Daphnia pulex obtusa*) and a migrating cladocran (*Daphnia galeata*). *Ergebnisse der Limnologie (Archiv für Hydrobiologie. Beiheft)*, **43**, 197–206.

Smith, K. C. & Macagno, E. R. (1990). UV photoreceptors in the compound eye of *Daphnia magna* (Crustacea: Branchiopoda). A fourth spectral class in single omatidia. *Journal of Comparative Physiology A*, **166**, 597–606.

Smith, R. C. & Baker, K. S. (1979). Penetration of UV-B biologically effective dose-rates in natural waters. *Photochemistry and Photobiology*, **29**, 311–23.

Sommaruga, R., Oberleiter, A. & Psenner, R. (1996). Effect of UV radiation on the bacterivory of a heterotrophic nanoflagellate. *Applied Environmental Microbiology*, **62**, 4395–400.

Sutherland, B. M. (1981). Photoreactivation. *BioScience*, **31**, 439–44.

Teai, T., Drollet, J. H., Bianchini, J. P., Cambon, A. & Martin, P. M. V. (1997). Widespread occurrence of mycosporine-like amino acid compounds in scleractinians from French Polynesia. *Coral Reefs*, **16**, 169–76.

Thomasson, K. (1956). Reflections on Arctic and Alpine Lakes. *Oikos*, **7**, 117–43.

Uchida, N., Mitani, H. & Shima, A. (1995). Multiple effects of fluorescent light on repair of ultraviolet-induced DNA lesions in cultured goldfish cells. *Photochemistry and Photobiology*, **61**, 79–83.

Vincent, W. F. & Roy, S. (1993). Solar ultraviolet-B radiation and aquatic primary production: damage, protection, and recovery. *Environmental Reviews*, **1**, 1–12.

Wahl, C. M., Mills, E. L., McFarland, W. N. & Degisi, J. S. (1993). Ontogenetic changes in prey selection and visual acuity of the yellow perch, *Perca flavescens*. *Canadian Journal of Fisheries and Aquatic Sciences*, **50**, 743–9.

Wang, K. S. & Chai, T. J. (1994). Reduction in omega-3 fatty acids by UV-b irradiation in microalgae. *Journal of Applied Phycology*, **6**, 415–21.

Watanabe, T., Kitajima, C. & Fujita, S. (1983). Nutritional values of live organisms used in Japan for mass propagation of fish: a review. *Aquaculture*, **34**, 115–43.

Williamson, C. E. (1995). What role does UV-B radiation play in freshwater ecosystems? *Limnology and Oceanography*, **40**, 386–92.

Williamson, C. E. (1996). Effects of UV radiation on freshwater ecosystems. *International Journal of Environmental Studies*, **51**, 245–56.

Williamson, C. E., Hargreaves, B. R., Orr, P. S. & Lovera, P. A. (1999). Solar UV radiation effects on higher trophic levels: the response of invertebrate and vertebrate predators in high and low DOC lakes and implications for lake acidification and recovery. *Limnology and Oceanography*, in press.

Williamson, C. E., Metzgar, S. L., Lovera, P. A. & Moeller, R. E. (1997). Solar ultraviolet radiation and the spawning habitat of yellow perch, *Perca flavescens*. *Ecological Applications*, **7**, 1017–23.

Williamson, C. E., Stemberger, R. S., Morris, D. P., Frost, T. M. & Paulsen, S. G. (1996).

Ultraviolet radiation in North American lakes: attenuation estimates from DOC measurements and implications for plankton communities. *Limnology and Oceanography*, **41**, 1024–34.

Williamson, C. E. & Zagarese, H. E. (eds.) (1994). *Impact of UV-B Radiation on Pelagic Freshwater Ecosystems*. Schweizerbart'sche Verlagsbuchhandlung, Stuttgart.

Williamson, C. E., Zagarese, H. E., Schulze, P. C., Hargreaves, B. R. & Seva, J. (1994). The impact of short-term exposure to UV-B radiation on zooplankton communities in north temperate lakes. *Journal of Plankton Research*, **16**, 205–18.

Yamazaki, H. & Squires, K. D. (1996). Comparison of oceanic turbulence and copepod swimming. *Marine Ecology Progress Series*, **144**, 299–301.

Yan, N. D., Keller, W., Scully, N. M., Lean, D. R. S. & Dillon, P. J. (1996). Increased UV-B penetration in a lake owing to drought-induced acidification. *Nature*, **381**, 141–3.

Yasuhira, S., Mitani, H. & Shima, A. (1991). Enhancement of photorepair of ultraviolet damage by preillumination with fluorescent light in cultured fish cells. *Photochemistry and Photobiology*, **53**, 211–15.

Yasuhira, S., Mitani, H. & Shima, A. (1992). Enhancement of photorepair of ultraviolet-induced pyrimidine dimers by preillumination with fluorescent light in the goldfish cell line. The relationship between survival and yield of pyrimidine dimers. *Photochemistry and Photobiology*, **55**, 97–101.

Young, A. R., O., B.L., Moan, J. & Nultsch, W. (1993). *Environmental UV Photobiology*. Plenum Press, New York. Zagarese, H. E., Cravero, W., Gonzalez, P. & Pedrozo, F. (1998a). Copepod mortality induced by fluctuating levels of natural solar radiation simulating vertical water mixing. *Limnology and Oceanography*, **43**, 169–74.

Zagarese, H. E., Cravero, W., Gonzalez, P. & Pedrozo, F. (1998) Copepod mortality induced by fluctuating levels of natural solar radiation simulating vertical water mixing. *Limnology and Oceanography*, **43**, 169–74.

Zagarese, H. E., Feldman, M. & Williamson, C. E. (1997a). UV-B induced damage and photoreactivation in three species of *Boeckella* (Copepoda, Calanoida). *Journal of Plankton Research*, **19**, 357–67.

Zagarese, H. E., Tartarotti, B., Cravero, W. & Gonzalez, P. (1998b). UV damage in shallow lakes: the implications of water mixing. *Journal of Plankton Research*, **20**, 1423–33.

Zagarese, H. E. & Williamson, C. E. (1994). Modeling the impacts of UV-B radiation on ecological interactions in freshwater and marine ecosystems. In *Stratospheric Ozone Depletion/UV-B Radiation in the Biosphere*, ed. R. H. Biggs & M. E. B. Joyner, pp. 315–28. Springer-Verlag, New York.

Zagarese, H. E., Williamson, C. E., Mislivets, M. & Orr, P. (1994). The vulnerability of *Daphnia* to UV-B radiation in the Northeastern United States. *Ergebnisse der Limnologie (Archiv für Hydrobiologie. Beiheft)*, **43**, 207–16.

Zagarese, H. E., Williamson, C. E., Vail, T. L., Olsen, O. & Queimaliños, C. (1997b). Long-term exposure of *Boeckella gibbosa* (Copepoda: Calanoida) to in situ levels of solar UVB radiation. *Freshwater Biology*, **37**, 99–106.

Zeeman, M. G. & Brindley, W. A. (1981). Effects of toxic agents upon fish immune systems: a review. In *Immunologic Considerations in Toxicology*, ed. R. P. Sharma, CRC Press, Boca Raton, FL.

Zellmer, I. D. (1996). The impact of food quantity on UV-B tolerance and recovery from UV-B damage in *Daphnia pulex*. *Hydrobiologia*, **319**, 87–92.

Zigman, S. & Rafferty, N. S. (1994). Effects of near UV radiation and antioxidants on the response of dogfish (*Mustelus canis*) lens to elevated H_2O_2. *Comparative Biochemistry and Physiology A – Physiology*, **109**, 463–7.

11

○ ○ ○ ○ ○ ○ ○ ○ ○ ○ ○ ○ ○ ○ ○ ○ ○ ○ ○

Implications of UV radiation for the food web structure and consequences on the carbon flow

Behzad Mostajir, Serge Demers*, Stephen J. de Mora, Robert P. Bukata and John H. Jerome

11.1 Introduction

Given the anticipated decreasing concentrations of stratospheric ozone, the concern regarding the effects of UV radiation (UVR) in the environment is very topical. The preceding chapters in this book provide a state-of-the-art review of UVR effects in the marine environment (see Figure 11.1). This last chapter synthesises briefly the potential UV-induced responses at community and ecosystem levels and gives an insight on how the global changes can interact with the direct impacts of UVR.

The increase of anthropogenic chlorofluorocarbons (CFCs) is responsible for the destruction of ozone in the upper atmosphere. This decrease in the ozone layer causes in an increase of UV-B radiation in the wavelength range 280 to 320 nm (Chapter 1). The past decade has been witness to a concentrated effort from the international scientific community to measure the increase of UVR and to understand its effects on different ecosystems. These efforts were supported by signatories to international agreements, such as the Montreal Protocol, to control and eliminate the production and emission of ozone-depleting substances. However, each spring, an 'ozone hole' larger than the size of Canada appears over the Antarctic region (Smith *et al.*, 1992). In the Arctic and into the north temperate zone, the ozone layer diminished by more than 45% during the 1997/8 winter, the most important decrease in 35 years (Wardle *et al.*, 1997). Notable ozone depletion has now been observed in both polar regions, as ozone loss rates in the Arctic region in recent years have reached values comparable to those recorded over the Antarctic (Rex *et al.*, 1997, and references therein). Moreover, during the spring of 1997 the ozone layer diminished by 7% below normal over the mid-latitude regions of Canada (Tarasick & Fioletov, 1997).

The ozone hole in Antarctica develops as a result of the formation of polar stratospheric clouds at very low temperatures. The Arctic condition of ozone depletion may at least in part be explained by increased greenhouse gases. In fact, an increase in atmospheric carbon dioxide causes stratospheric cooling and thereby accelerates the destruction of stratospheric ozone (Shindell, Rind & Lonnergan, 1998). This links directly the ozone depletion phenomenon to the planetary global warming problem. Similarly, Schindler *et al.* (1996), using data from 20 years of whole-lake acidification studies at the Experimental Lakes Area (ELA) in northwestern Ontario, observed that both climatic warming and lake acidification have produced significant declines in the columnar dissolved organic matter (DOM) of boreal lakes. Since a decline in DOM concentration allows deeper UV penetration (Chapters 2, 4 and 5), the possibility of UV-induced chemical and/or biological reactions increases in the aquatic environment. This increase in UV penetration may have been responsible for changes to the aquatic ecosystems (e.g. community structure) that have hitherto been attributed to acidification processes. The environmental impact of this predicted rise in solar UV-B has recently become a source of much concern and speculation in the popular press as well as in scientific literature.

11.2   UV radiation and biological systems

The impacts of UVR on aquatic ecosystems are linked directly through photobiological and photochemical processes to aquatic carbon cycle dynamics, which is itself intimately influenced by the interaction between trophic levels. The effectiveness of UVR, an agent of photobiological and photochemical change, is, in addition to the irradiance intensity of the impinging radiation field, dependent upon:

1. The depth to which the UV radiation can penetrate the water body (UV photic zone).
2. The times (short-term and cumulative) that aquatic organisms are exposed to UV radiation (determined by the duration of time and periodicity that aquatic organisms spend in and out of the UV photic zone; thereby representing the balance between organism damage and organism repair or reprieve).
3. Any parameter or process that controls either of the above determinants.

Solar UV-B radiation is known to have a wide range of harmful effects (manifest as reduced productivity) on freshwater and marine organisms

(Chapters 3 and 6), including bacterioplankton (Chapter 8) and phytoplankton (Chapter 9). Analogous studies on zooplankton and on the early life history stages of fishes, although rare, indicate that exposure to relatively low levels of UV-B deleteriously affects these groups as well (Chapter 10). All plant, animal and microbial groups appear to be susceptible to UV-B, but to a highly variable extent that depends on the individual species, its environment and their strategies to minimise UV-induced damage (Chapter 7).

In order to evaluate the potential influence of enhanced UV-B on ecosystems, it is necessary to observe and model the effects of natural and experimentally altered UV radiation on several different biological and chemical systems. This requires accurate measurement of spectral irradiance during experiments and also in the field, where responses must be eventually predicted. Because biological and chemical effects of UV are strongly dependent on wavelength, spectral weighting functions (action spectra) must be applied to relate UV exposure to response (Chapter 3). Spectral weighting functions have been determined for many processes, such as damage to DNA (Setlow, 1974; Quaite, Sutherland & Sutherland, 1992), inhibition of partial reactions of photosynthesis in chloroplasts (Jones & Kok, 1966), and inhibition of photosynthesis in higher plants (Caldwell *et al.*, 1986) and in marine phytoplankton (Cullen, Neale & Lesser, 1992; Helbling *et al.*, 1992; Neale, Lesser & Cullen, 1994). The spectra have different shapes, and many extend well into the UV-A, showing that it is not possible to quantify the potential for ecological effects with any one weighting function, or with unweighted measurements of UV-B radiation. Thus, accurate radiometry and evaluation of weighting functions are essential to environmentally relevant studies of UV effects on ecosystems. This is especially true for studies aimed at elucidating the effects of UV radiation at multiple trophic levels.

11.3   Herbivorous versus microbial food web: implication of the UV-B radiation

Biological oceanographers generally recognise that the biogenic flux of carbon from autotrophic to heterotrophic organisms in the pelagic marine environment follows two contrasting trophic pathways: the herbivorous and the microbial food webs (Legendre & Rassoulzadegan, 1995; Azam, 1998). As Figure 11.1 illustrates, a herbivorous food web is characterised by an ample NO_3^- supply, allowing the development of large phytoplankton (e.g. diatoms) that are linked to fish directly or indirectly via large zooplankton (e.g. copepods). The herbivorous food

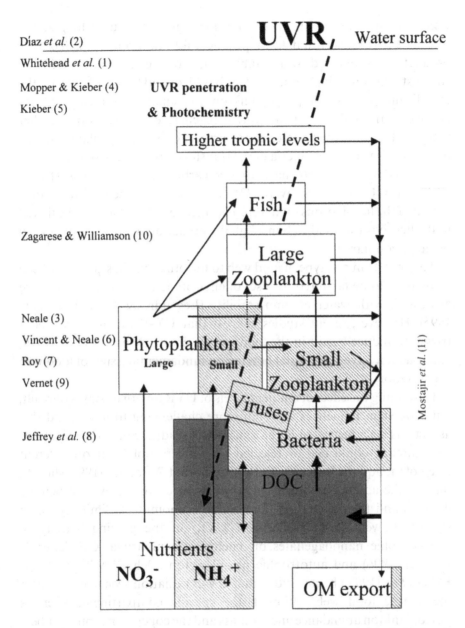

Díaz *et al.* (2)

Whitehead *et al.* (1)

Mopper & Kieber (4)

Kieber (5)

Zagarese & Williamson (10)

Neale (3)

Vincent & Neale (6)

Roy (7)

Vernet (9)

Jeffrey *et al.* (8)

UVR, Water surface

UVR penetration
& Photochemistry

Higher trophic levels

Fish

Large Zooplankton

Phytoplankton
Large   Small

Small Zooplankton

Viruses

Bacteria

DOC

Nutrients
NO$_3^-$ NH$_4^+$

OM export

Mostajir *et al.* (11)

Figure 11.1. Schematic representation of the simplified pelagic food web structure. The authors and the corresponding chapter in this book (in parenthesis) describing the effect of UV radiation are marked in front of each compartment. The open boxes illustrate the herbivorous food web compartments and shaded boxes indicate the principal components of microbial food webs. The organisms are supported by the nutrients (NO$_3^-$ and NH$_4^+$) and dissolved organic matter (DOC) shown in the grey box (for details see the text). Boxes and arrows are not drawn to scale. OM, organic matter.

web is encountered in productive marine regions (e.g. upwelling zones, coastal environment) and large phytoplanktonic production (i.e. new production) is followed by a notable export of organic matter (OM). In contrast, the microbial food wed is based on NH_4^+ and leads to the establishment of small phytoplankton (i.e. eukaryotic algae and cyanobacteria). Small phytoplankton (i.e. regenerated production), together with the heterotrophic bacteria, are grazed by small zooplankton (e.g. protozoa). The role of bacteria in the transformation of dissolved organic carbon (DOC) into particulate organic carbon (POC) in this system is noteworthy. However, the overall OM export of this system is low relative to that of the herbivorous food web. The microbial food web can be linked to the herbivorous food web through the predation of microzooplankton by large zooplankton.

The shift from one type of food web to the other one has generally been recognised to be favoured by a set of environmental conditions that are associated with water column stability (Legendre & Rassoulzadegan, 1995). However, recent studies suggest that UV-B radiation can be a trigger to induce a shift in the food web structure (Mostajir *et al.*, 1999) and thus can have substantial bearing on fundamental issues of food web organisation.

Describing the effect of ambient levels of UVR on organisms is difficult, but ecosystem influences are even more challenging to assess and this aspect has been neglected in the majority of studies. Some investigations have already confirmed or rejected the influence of UVR on different levels of the aquatic food web. Bothwell, Sherbot & Pollock (1994) showed that, in the aquatic environment, UVR reduced the growth of benthic diatoms and their predators (Diptera: Chironomidae). Similarly, some studies showed harmful effects of UVR on the grazing activity of heterotrophic nanoflagellates on bacteria (Sommaruga, Oberleiter & Psenner, 1996) and autotrophic picoplankton (Ochs, 1997). However, Keller *et al.* (1997a,b) concluded that UV-B radiation seems not to affect the marine food web, although the specific compartments such as phytoplankton abundance and biomass and the copepod nauplii numbers are significantly reduced under UV-B stress in a stratified coastal system. They concluded that, despite the ten-fold increase in biologically damaging UV-B applied in their experiment, effects were not seen at the higher trophic levels. They suggested that it was most likely because of the rapid extinction of UV-B in the highly coloured coastal water. Such variability in responses has resulted in a great deal of controversy and confusion about the implications of increasing UV-B radiation for specific ecosystems.

The effects of UV-B on heterotrophs (bacteria and zooplankton) and on the whole pelagic food web are not well understood and consequently difficult to predict. Moreover, viruses, bacterioplankton, phytoplankton, zooplankton and the eggs and larvae of commercially important fishes are usually more concentrated in, or restricted to, the upper 25 m of the water column. Therefore, they are particularly susceptible to fluctuations in incident UVR. The lack of information on levels of UVR in the water column, and on the sensitivity of aquatic and marine ecosystem productivity to fluctuations in UVR, is a serious deficiency.

As pointed out by Mostajir et al. (1999), the nature of pelagic ecosystem responses to increased UV-B radiation, via its effect on trophic interactions, is to date the most elusive ecological question related to UV-B impact studies. The major problem in evaluating the UV-B effect at the ecosystem level seems to be methodological. There are two major obstacles to studying a wide size-range of organisms from viruses to fish larvae: time and space scales. For the microbiologists, a micro and meso time scale (hours, days) to evaluate the changes (physiological or at the community level) in several generations of pico-, nano- and micro-organisms (viruses, bacteria and protozoa) is usually sufficient. In contrast, to study these changes for larger organisms (zooplankton, fish larvae and fishes), a much longer time scale (days, weeks and months) is necessary. Similarly, the microbiologists can effectively study samples in volumes of millilitres or litres but for the investigation of larger organisms, larger volumes of samples are required. Therefore, to evaluate the UV-B effect (as well as effects of the other stress parameters) at the ecosystem level (at least from viruses to large zooplankton), a large sample volume, which can be sampled during several days or weeks, is indispensable. A mesocosm approach can offer these possibilities as proposed by Belzile et al. (1998). To extend the results from laboratory and mesocosm studies to whole ecosystems, some fundamental experimental design criteria must be considered (Mostajir et al., 1999):

1. The organisms used in the UV-B experiment must be representative of natural conditions.
2. The results of UV-B experiments using a single trophic level cannot be applied directly to the natural environment with a complex web of trophic interactions.
3. Experimentally enhanced UV-B doses must be representative of plausible natural conditions and natural UV-B controls also are necessary for comparison.

4. Changes in the dynamics of environmental variables such as nutrients and temperature should be also investigated in enhanced UV-B studies.
5. Linkage between receiving UVR by planktonic organisms and mixing dynamics of the water column in the pelagic system should be considered as well. The euphotic zone of the pelagic system is heterogeneous with respect to UV-B response, permitting periodic damage at the surface as well as periodic repair in deeper waters. Good experimental design will ensure that mixing is uniform, reproducible and representative of the natural conditions.

Using a pseudo-ecosystem (mesocosm) approach, Mostajir et al. (1999) observed a decline of 66% in the ciliate population (15–35 μm) and 63% for large phytoplankton communities (5–20 μm) under high enhanced UV-B relative to natural conditions. This decrease in the ciliate population allowed the development of their potential prey, namely heterotrophic bacteria, heterotrophic flagellates (2–10 μm) and small phytoplankton (< 5 μm). This reduction of ciliate and large phytoplankton communities would strongly limit upward transfer of mass and energy. In fact, as ciliates can be the most important means of transferring production from the lower trophic levels to higher ones, this 66% deficit of carbon will perturb the transfer of energy from primary producer to herbivorous consumers. In the same experiment, Chatila et al. (1999) showed similar results concerning the grazing activity of heterotrophic nanoflagellates (2–10 μm), which seemed higher under enhanced UV-B than that of ciliates. In parallel, the phaeopigment concentration as an indicator of the herbivorous grazing activity was significantly reduced under high UV-B levels relative to natural conditions (Walsh et al., unpublished data). The conclusion was that the adverse effect of enhanced UV-B on ciliates and the consequently lower grazing pressure on their prey could interrupt the trophic transfer of matter and energy, thereby channelling more carbon into the microbial food web within the pelagic ecosystem. Moreover, the reduced number of large phytoplankton (i.e. diatoms) under enhanced UV-B could hinder the development of the herbivorous food web. The shift in the ecosystem from a herbivorous to a microbial food web under enhanced UV-B radiation would favour the trapping of carbon in small organisms, which can induce changes in the structure and dynamics of the pelagic food web.

The above illustration of the interdependent and non-uniform responses of the individual members of an aquatic ecosystem to not only a varying

UVR field but also to a suite of atmospheric stressors (stressors that may appear to be independent variables but which are in reality dependent variables) is consistent with the work of Bukata & Jerome (1997), who utilised replicate microcosms in lake water to illustrate ecosystem responses to reduced, ambient and enhanced downwelling fluxes of UVR. They observed that an ecosystem exposed to 'normal stress' (i.e. to a suite of ambient stressors to which the ecosystem has adapted) displays maximum degrees of freedom in its evolutionary development and, consequently, several 'futures' are available to that ecosystem. However, an ecosystem exposed to departures from that 'normalcy' (i.e. the increase, decrease, or complete removal of one of the stressors from the suite to which it had adapted) loses those degrees of freedom in its evolutionary development and a restricted 'future' protocol is selfinvoked (possibly predictable under certain conditions), one predicated upon the interdependent and non-uniform synergistic responses of the ecosystem members. The observation and study of such protocols under quasicontrolled conditions might provide the linkages between laboratory and field determinations of individual and collective organism response(s) to UVR.

11.4 UV-B and the carbon flux

The implication of modification to the community structure induced by the UVR could seriously affect the role of the biological CO_2 pump in the ocean (Volk & Hoffert, 1985). The concept of the biological CO_2 pump relies on the photosynthetic incorporation of inorganic carbon into organic molecules by large algae ($> 5 \mu m$). A fraction of these large algal cells sinks to deep waters (as intact cells, faecal pellets, marine snow) or is actively transported by vertically migrating organisms. According to Legendre & Le Fèvre (1991), the phytoplankton production and the community structure influence the pathways of carbon export and sequestration as well as the renewable marine resources. Whether bacteria or phytoplankton and detritus dominate the food for grazers in the euphotic zone is likely to influence the food web structure, nutrient cycling, and sinking flux (Cho & Azam, 1990). Production by large phytoplankton ($> 5 \mu m$), is generally thought to favour rapid export (Legendre & Le Fèvre, 1989), whereas smaller cells (ultraplankton, $< 5 \mu m$; Sverdrup, Johnson & Fleming, 1942) and bacteria enter the food web and are either not exported or are exported via large heterotrophs on a longer time scale.

The size abundance spectrum of plankton in a given water column is the result of a combination of diverse processes at the level of the individual (growth, respiration, exudation, natural mortality, sinking), the interactions with other individuals (competition, predation) and the physical environment (coagulation, sinking, dissolution, horizontal transport, vertical mixing, turbulence), and occur on particular spatial and temporal scales (for a review, see Rodriguez, 1994). However, the UVR has never been invoked as a factor that could influence the size structure and the dynamics of the ecosystem. A system dominated by the microbial web is portrayed as a recycling loop from which there is very little export. UV-B could favour the development of a planktonic community dominated by small cells included in the microbial food web (Mostajir et al., 1999), which could seriously influenced the role of the biological CO_2 pump in the pelagic marine ecosystems.

Acknowledgements

This contribution was supported by NSERC (Natural Sciences and Engineering Research Council of Canada), the Fonds FCAR (Formation de Chercheurs et Aides à la Recherche du Québec), FODAR (Fonds pour le Développement et l'Avancement de la Recherche) and the Inter-American Institute (IAI) for global change. This is a contribution to the research programme of the Groupe de Recherche en Environnement Côtier.

References

Azam, F. (1998). Microbial control of oceanic carbon flux: the plot thickens. *Science*, **280**, 694–6.

Belzile, C., Demers, S., Lean, D. R. S., Mostajir, B., Roy, S., de Mora, S. J., Bird, D., Gosselin, M., Chanut, J.-P. & Levasseur, M. (1998). An experimental tool for the study of the effects of ultraviolet radiation on planktonic communities: a mesocosm approach. *Environmental Technology*, **19**, 667–82.

Bothwell, M. L., Sherbot, D. M. J. & Pollock, C. M. (1994). Ecosystem response to solar ultraviolet-B radiation: influence of trophic-level interactions. *Science*, **265**, 97–100.

Bukata, R. P. & Jerome, J. H. (1997). A (non-linear) perspective on ultraviolet radiation and freshwater ecosystems. *Proceedings of the Ontario Climate Advisory Committee Workshop on Atmospheric Ozone*, pp. 136–65.

Caldwell, M. M., Camp, L. B., Warner C. W. & Flint, S. D. (1986). Action spectra and their key role in assessing biological consequences of solar UV-B radiation change. In *Stratospheric Ozone Reduction, Solar Ultraviolet Radiation, and Plant Life*, ed. R. C. Worrest & M. M. Caldwell, pp. 87–111. Springer-Verlag, Berlin.

Chatila, K., Demers, S., Mostajir, M., Gosselin, M., Chanut, J.-P. & Monfort, P. (1999). Bacterivory of a natural heterotrophic protozoan population exposed to different intensities of ultraviolet-B radiation. *Aquatic Microbial Ecology*, in press.

Cho, B. C. & Azam, F. (1990). Biogeochemical significance of bacterial biomass in the ocean's euphotic zone. *Marine Ecology Progress Series* **63**, 253–9.

Cullen, J. J., Neale, P. J. & Lesser, D. M. P. (1992). Biological weighting function for the inhibition of phytoplankton photosynthesis by ultraviolet radiation. *Science*, **258**, 646–50.

Helbling, E. W., Villlafañe, V. E., Ferrario, M. & Holm-Hansen, O. (1992). Impact of natural ultraviolet radiation on rates of photosynthesis and on specific marine phytoplankton species. *Marine Ecology Progress Series*, **80**, 89–100.

Jones, L. W. & Kok, B. (1966). Photoinhibition of chloroplast reactions. II. Multiple effects. *Plant Physiology*, **41**, 1044–9.

Keller, A., Hargraves, P., Jeon, H., Klein-Macphee, G., Klos, E., Oviatt, C. & Zhang, J. (1997a). Ultraviolet-B radiation enhancement does not affect marine trophic levels during a winter-spring bloom. *Éco-science*, **4**, 129–39.

Keller, A., Hargraves, P., Jeon, H., Klein-Macphee, G., Klos, E., Oviatt, C. & Zhang, J. (1997b). Effects of ultraviolet-B enhancement on marine trophic levels in a stratified coastal system. *Marine Biology*, **130**, 277–87.

Legendre L. & Le Fèvre, J. (1989). Hydrodynamical singularities as controls of recycled versus export production in oceans. In *Productivity of The Ocean: Present and Past*, ed. H. Berger, V. S. Smetacek & G. Wefer, pp. 49–63. W. J. Wiley and Sons Ltd., Chichester.

Legendre, L. & Le Fèvre, J. (1991). From individual plankton cells to pelagic marine ecosystems and to global biogeochemical cycles. In *Particle Analysis in Oceanography*. NATO ASI Series, G 27, ed. S. Demers, pp. 261–300. Springer-Verlag, Berlin.

Legendre, L. & Rassoulzadegan, F. (1995). Plankton and nutrient dynamics in marine waters. *Ophelia*, **41**, 153–72.

Mostajir, B., Demers, S., de Mora, S., Belzile, C., Chanut, J.-P., Gosselin, M., Roy, S., Villegas, P. Z., Bouchard, J., Bird, D., Monfort, P. & Levasseur, M. (1999). Experimental test of the effect of UV-B radiation on a planktonic community. *Limnology and Oceanography*, **44**, 586–96.

Neale, P. J., Lesser, M. P. & Cullen, J. J. (1994). Effects of ultraviolet radiation on the photosynthesis of phytoplankton in the vicinity of McMurdo station, Antarctica. In *Ultraviolet Radiation in Antarctica: Measurements and Biological Effects*, Antarctic Research Series **62**, ed. C. S. Weiler & P. A. Penhale, P. A. pp. 125–42. American Geophysical Union, Washington DC.

Ochs, C. A. (1997). Effects of UV radiation on grazing by two marine heterotrophic nanoflagellates on autotrophic picoplankton. *Journal of Plankton Research*, **19**, 1517–36.

Quaite, F. E., Sutherland, B. M. & Sutherland, J. C. (1992). Action spectrum for DNA damage in alfafa lowers predicted impact of ozone depletion. *Nature*, **358**, 576–8.

Rex, M., Harris, N. R. P., van der Gathen, P., Lehmann, R., Braathen, G. O., Reimer, E. *et al.* (1997). Prolonged stratospheric ozone loss in the 1995–96 Arctic winter. *Nature*, **389**, 835–8.

Rodriguez, J. (1994). Some comments on the size-based structural analysis of the pelagic ecosystem. In *The Size Structure and Metabolism of the Pelagic Ecosystem*, ed. J. Rodriguez & W. K. W. Li, pp. 1–10.

Schindler, D. W., Curtis, J. P., Parker, B. R. & Stainton, M. P. (1996). Consequences of climate warming and lake acidification for UV-B penetration in North American boreal lakes. *Nature*, **379**, 706–8.

Setlow, R. B. (1974). The wavelengths in sunlight effective in producing skin cancer: a theoretical analysis. *Proceedings of the National Academy of Sciences, USA*, **71**, 3363–6.

Shindell, D. T., Rind, D. & Lonnergan, P. (1998). Increased polar stratospheric ozone losses and delayed eventual recovery owing to increasing greenhouse-gas concentrations. *Nature*, **392**, 589–92.

Smith, R. C., Prézelin, B. B., Baker, K. S., Bidigare, R. R., Boucher, N. P., Coley, T., Karentz, D., MacIntyre, S., Matlick, H. A., Menzies, D., Ondrusek, M., Wan, Z. & Waters, K. J. (1992). Ozone depletion: ultraviolet radiation and phytoplankton biology in Antarctic waters. *Science*, **255**, 952–9.

Sommaruga, R., Oberleiter, A. & Psenner, R. (1996). Effect of UV radiation on the bacterivory of a heterotrophic nanoflagellate. *Applied and Environmental Microbiology*, **62**, 4395–400.

Sverdrup H. U., Johnson, M. W. & Fleming, R. H. (1942). *The Oceans. Their Physics, Chemistry and General Biology*. Prentice-Hall, Englewood Cliffs, NJ.

Tarasick, D. W. & Fioletov, V. E. (1997). The distribution of ozone and ozone-depleting substances in the atmosphere and observed changes. In *Ozone Science: A Canadian Perspective on the Changing Ozone Layer*, ed. D. I. Wardle, J. B. Kerr, C. T. McElroym & D. R. Francis, pp. 9–14. Environment Canada (printed and bound in Canada by the University of Toronto Press, Toronto).

Volk, T. & Hoffert, M. I. (1985). Ocean carbon pumps: analysis of relative strengths and efficiencies in ocean driven CO_2 changes. In *The Carbon Cycle and Atmospheric CO_2: Natural Variations Archean to Present*, AGU Monograph 32, ed. E. T. Sundquist & W. S. Broecker, pp. 99–110. American Geophysics Union, Washington DC.

Wardle, D. I., Kerr, J. B., McElroy, C. T. & Francis, D. R. (eds.) (1997). *Ozone Science: A Canadian Perspective on the Changing Ozone Layer*. Environment Canada (printed and bound in Canada by the University of Toronto Press, Toronto).

Index